Mechanically Interlocked Materials

Mechanically Interlocked Materials

Polymers, Nanomaterials, MOFs, and more

Edited by
Emilio M. Pérez

WILEY-VCH

Editor

Prof. Emilio M. Pérez
IMDEA Nanoscience
c/ Faraday 9
28049 Madrid
Spain

Cover Image: Courtesy of Patricia Bondia

Library of Congress Card No.: applied for

British Library Cataloguing-in-Publication Data
A catalogue record for this book is available from the British Library.

Bibliographic information published by the Deutsche Nationalbibliothek
The Deutsche Nationalbibliothek lists this publication in the Deutsche Nationalbibliografie; detailed bibliographic data are available on the Internet at <http://dnb.d-nb.de>.

Print ISBN: 978-3-527-34793-3
ePDF ISBN: 978-3-527-82895-1
ePub ISBN: 978-3-527-82896-8
oBook ISBN: 978-3-527-82897-5

Typesetting Straive, Chennai, India
Printing and Binding CPI Group (UK) Ltd, Croydon, CR0 4YY

C9783527347933_130324

Contents

Preface

Dear reader,

Welcome to the world of mechanically interlocked materials! You are about to discover the way that one of the most interesting tools of chemistry, the mechanical bond, is influencing the world of materials science.

Synthetic chemists started wondering about the possibility of connecting molecular fragments mechanically, like the links on a chain, very early on. The first monograph on the subject was written by Schill more than 50 years ago [1] and mentions that the earliest known discussion of such compounds, attributed to Willstätter, dates back somewhere between 1900 and 1912. A [2]catenane (two interlocked rings) was first intentionally made by Wasserman using statistical approaches in 1960 [2, 3], followed seven years later by the synthesis of a rotaxane by Harrison and Harrison [4]. In the meantime, Schill had started exploring a covalent template approach and succeeded in making catenanes in 1964 [5, 6]. The advent of noncovalent templated synthetic methods, where weak intermolecular interactions are used to place functional groups in the right positions, allowed chemists to make interlocked molecules in meaningful yields. This approach was pioneered in the early 1980s by Jean-Pierre Sauvage, with the use of metal–ligand interactions [7, 8] and later expanded to π–π and charge–transfer interactions by Sir J. Fraser Stoddart [9] and to H-bonding by David Leigh [10, 11]. At approximately the same time, Fujita used dynamic metal–ligand interactions to promote the formation of a [2]catenane from two separate macrocycles [12]. These are some of the most prominent milestones and groups that enabled the synthesis of mechanically interlocked molecules (MIMs), and, therefore, the study of the unique features of the mechanical bond. We now know that the mechanical bond is strong (it cannot be taken apart without breaking covalent bonds within the components), but it is also respectful of the structural integrity of each part of the ensemble, combining the best of covalent and supramolecular chemistries [13]. But the star of the show is perhaps the dynamic nature of the mechanical bond: the components of MIMs can move with respect to each other. This has allowed the construction of fascinating synthetic molecular machines, which account for two-thirds of the 2016 Nobel Prize in Chemistry awarded to Stoddart [14] and Sauvage [15] with the work of Ben Feringa on molecular machines that move around covalent linkages claiming the remaining third [16].

While the chemistry of MIMs is a relatively recent endeavor, materials science is one of the oldest and most decisive of human interests: materials have literally defined ages and civilizations. This is so because making tools and shaping objects to perform a specific function is perhaps the key defining characteristic of humans [17] – although several other animals can *use* tools, their manufacturing abilities are typically much more limited [18]. Of course, the function of the tool is directly related to the material it is made of, which makes materials the cornerstone of civilization [19]. For example, the three-age system that divides human prehistory into Stone Age, Bronze Age, and Iron Age is attributed to Christian Jürgensen Thomsen c. 1836 and is still widely accepted [20]. The history of steel is a most fascinating one, including its serendipitous discovery (most likely iron was left for too long in coal ovens), the famous Damascus sabres that are often cited as one of the earliest examples of nanotechnology [21], and its key role in the Second Industrial Revolution, after mass production of steel was made possible by the Bessemer and the Siemens-Martins processes [22]. The invention of modern (Portland) cement by Joseph Aspdin [23] re-shaped the world to such an extent that its production has been traditionally used as one of the most reliable indicators of economic development [24]. It is such a central material that its eco-friendly use is currently a focal point in the battle against climate change [25]. Introduced by Alexander Parkes at the London International Exhibition in 1862, Parkesine is the first known example of a man-made plastic [26]. However, it is Baekeland's bakelite that inaugurated the field of synthetic polymers [27], that would be officially born with Staudinger's famous 1920 article "Über polymerisation" [28]. The influence that synthetic polymers have had on modern life cannot be overstated. On the positive side, they have increased our quality of life tremendously. Their role as enablers of other technologies is unmatched by any other family of materials. Hi-tech or low-tech, everything and anything is made of (or contains a bit of) plastic: the keyboard I am writing on, the screen you are looking at, the helmet I wear when I ride my bike, the plane that you flew in for your last holiday, the clothes you are wearing, or the wrapping of the frozen pizza you plan to have for dinner… just look around! [29]. This super-intensive use of plastics also has its dark side, of course, because synthetic polymers are typically not biodegradable, and pollution by (micro)plastics has become a major concern, that will require an entirely new approach to the way we make and use polymers [30]. And we still have to talk about semiconductors and the way that electronics have changed our way of living! The invention of the transistor by Shockley, Bardeen and Brattain [31] was recognized with the Nobel Prize in Physics in 1956. The development of electronics has indeed been mostly a physics endeavor, but one where the purity of silicon (and other semiconductors) has played a major role [32]. Once more, materials enable technology.

As the chemistry of the mechanical bond progressed, it was soon apparent that its characteristics made it a very interesting tool for materials chemistry. For example, strength and adaptability are often sought after together, but far from easy to achieve within the same material. In principle, the dynamic nature of the mechanical bond makes it a good candidate for adaptable materials, while its strength is typically comparable to that of a covalent bond. But there are many other opportunities, as you will see!

Each chapter of the book is completely self-standing. Our contributors have done an excellent job of putting their subject into perspective, summarizing the main advances, and providing a far-reaching outlook for their specific field. So, if you are only (or particularly) interested in a certain subject, by all means, dive into the corresponding chapter directly. If you prefer to get a global view of the subject, we have tried to help you by organizing the book in the following way:

The first section of the book is dedicated to understanding at the fundamental level how the dynamic character of MIMs can contribute to the dynamic and mechanical properties of materials. To understand, we need to measure in detail how individual MIMs respond under mechanical stress. We will begin with a chapter on single-molecule force spectroscopy of MIMs by James Ormson, Anne-Sophie Duwez, and Guillaume De Bo. Next, we start moving from the purely molecular to the materials world step by step. We will first see how MIMs behave on very small and essentially 0D materials. Euan Kay will explain how to interface MIMs and nanoparticles and the opportunities that arise from the combination of these two very hot research topics. The following contribution is by Alejandro López-Moreno and myself, and will take us to 1D materials. It is focused on the making and studying of mechanically interlocked derivatives of single-walled carbon nanotubes. We will progress one step further to the second dimension by seeing how molecular motors behave when they are supported on surfaces in a chapter prepared by Monika Schied. Finally, we complete this first stage of our journey with the third dimension, in the chapter prepared by Benjamin Wilson and Steve Loeb. They write about how to organized MIMs and study their motion within 3D porous crystalline solids.

The next block is dedicated to biopolymers featuring mechanical bonds. Proteins are the paradigmatic functional biomolecule. We will start with a chapter on mechanically interlocked proteins by Yu-Xiang Wang, Wen-Hao Wu, and Wen-Bin Zhang. The authors will take you through both natural and synthetic peptides and proteins featuring mechanical bonds. You might be surprised to learn that up to 6% of the structures deposited in the Protein Data Bank contain some form of interlocking! The following chapter is by Yinzhou Ma, Ze Yu, and Julián Valero, who will give us an overview of DNA-based mechanically interlocked structures. Besides being the information material of the cell, DNA is one of the most versatile building blocks for nanotechnology, so get ready to be amazed with intricate structures and examples of advanced DNA-based molecular machines.

We will finish the book on a high note with two chapters on purely synthetic mechanically interlocked materials. First, we will learn about oligo- and polycatenanes in the chapter contributed by Sougata Datta, Atsushi Isobe, and Shiki Yagai. These were perhaps the structures that started it all, as the conceptual move from two interlocked rings to a full chain of them was immediate. Not so with the synthesis, though. Making and characterizing polycatenanes is a grand challenge, and you will see how different groups have tackled and succeeded at this problem. The final chapter of the book is on mechanically interlocked polymers and is authored by Leonie Braks and Ali Coskun. Among these structures are some of the most advanced mechanically interlocked materials that have already demonstrated significant improvements over traditional structures. Accordingly, it includes specific sections on potential applications in fields from composites to biomedicine to batteries.

I hope you find the complete journey instructive and that you will be amazed by some of the advances we have already achieved. I hope that you will learn plenty; I certainly have. But, more than anything, I hope you will find this book inspiring, and that reading what has been done will encourage you to set your gaze on all that is yet to come. And if you choose to set your sail towards that Ithaka, I hope that your road is a long one, full of adventure, full of discovery.

I want to conclude this preface by sincerely thanking all authors of the chapters of this book. It has been a long journey for us, too! Thank you very much for your contributions. I am sure you will forgive the many emails, the constantly moving deadlines, and the many delays… and that you will agree with me that the final result was worth the effort.

Madrid, Spain *Emilio M. Pérez*
February 2024

References

1 Schill, G. (1971). *Catenanes, Rotaxanes and Knots*. New York: Academic Press.
2 Wasserman, E. (1960). The preparation of interlocking rings: a catenane. *J. Am. Chem. Soc.* 82: 4433–4434.
3 Baluna, A.S., Galan, A., Leigh, D.A. et al. (2023). In search of Wasserman's catenane. *J. Am. Chem. Soc.* 145 (17): 9825–9833.
4 Harrison, I.T. and Harrison, S. (1967). Synthesis of a stable complex of a macrocycle and a threaded chain. *J. Am. Chem. Soc.* 89 (22): 5723–5724.
5 Schill, G. and Lüttringhaus, A. (1964). The preparation of catena compounds by directed synthesis. *Angew. Chem. Int. Ed. Engl.* 3 (8): 546–547.
6 Brückner, R. (2019). Pioneering work on catenanes, rotaxanes, and a knotane in the University of Freiburg 1958–1988. *Eur. J. Org. Chem.* 2019 (21): 3289–3319.
7 Dietrich-Buchecker, C.O., Sauvage, J.P., and Kintzinger, J.P. (1983). New family of molecules: the metallo-catenanes. *Tetrahedron Lett.* 24 (46): 5095–5098.
8 Dietrich-Buchecker, C.O., Sauvage, J.P., and Kern, J.M. (1984). Templated synthesis of interlocked macrocyclic ligands: the catenands. *J. Am. Chem. Soc.* 106 (10): 3043–3045.
9 Ashton, P.R., Goodnow, T.T., Kaifer, A.E. et al. (1989). A [2]catenane made to order. *Angew. Chem. Int. Ed. Engl.* 28 (10): 1396–1399.
10 Johnston, A.G., Leigh, D.A., Nezhat, L. et al. (1995). Structurally diverse and dynamically versatile benzylic amide [2]catenanes assembled directly from commercially available precursors. *Angew. Chem. Int. Ed. Engl.* 34 (11): 1212–1216.
11 Johnston, A.G., Leigh, D.A., Pritchard, R.J., and Deegan, M.D. (1995). Facile synthesis and solid-state structure of a benzylic amide [2]catenane. *Angew. Chem. Int. Ed. Engl.* 34 (11): 1209–1212.

12 Fujita, M., Ibukuro, F., Hagihara, H., and Ogura, K. (1994). Quantitative self-assembly of a [2]catenane from two preformed molecular rings. *Nature* 367 (6465): 720–723.

13 Stoddart, J.F. (2009). The chemistry of the mechanical bond. *Chem. Soc. Rev.* 38 (6): 1802–1820.

14 Stoddart, J.F. (2017). Mechanically interlocked molecules (MIMs)—molecular shuttles, switches, and machines (nobel lecture). *Angew. Chem. Int. Ed.* 56 (37): 11094–11125.

15 Sauvage, J.-P. (2017). From chemical topology to molecular machines (nobel lecture). *Angew. Chem. Int. Ed.* 56 (37): 11080–11093.

16 Feringa, B.L. (2017). The art of building small: from molecular switches to motors (nobel lecture). *Angew. Chem. Int. Ed.* 56 (37): 11060–11078.

17 Stout, D. and Chaminade, T. (2007). The evolutionary neuroscience of tool making. *Neuropsychologia* 45 (5): 1091–1100.

18 Seed, A. and Byrne, R. (2010). Animal tool-use. *Curr. Biol.* 20 (23): R1032–R1039.

19 Sass, S.L. (1998). *The Substance of Civilization: Materials and Human History from the Stone Age to the Age of Silicon*. Arcade Publishing.

20 Rowley-Conwy, P. (2007). *From Genesis to Prehistory: the Archaeological Three Age System and Its Contested Reception in Denmark, Britain, and Ireland*. Oxford University Press.

21 Reibold, M., Paufler, P., Levin, A.A. et al. (2006). Carbon nanotubes in an ancient Damascus sabre. *Nature* 444 (7117): 286.

22 Mathews, J.A. (1926). The steel age—1876 to 1926. *Ind. Eng. Chem.* 18 (9): 913–914.

23 Ryan, J.F. (1929). The story of Portland cement. *J. Chem. Educ.* 6 (11): 1854.

24 Daugherty, K.E. (1973). Cement as an indicator of economic development. *Cem. Concr. Res.* 3 (2): 163–183.

25 Damineli, B.L., Kemeid, F.M., Aguiar, P.S., and John, V.M. (2010). Measuring the eco-efficiency of cement use. *Cem. Concr. Compos.* 32 (8): 555–562.

26 Parkes, A. (1865). On the properties of parkesine, and its application to the arts and manufactures. *J. Soc. Arts* 14: 81.

27 Baekeland, L.H. (1909). The synthesis, constitution, and uses of bakelite. *J. Ind. Eng. Chem.* 1 (3): 149–161.

28 Staudinger, H. (1920). Über polymerisation. *Berichte der deutschen chemischen Gesellschaft (A and B Series).* 53 (6): 1073–1085.

29 Geyer, R. (2020). Chapter 2 - Production, use, and fate of synthetic polymers. In: *Plastic Waste and Recycling* (ed. T.M. Letcher), 13–32. Academic Press.

30 Haque, F.M., Ishibashi, J.S.A., Lidston, C.A.L. et al. (2022). Defining the macromolecules of tomorrow through synergistic sustainable polymer research. *Chem. Rev.* 122 (6): 6322–6373.

31 Bardeen, J. and Brattain, W.H. (1948). The transistor, a semi-conductor triode. *Phys. Rev.* 74 (2): 230–231.

32 Ross, I.M. (1997). The foundation of the silicon age. *Bell Labs Tech. J.* 2 (4): 3–14.

1

Force-Promoted Transformations in Mechanically Linked Molecules

James Ormson[1], Anne-Sophie Duwez[2], and Guillaume De Bo[1]

[1] *University of Manchester, Department of Chemistry, Oxford Road, Manchester M13 9PL, UK*
[2] *University of Liège, Molecular Systems, Department of Chemistry, Campus of Sart Tilman, Allee du Six aout, Liège 4000, Belgium*

1.1 Introduction

Mechanical bonds present useful dynamic properties that have been instrumental in the development of molecular machines and smart materials, as their subcomponents can undergo large internal displacement (e.g. shuttling of a macrocycle along the axle of a rotaxane) [1]. When coupled with a directional elongation force, such a large amplitude movement can trigger a useful mechanochemical transformation [2]. As an external force is applied, molecules respond by bond rotation and bending before cleavage of covalent bonds occurs at high force. Mechanical bonds follow a similar two-stage response to external force. If the macrocycle of a mechanical bond is forced away from its equilibrium position, the molecule will go through a stage of deformation (Figure 1.1, dashed arrows) before the eventual cleavage of a covalent bond (Figure 1.1, plain arrows). An alternative non-covalent dissociation pathway is accessible if the macrocycle can pass over the stopper of a rotaxane (dethreading).

Different experimental techniques can be used to study the two force ranges involved in the application of force to mechanically interlocked molecules. Single-molecule force spectroscopy (SMFS) has been used to probe non-covalent (e.g. elongation, dethreading) and dynamic (e.g. shuttling) processes in interlocked molecules. These phenomena typically occur in the pN regime, and SMFS provides quantitative information on the force and elongation distance involved. Ultrasound activation generates forces in the nN regime where the scission of covalent bonds occurs. This technique has been used to activate mechanically interlocked molecules and elucidate their mechanism of dissociation under tension. This chapter will cover the recent advances in the study of mechanical bonds under external force moving from low force (SMFS, Section 1.2) to high force (ultrasonication in solution, Section 1.3) processes. The final section presents the recent developments in the use of interlocked molecules in mechanoresponsive materials (Section 1.4).

Mechanically Interlocked Materials: Polymers, Nanomaterials, MOFs, and more, First Edition.
Edited by Emilio M. Pérez.

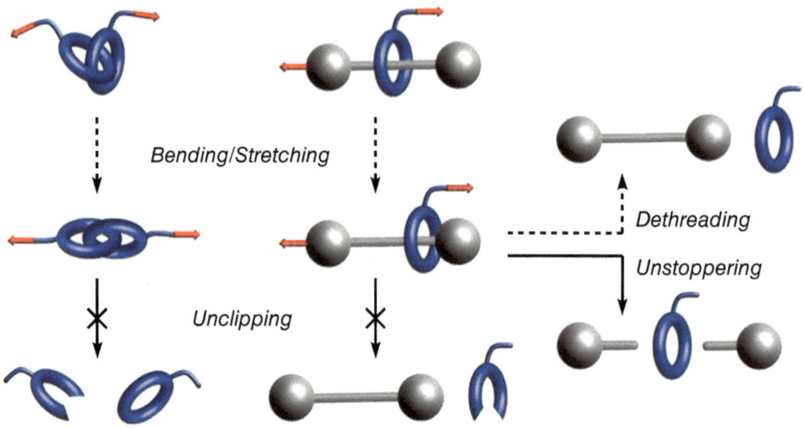

Figure 1.1 Response of catenanes and rotaxanes to external force.

1.2 SMFS in the Study of Non-covalent Interactions

SMFS consists of trapping and stretching a molecule between an atomic force microscopy (AFM) tip and a surface, or in a laser trap or between magnetic beads, to probe molecular processes *in situ* and real-time through the application of mechanical forces. For more than two decades, SMFS has proved efficacious in deciphering mechanistic information of individual biomolecules and in quantifying their force response to external stress. Such experiments have provided unprecedented insights into the structure and function of many biological systems, including DNA, proteins, enzymes, and biomolecular machines [3–15]. In addition to its widespread use with biological macromolecules, the technique has been adapted for the study of intra-molecular processes in synthetic macromolecules, including polymers [16], mechanophores [17–20], and interlocked systems [21, 22]. However, only a few investigations on intra-molecular processes and single-molecule mechanics have been realized on small molecules (1–5 nm), with successful examples including molecular recognition pairing [23], helical structures [24], and artificial molecular machines prototypes [25–30]. The rarity of such studies stems from the difficulty in developing proper tools and preparing appropriate molecules that can be interfaced with SMFS techniques, especially when one wants to probe sub-molecular motions.

1.2.1 Rotaxanes

Stoddart, Houk, Ho, and coworkers have pulled the macrocycle over the bulky end groups of a redox-active rotaxane with an AFM tip [31]. They compared the large forces required to de-thread oxidized and unoxidized molecules. They have used the switchable, bistable, [2]rotaxane $\mathbf{R}^{4+}$ shown in Figure 1.2, whose mechanism is based on the redox-controlled mechanical shuttling of the macrocycle along its axis. The molecule was specifically designed with a short thioctic acid linker tethered to the ring for tight attachment to a gold AFM tip and with a hydroxymethyl group

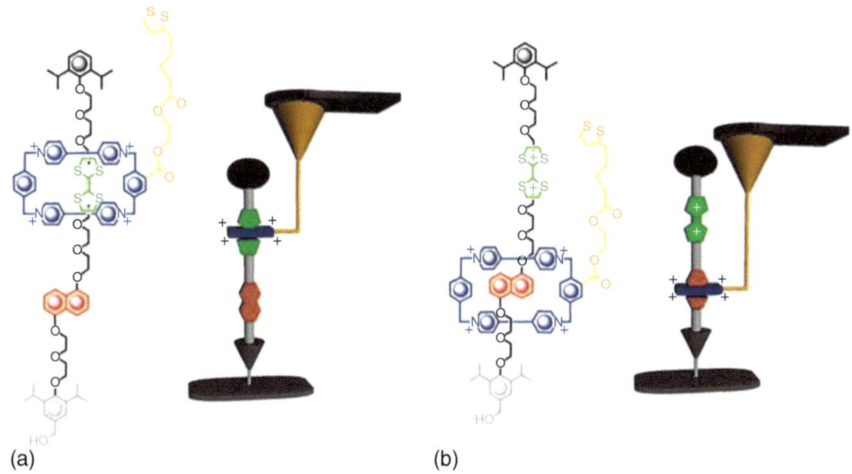

Figure 1.2 Chemical structure and schematic representation of the AFM force spectroscopy experiment of the bistable [2]rotaxane $\mathbf{R}^{4+}$, in which an electron-poor cyclobis(paraquat-*p*-phenylene) ($CBPQT^{4+}$) ring is confined to an axis containing two electron-rich recognition sites, tetrathiafulvalene (TTF) and 1,5-dioxynaphthalene (DNP), by the presence of bulky 2,6-diisopropylphenyl ether stoppers at each end. The stopper close to the DNP site carries a hydroxymethyl group for subsequent attachment to silicon wafers. (a) The $CBPQT^{4+}$ ring, which carries a very short tether terminated by a thioctic acid ester for attachment to a gold-coated AFM tip, displays a stronger interaction with TTF than with DNP and thus resides selectively on the former. (b) Chemical oxidation of TTF to TTF^{2+} results in a strong charge–charge repulsion between the $CBPQT^{4+}$ ring and TTF^{2+}, a situation that causes the $CBPQT^{4+}$ ring to shuttle to DNP in the oxidized [2]rotaxane $\mathbf{R}^{6+}$. Source: Brough et al. [31] / National Academy of Science.

on one of the stoppers for attachment to SiO_2 substrates via covalently bound monolayers of isocyanatopropyl linkers. They used force spectroscopy to probe the steric and electrostatic interactions present in the ground and oxidized states of $\mathbf{R}^{4+}$ (Figure 1.2). Control measurements consisted of probing molecules to measure the repulsive steric interactions between the cyclobis(paraquat-*p*-phenylene) ($CBPQT^{4+}$) ring and the diisopropylphenyl ether stopper. The probing of the $\mathbf{R}^{6+}$ molecules in an oxidizing solution was used to measure the repulsive force between the $CBPQT^{4+}$ ring and an oxidized tetrathiafulvalene (TTF^{2+}), which is responsible for molecular actuation. On the basis of the differences in the force required to pass the ring over the stopper in the presence and absence of the $Fe(ClO_4)_3$ oxidizing agent, they estimated the repulsive electrostatic barrier, which is ultimately responsible for the molecule's actuation.

By using the measured force spectroscopy values of 66 pN for the dethreading barrier under nonoxidative conditions and 145 pN for the oxidant-induced electrostatic repulsion barrier, along with the distance of 8.6 Å of the energy barrier produced from molecular simulations for the dethreading of the $CBPQT^{4+}$ ring over the bulky stopper, and 13.0 Å as the distance necessary for the $CBPQT^{4+}$ ring to travel from the 1,5-dioxynaphthalene (DNP) recognition site to the TTF^{2+} dication, the difference in interaction energies can be calculated as 19 kcal mol^{-1}. By combining the 19 kcal mol^{-1} difference in oxidative and nonoxidative interaction energies with the theoretically determined value of 46 kcal mol^{-1} for the energy associated with ground-state dethreading, a final value of 65 kcal mol^{-1} was obtained, representing the amount of repulsive actuation energy between $CBPQT^{4+}$ and TTF^{2+} produced by the oxidation of $\mathbf{R}^{4+}$. As estimated by this approach, 65 kcal mol^{-1} represents the upper limit of the total actuation energy.

Unfortunately, the irreversible rupture of molecules due to the dethreading required by this experiment prohibits the realization of a relaxing step to complete the cycle required to directly probe the mechanical work generated against the external load and to verify the estimation of 65 kcal mol^{-1}.

In 2011, Duwez, Leigh, and coworkers succeeded in detecting sub-molecular movements in a hydrogen-bonded rotaxane and in directly measuring the work generated by this molecule against a load [25]. They designed a hydrogen-bonded rotaxane with a poly(ethylene oxide) (PEO) tether attached to the ring to track its motion by an AFM cantilever (Figure 1.3). The [2]rotaxane consists of a ring mechanically locked onto a thread by two bulky groups situated at either end of the axle. The thread bears a fumaramide and a succinic amide-ester site, each of which can bind to the macrocycle through four intercomponent hydrogen bonds [32]. The ring predominantly resides over the fumaramide site (green in Figure 1.3), the occupancy ratio being higher than 95 : 5, compared to the succinimide ester (orange in Figure 1.3). A chain, suitable for binding to an AFM tip, was attached to the macrocycle. Close to the fumaramide binding site, a 1,2-dithiolane ring was introduced to enable the grafting of the rotaxane onto a gold surface.

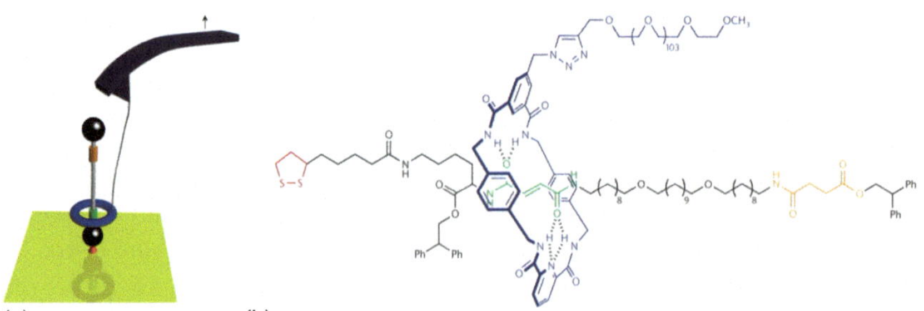

(a) (b)

Figure 1.3 (a) Schematics of the single-molecule force spectroscopy experiment of the rotaxane. The rotaxane is grafted onto gold, and the PEO tether is caught by the AFM tip. (b) Chemical structure of the rotaxane molecule. The rotaxane consists of a benzylic amide molecular ring (in blue) mechanically locked onto an axle by bulky diphenylethyl ester groups situated at either end. The axle bears a fumaramide group (in green) and a succinic amide-ester group (in orange), either of which can act as a binding site for the ring through up to four intercomponent hydrogen bonds. The affinity of the ring for the fumaramide site is much higher than for the succinic amide-ester site so the fumaramide: succinic amide-ester occupancy ratio is higher than 95 : 5. Next to the fumaramide binding site, a disulfide group (in red) was introduced to enable the grafting of the molecule onto gold substrates. A PEO tether (in blue) is attached to the ring in order to link the molecule to the AFM probe and track the motion of the ring along the axle. Source: Lussis et al. [25] / Springer Nature.

They used the cantilever of an AFM microscope to catch the tether, then applied a mechanical load to the ring of the rotaxane and followed its movement. The caught molecules were stretched in a controlled manner by moving the tip away from the substrate at a fixed pulling rate, and the force–extension profiles were measured. Thanks to the long tether, the detachment of the macrocycle from its station is directly observable in the stretching profile. Indeed, the progressive stretching of the PEO tether gives the characteristic parabolic profile of a random coil, which loses entropy on stretching, inducing a restoring force. Once the force exerted on the tether exceeds the strength of the hydrogen bonds between the macrocycle and the fumaramide station, the hydrogen bonds break, which results in the appearance of a small peak in the main force profile. This peak appears at about 27 pN in dimethylformamide and 45 pN in tetrachloroethane (TCE), reflecting the relative strengths of solute–solute hydrogen bonding in the two solvents. They fitted the pulling profiles with the worm-like chain model (an entropic elasticity model that predicts the relationship between the extension of a linear polymer and the entropic restoring force). The fits of the profile before and after the rupture peak give the increase in length (ΔL_c) of the molecule after the rupture of the inter-component hydrogen bonds of about 4 nm, which is consistent with the theoretical length of the fully extended thread between the two binding sites.

Pulling–relaxing cycles were also recorded, and they observed a peak, which is the signature of force acting on the tether. This shows that the macrocycle has traveled back from the succinic amide-ester site to the fumaramide site and that it is able to generate force when rebinding against the external load exerted by the cantilever. It means that if we trap the ring and then slightly decrease the force on the tether while maintaining the pulling, the ring is capable of pulling more strongly to recover its preferred position. In TCE, the macrocycle is able to travel back against an external load of 30 pN. The mechanical work produced by this sub-molecular motion is about 6 kcal mol^{-1}.

They also used the Crooks fluctuation theorem to estimate the free energy driving the macrocycle to bind to the fumaramide binding site at zero force. The work done on/by the molecules during pulling/relaxing is given by the areas below the corresponding force–extension curves. The rupture and rebinding work distributions overlap over a large range of work values and cross at a value of the work ΔG of $9.3 \pm 2.3\ k_B T = 5.5 \pm 1.3$ kcal mol^{-1}. The work value for a loading rate of 500 pN s^{-1} is close to this binding-energy difference between the two hydrogen-bonding motifs, indicating that under these conditions, the rotaxane is able to make use of almost all the energy available from hydrogen bonding to perform work along the direction of the applied load.

This example was the first quantitative measurement of the work done by a single synthetic molecular machine. It shows that the biased Brownian motions caused by thermal energy can be harnessed within a single synthetic small molecule to generate significant directional forces of similar magnitude to those generated by natural molecular machines, which are a lot bigger.

More recently, they detected the real-time fluctuations in the force experiments on these rotaxanes and revealed an intermediate weak binding state during shuttling [30]. They mechanically triggered the translocation of the ring between the two principal binding sites on the axle. Equilibrium fluctuations reveal another interacting site involving the two oxygen atoms in the middle of the thread. They characterized the ring occupancy distribution over time, which confirms the intermediate in both shuttling directions. The study provides evidence of weak hydrogen bonds that are difficult to detect using other methods and shows how the composition of the thread can significantly influence the shuttling dynamics by slowing down the ring motion between the principal binding sites. More generally, the study illustrates the utility that single-molecule experiments, such as force spectroscopy, can offer for elucidating the structure and dynamics of synthetic molecular machines.

Ibarra and coworkers studied the dynamics of a related rotaxane, modified to contain a very long thread so that the molecule could be investigated using optical tweezers in an aqueous environment (Figure 1.4) [27]. Force-clamp experiments under low external forces were made possible due to the high stability of the trapping, which allowed them to measure the dynamics of the rotaxane in an aqueous medium. They observed a lower rupture force of 8.5 pN for the breaking of hydrogen bonds between the macrocycle and the fumaramide station, in agreement with the relative strengths of hydrogen bonding in water and organic solvents, intercomponent hydrogen bonds being stronger in less polar solvents. Hopping events were also evidenced between the fumaramide and succinic amide-ester stations, but no intermediate states were observed. Since these experiments were performed in aqueous conditions, we suggest that the formation of weak hydrogen bonds between the macrocycle and the polyether thread would be much less favorable, and so, under those experimental conditions, the intermediate states may not be significant. Furthermore, the detection of such close intermediate states is difficult, given the spatial resolution of optical tweezers.

Their experimental set-up allowed very stable pulling–relaxing experiments to probe real-time shuttling events. Under these conditions, hundreds of shuttling events of the macrocycle between the two stations were monitored in real-time during minutes. From these data, they obtained crucial information about the dynamics of shuttling and a detailed picture of the energy landscape at different forces.

In 2018, Duwez, Sluysmans, Stoddart, and coworkers investigated the mechanochemical properties of donor–acceptor oligorotaxane foldamers, in which interactions between the mechanically interlocked component parts dictate the single-molecule assembly into a folded secondary structure (Figure 1.5) [28].

They used AFM-based SMFS to mechanically unfold molecules made of oligomeric dumbbells incorporating DNP units encircled by $CBPQT^{4+}$ rings. Real-time capture of fluctuations between unfolded and folded states revealed that the molecules exert forces of up to 50 pN against a mechanical load of up to 150 pN and displayed transition times of less than 10 μs. They showed that the folding is at least as fast as that observed in proteins and remarkably more robust, thanks to the mechanically interlocked structure.

The mechanical breaking of the donor–acceptor interactions responsible for the folded structure was measured over a broad range of loading rates [29]. The

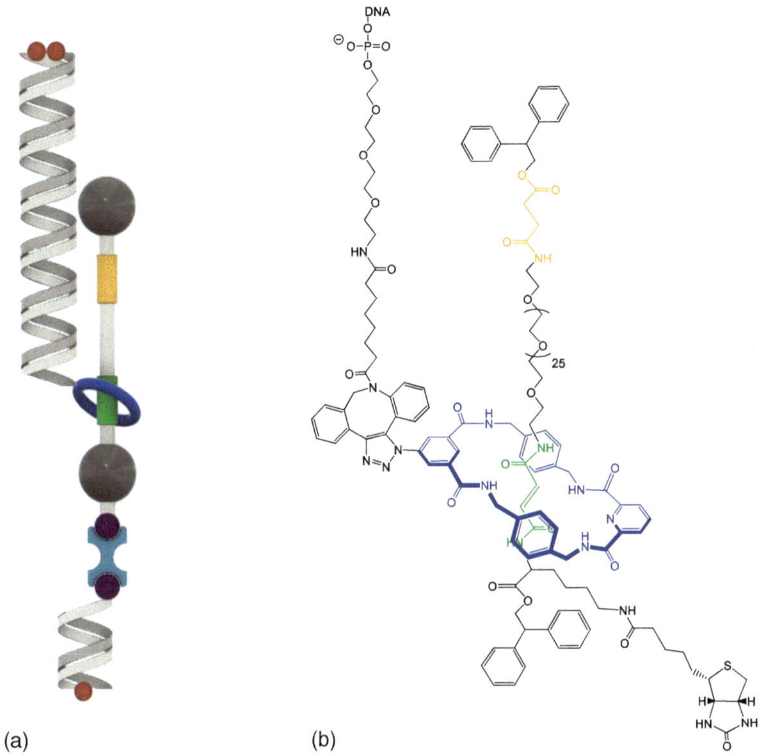

Figure 1.4 (a) Schematics of the OT-based single-molecule force spectroscopy experiment of the rotaxane. The rotaxane is attached to two polystyrene beads (diameter ~3 μm) through two dsDNA molecules: the 2686 bp dsDNA molecule connects the macrocycle to the bead via digoxigenin–antidigoxigenin (Dig–αDig) connections. The 830 bp dsDNA molecule connects the shuttle to a bead held by suction on a micropipette, via biotin–streptavidin connections at one end and Dig–αDig at the other end. Load is applied to the system by moving the optical trap away from the micropipette. At constant load, the shuttling dynamics of the macrocycle are inferred by measuring the motions of the bead. (b) Chemical structure of the rotaxane molecule modified to be trapped in a laser in an aqueous medium. The axle contains fumaramide (green) and succinic amide-ester (orange) stations separated by a very long oligoethyleneglycol spacer. A DNA oligonucleotide was covalently attached to the macrocycle to allow the manipulation in the optical trap.

results showed a high constant rupture force over a leading rate covering three orders of magnitude. In comparison with dynamic force spectroscopy performed during the past 20 years on various (bio)molecules, the near-equilibrium regime of oligorotaxanes persists at much higher loading rates, at which biomolecules have reached their kinetic regime, illustrating the high-speed dynamics and remarkable rebinding capabilities of the intramolecular donor–acceptor interactions. The authors investigated single interactions at a time in pulling–relaxing cycles and evidenced the near-equilibrium stochastic behavior. Indeed, even if the pulling experiments are performed near thermodynamic equilibrium, small molecules are always being submitted to random thermal fluctuations that constantly modify their unfolding and refolding paths. Using the Crooks fluctuation theorem, they measured the mechanical work produced during the breaking and rebinding to

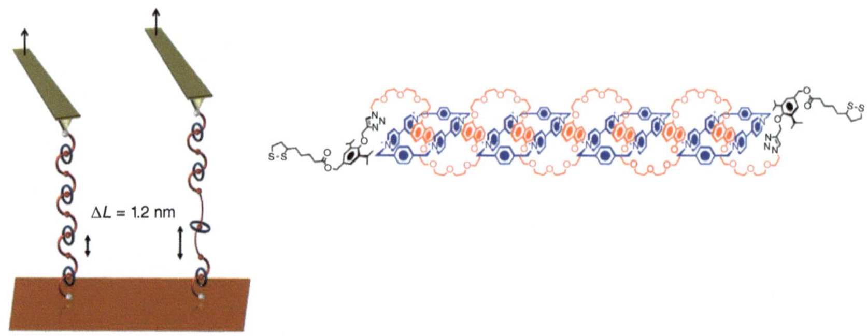

(a) (b)

Figure 1.5 (a) Schematics of the single-molecule force spectroscopy experiment of the oligorotaxane. The theoretical variation of length after breaking one interaction between a free DNP and the contiguous ring is about 1.2 nm, on the basis of crystallographic data. (b) Chemical structure of the oligorotaxane molecule made of oligomeric dumbbells incorporating 1,5-dioxynaphthalene units encircled by cyclobis(paraquat-*p*-phenylene) rings.

determine a free-energy difference, ΔG, of 6 kcal mol^{-1} between the two local conformations around a single bond.

1.2.2 Poly-(pseudo)rotaxanes

In contrast to rotaxanes, pseudorotaxanes are kinetically labile as their macrocycle is able to dethread spontaneously (i.e. the stoppers are too small to retain the macrocycle). This ability has been studied experimentally [33–40] and computationally [33, 35, 41], and subsequent mechanochemical studies have provided information of the forces needed to overcome steric groups and distinguish rotaxanes from pseudrotaxanes. Cyclodextrins (CD) spontaneously form pseudorotaxanes by threading onto polyethyleneglycol (PEG) chains [42], and the dethreading behaviour of these systems has been studied by Round and co-workers in 2008 [43]. Polyrotaxanes of varied lengths were probed by AFM (Figure 1.6). Pulling experiments at loading rate in the range of 7–12 nN s^{-1} indicate a rupture force around 114 pN with a rupture length corresponding to length of the PEG axle and the length of the tip tether. This observation, and the fact that the detachment of the of the polymer from the tip occurs at a higher force, suggests that the rotaxane disassembles with the dethreading of CD macrocycle over the bulky tricarboxylic acid benzoyl stopper.

The same authors have used a CD-based pseudorotaxane to measure the strength of a calcium-mediated eggbox junction (a cross-link between sequences of oligoguluronic acids [oligoGs]) [44]. After some equilibration time, an AFM tip was used to pick an α-CD in order to unzip the eggbox junction. The authors determined how much force is required to slide along a single strand of oligoGs. Slipping over one guluronic acid requires about 47 pN, while unzipping requires 58, 112, and 141 pN (at 6–20 nN s^{-1}) for oligoGs containing 2, 4, and 8 Ca^{2+}, respectively. The strength of the first cross-link depends on the number of subsequent cross-links, and inspection of the force–extension curves of the different oligomers show that the CD does not break one complex at a time but rather disrupts a sequence of four Ca^{2+} at once.

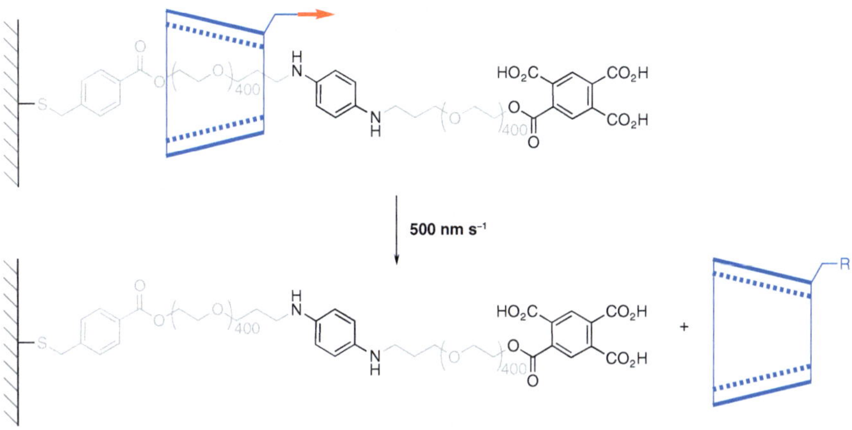

Figure 1.6 Single-molecule force spectroscopy experiment of the deslipping behavior of α-CD over a tricarboxylic acid benzoyl stopper on a PEG track. The red arrow indicates the direction of force.

The same group used a similar design measure the strength of a calcium-mediated eggbox junction (a cross-link between sequences of oligoguluronic acids [oligoGs]) [44]. After some equilibration time, an AFM tip was used to pick an α-CD to unzip the eggbox junction. They found that this macrocycle requires ~47 pN to slip over one guluronic acid, while unzipping requires 58, 112, and 141 pN (at 6–20 nN s^{-1}) for oligoGs containing 2, 4, and 8 Ca^{2+}, respectively. Inspection of the force-extension curves showed that the first cross-link is influenced by the subsequent cross links causing 4 Ca^{2+} units to break in a single event.

They also exploited the sliding and deslipping behavior of α-CD to map the position of specific groups along a polymer chain and called their approach "sliding contact force spectroscopy" [45].

1.2.3 Catenanes

In 2014, Duwez, Fustin, Leigh, and coworkers probed the mobility of the rings in a [2]catenane of the benzylic amide family [26]. They selected two catenanes that differ only by methylation of the amide groups and the corresponding ability to form inter-component hydrogen bonds (Figure 1.7). Catenane CatNH has secondary amide groups and can thus form hydrogen bonds between the two macrocycles, restricting their mobility. In catenane CatNMe, the amide groups are methylated, and the rings are relatively free to rotate. PEO chains were grafted at each side of the catenane for pulling with the AFM tip. The experiment consisted of performing approach-retraction cycles with the AFM tip on the molecules adsorbed on the substrate in both TCE and DMF. The persistence length of each molecule was extracted by fitting the force–extension curves with a worm-like chain model.

Catenane CatNMe, which is unable to form any internal hydrogen bonds, showed the same persistence length (L_p) in both solvents (0.45 nm). In contrast, catenane CatNH showed a large increase in L_p from DMF (0.5 nm) to TCE (1 nm), due to strong hydrogen bond interactions between the two macrocycles that transform the

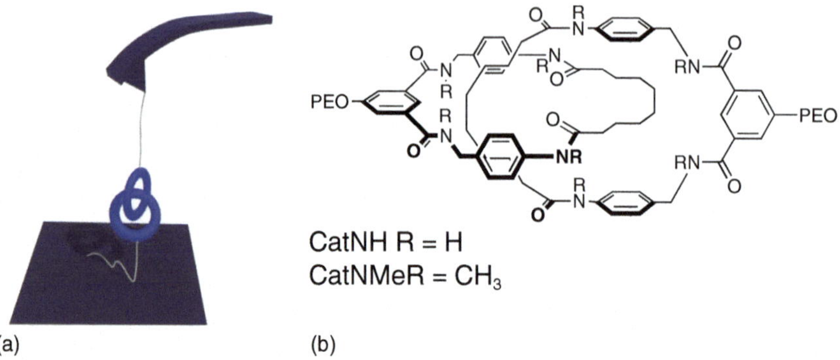

Figure 1.7 (a) Schematics of the single-molecule force spectroscopy experiment of the catenane. (b) Chemical structure of the [2]catenane of the benzylic amide family.

catenane in a long rigid segment. This result showed that SMFS is able to address the mobility of the rings of a single catenane unit, and demonstrates the impact that a single mechanical bond can have on the mechanical properties of polymers.

In 2019, Zhang, Huang, and coworkers investigated a poly[2]catenane made of the same catenane units of the benzylic amide family with inter-component hydrogen bonds (Figure 1.8) [22]. The catenane units are separated by short alkane tethers. When pulling on these polymers, they observed force–extension curves with a regular sawtooth pattern containing evenly spaced peaks. The contour

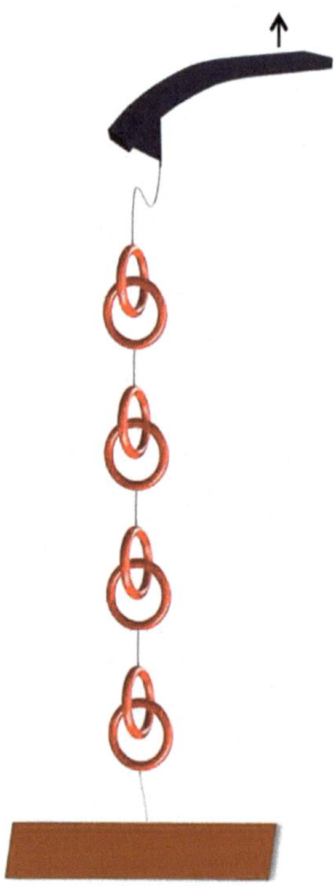

Figure 1.8 Schematics of the AFM-based single-molecule force spectroscopy experiment of a poly[2]catenane.

length increments, 1.68 ± 0.12 nm, were in good agreement with the length change expected (1.5 nm) after the rupture of the inter-component hydrogen bonds from the theoretical modeling of the structures. They performed the pulling experiments at eight different pulling rates and determined the characteristic force rupture for the inter-component hydrogen bonds. They could observe both the equilibrium (below ~1500 pN s^{-1}) and kinetic regimes (above ~1500 pN s^{-1}). Kinetic and thermodynamic parameters of the dissociation were extracted from the data with the Bell–Evans model [46]. The spontaneous unbinding rate in the absence of external force was estimated at about 0.07 s^{-1}, and the distance between the bound state and the transition state was estimated at about 2.82 Å. Since the near-equilibrium regime is probed, it would be interesting to re-analyze the data with the Friddle–Noy–De Yoreo model that consists of a nonlinear fit including both the near-equilibrium (loading rate-independent) and the kinetic regimes where the force is proportional to the logarithm of the loading rate [47]. Cycles of one pulling and one relaxing experiment were also realized and showed that the hydrogen bonds can reform between the macrocycles of the catenanes during the relaxing step. The rebinding force was determined at the different loading rates. The spontaneous rebinding rate was estimated at 3.6×10^5 s^{-1}, and the distance between the unbinding state and the transition state was estimated at 3.51 Å, using the Bell–Evans model.

Janshoff, Marszalek, and coworkers designed an oligomer system made of interlocked calix[4]arenes dimers [21]. The dimers are held together by 16 hydrogen bridges, and the interlocking limits how far the dimers can be separated after the rupture of the H bonds (Figure 1.9). The design prevents the bond rupture from being irreversible and thus enables the study of the rebinding of the calix[4]arenes dimers. Oligomers were prepared by tethering several of these dimers together. The force–extension curves obtained by pulling the molecule displayed a regular sawtooth pattern indicative of the successive rupture of the calixarene dimers.

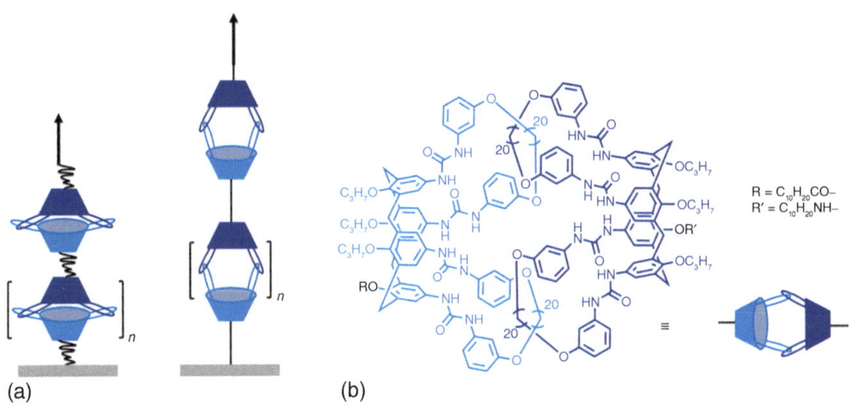

Figure 1.9 (a) Schematics of the AFM-based single-molecule force spectroscopy experiment of interlocked calix[4]arenes dimeric capsules. (b) Chemical structure of the capsules bound together by 16 hydrogen bonds.

The assignment of the peaks was unambiguous in light of their equidistance, which corresponds to the theoretical distance for the dimer separation. The strength and reversibility of the hydrogen bonds were measured at various loading rates, covering both the equilibrium regime and out-of-equilibrium regime, corresponding to pulling rates close to the spontaneous bond dissociation rate or much higher than this rate, respectively. The results, together with stochastic modeling using a three-well potential under external load, allowed the reconstruction of the energy landscape of the system.

1.3 Strength of Mechanical Bonds

In polymer mechanochemistry, polymer chains are used to activate mechanophores (force-sensitive molecules) in a way reminiscent of a tug of war. This can be achieved in solution with ultrasound-induced cavitation, where acoustic waves cause the nucleation, growth, and collapse of microbubbles. As bubbles collapse, a high gradient elongational flow is created in the surroundings, causing a nearby macromolecule to be stretched until covalent bond scission occurs in the central region of the chain [48]. This technique has been used to activate a variety of mechanophores [49], sometimes along unusual reaction pathways [50–52], prepare functional materials [53], carry out *in situ* catalysis [54, 55], or induce the release of small molecules [56], among others. Similarly, a mechanical bond at the center of a polymer chain will be subjected to a much higher strain rate than what can be achieved by SMFS, allowing for the dissociation pathways to be explored. These studies have revealed that dissociation via unclipping is unlikely since macrocycles can spread tensional stress around their circumference (Figure 1.1). As a result, rotaxanes break preferentially by unstoppering, as bonds in the axle are weakened, while catenanes act as a mechanochemical protecting group, as tensional stress accumulates in the linear segments outside the catenane (see below).

1.3.1 Polymers Containing a Rotaxane

The first investigation into the strength of the mechanical bond was carried out by Stoddart and coworkers in 2011, with the activation of polymers containing rotaxane mechanophores at the center by sonication [57]. In this technique, macromolecules are elongated by solvodynamic shear in the vicinity of collapsing ultrasound-induced cavitation bubbles [58]. The activation of polymers containing $CBPQT^{4+}$-derived [2]-rotaxanes (Figure 1.10) was compared to the linear and rotaxane end group polymers as controls. Diminution of the charge-transfer UV–vis absorption, associated with the interaction between the ring and the electron-rich DNP unit, was higher for the rotaxane-centered polymer compared to the controls as the applied force pulls the macrocycle away from the DNP unit. Although the authors provided no further details of the possible dissociation mechanism of the rotaxane, an earlier AFM study by the same group showed how a similar rotaxane was disassembled

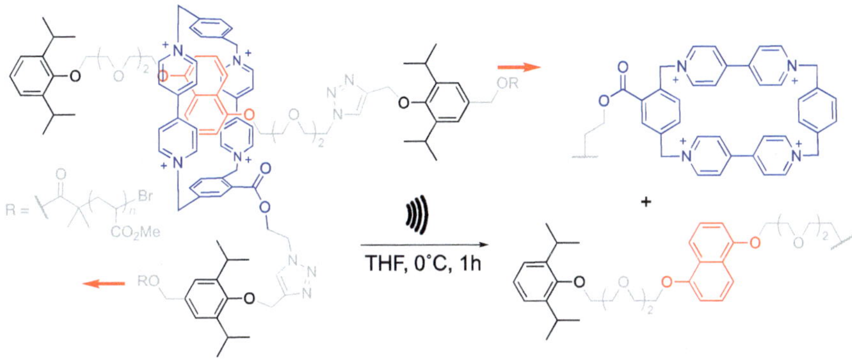

Figure 1.10 Mechanical activation of Stoddart's rotaxane in the center of a poly(methyl acrylate) backbone, red arrows indicate the direction of force, counter ions not shown for clarity.

by dethreading of a $CBPQT^{4+}$ macrocycle over the same diisopropylphenyl stopper at a force much lower than that needed for the cleavage of a covalent bond [31]. This is further supported by recently constrained geometries simulate external force (CoGEF) calculations, which predict the dethreading of the macrocycle at a force of 0.96 nN [59].

To shed light on how entanglement can affect the mechanical reactivity of a covalent bond, Zhang and De Bo studied the activation of a Diels–Alder (DA) mechanophore placed in the axle of a rotaxane [60], where a benzo-21-crown-7 (B21C7) macrocycle is stoppered on one side by a mechanically active *proximal-exo* DA adduct (i.e. an DA adduct of *exo* configuration where the polymer handle is close to the furan/maleimide junction) [61] and on the other side by a large stopper that prevents any dethreading even under tension (**1**, Figure 1.11). Upon elongation of the rotaxane, the macrocycle is pulled away from the ammonium binding station until it reaches the large stopper. As it is unable to cross the steric obstacle, tension increases in the macromolecule until the dissociation of the DA mechanophore.

The mechanical activation was performed using high-intensity ultrasound (20 kHz, 11.5 W cm^{-2}, 1 s ON/2 s OFF). The dissociation of the mechanophore and the generation of rotaxane **2** (the resulting maleimide is a stopper for B21C7) was confirmed by ^{1}H NMR. The mechanical activation of rotaxane **1** was slower than its force-uncoupled rotaxane **3** (in which the macrocycle is not covalently linked to the polymer backbone and hence not coupled to the force) and axle **4** counterparts. In other words, the mechanical activation of a DA mechanophore is slower when a rotaxane is used as a force actuator. The difference in reactivity originates from the constriction of the axle by the stretched macrocycle, which induces a significant amount of deformation in the chain and results in the accumulation of tensile, bending, and torsional stress in this region of the rotaxane. The presence of a competing high-stress region effectively decreases the mechanical coupling (i.e. how much the bond elongates as the mechanophore is stretched) of the scissile bond in the mechanophore, which in turn reduces the rate of activation. This region is unique to the rotaxane architecture, and the replacement of the *proximal-exo*

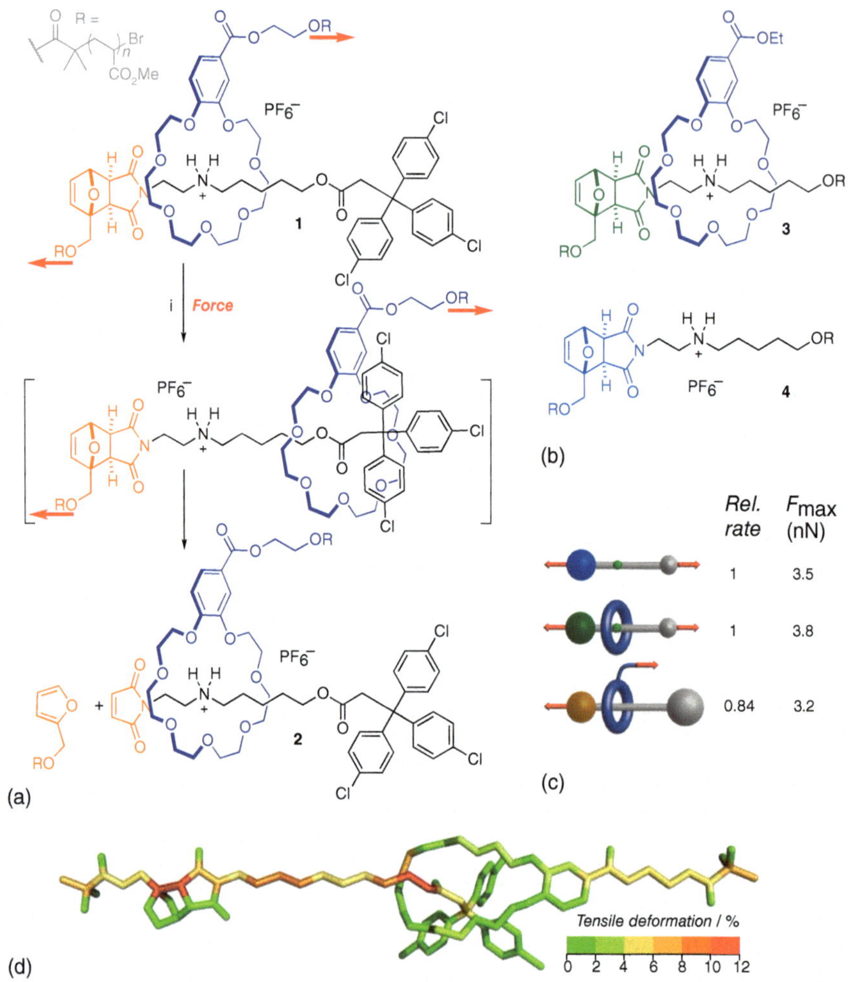

Figure 1.11 Rotaxane as a force actuator in a Diels–Alder reaction. (a) Mechanical activation of rotaxane **1**. (i) US (20 kHz, 13.0 W cm^{-2}, 1 s ON/2 s OFF) THF, 5–10 °C, 240 min. Red arrows indicate the direction of force. (b) Uncoupled rotaxane and linear mechanophore were used as controls. (c) Relative rates of dissociation and maximum force before cleavage of mechanophores. (d) Distribution of tensile stress in rotaxane mechanophore obtained from CoGEF calculations (DFT B3LYP/6-31G*).

DA mechanophore by a mechanically inert *distal-exo* DA adduct suggests the possibility of a selective cleavage of the terminal bulky stopper [62]. In this case, CoGEF calculations show the presence of a unique high-stress region centered on the axle-stopper junction (Figure 1.12b). Sonication of rotaxane **5** confirmed the predicted unstoppering dissociation pathway, as carboxylic acid **6** was isolated from the post-sonication mixture along with macrocycle- and amine-terminated polymers **7** and **8** respectively (as the linker joining the stopper to the ammonium station is eliminated in the process). These results suggest that the rotaxane cleaves via an unstoppering pathway, which is initiated by the homolytic scission of the

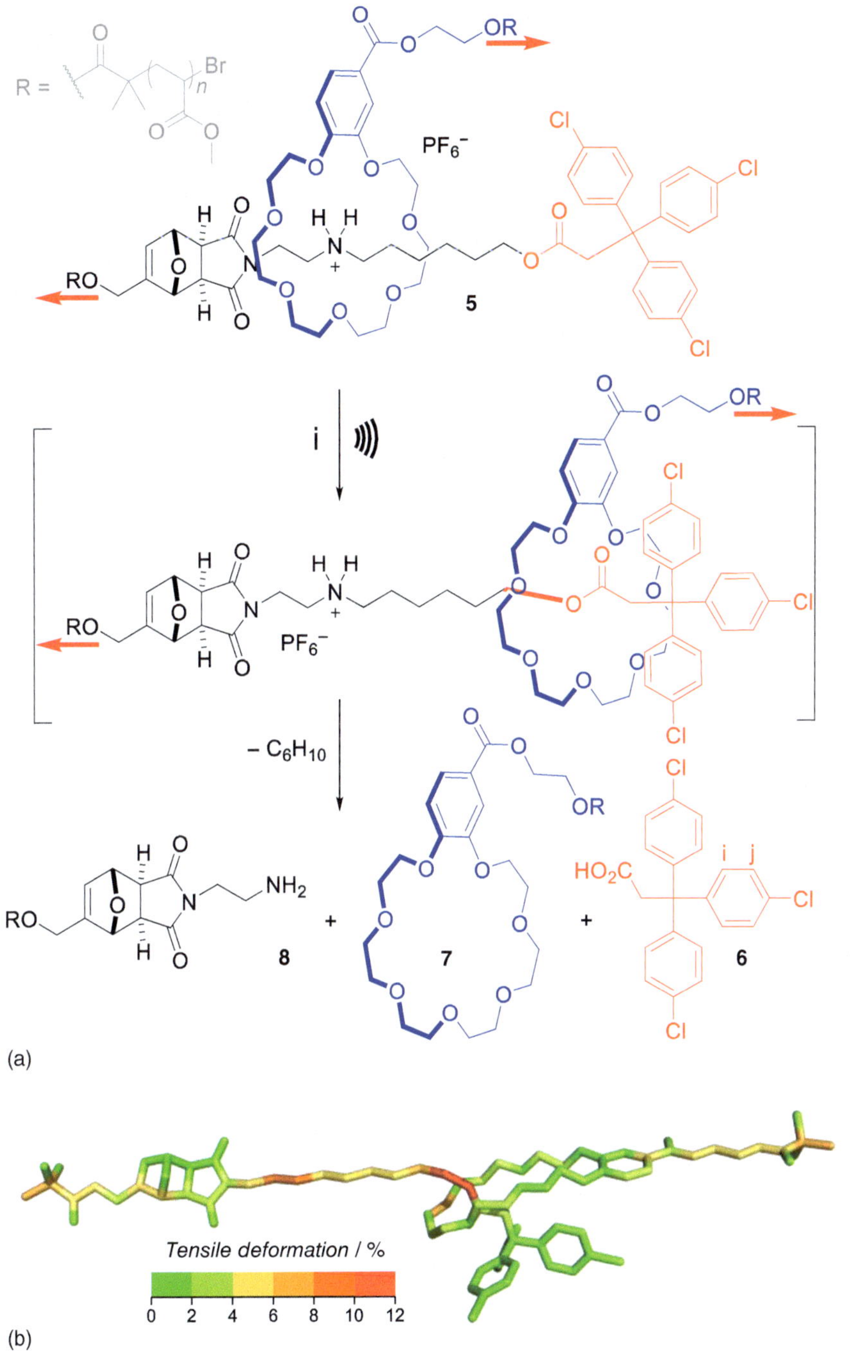

Figure 1.12 (a)Mechanical dissociation of rotaxane mechanophore **5** by an unstoppering process. (i) US (20 kHz, 11.5 W cm^{-2}, 1 s ON/2 s OFF), CH_3CN, 5–10 °C, 240 min. Red arrows indicate the direction of force. (b) Distribution of tensile stress in rotaxane mechanophore obtained from CoGEF calculations (DFT B3LYP/6-31G*).

C—O bond at the axle-stopper junction. This bond is only observed to cleave in rotaxane **5**, as its force-uncoupled rotaxane and axle counterparts cleave in the poly(methyl acrylate) backbone. This selectivity originates from the constriction of the axle by the stretched macrocycle, resulting in the accumulation of high tensile, bending, and torsional stress that active covalent bonds that are otherwise unreactive. In other words, a rotaxane actuator enhances the mechanical reactivity of the covalent bond in the axle. This finding has significant implications for the future development of slide-ring materials (a network of polyrotaxanes connected by their macrocycles) [63]. They display a remarkable ability to absorb mechanical energy at low extension due to "pulley effect" of their mobile cross-links, but this finding suggests that their rotaxane architecture could accelerate the failure of this material at maximal elongation (when the macrocyclic crosslinks are forced against the terminal stoppers of the polymer axles).

1.3.2 Polymers Containing a Catenane

As rotaxane actuators are able to induce a substantial amount of deformation in the axle (Figure 1.12b), it is noteworthy that the macrocycle itself is virtually free from

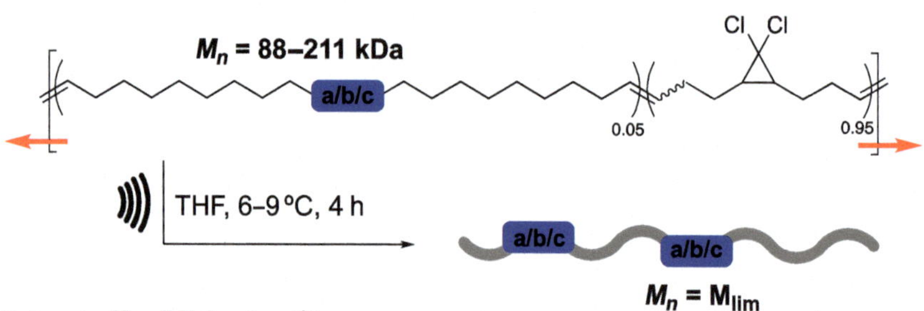

Polym.	M_{lim} (kDa)	L_{lim} (Å)
a	~30	~1400
b	~28	~1430
c	~29	~1560

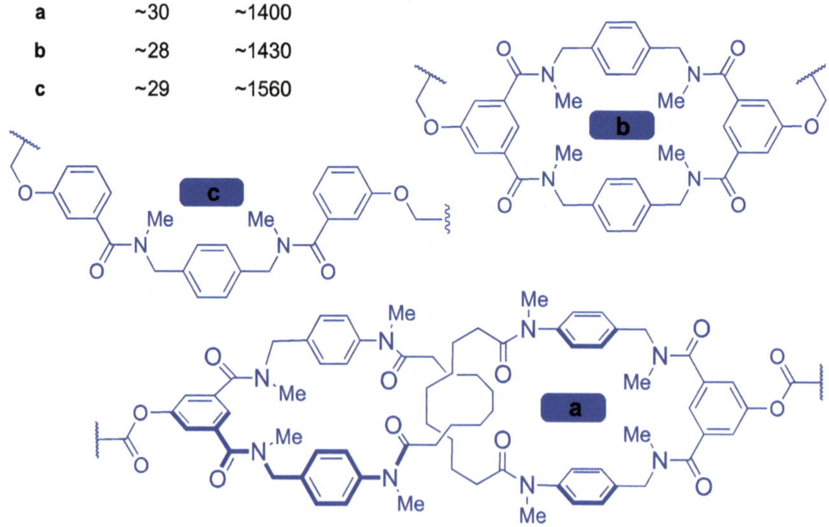

Figure 1.13 Sonication data for Craig' s polymers. Limiting mass and length of polymers containing a catenane (a), compared to macrocyclic (b), and linear (c) controls. Red arrows indicate the direction of force.

deformation. Indeed, because tension can spread over the entire circular backbone of macrocycles, they are less susceptible to elongational deformation than linear segments.

In an early study, Craig and coworkers have shown that copolymers containing [2]catenane junctions were as strong as their non-interlocked and linear counterparts. Multi-mechanophore gem-dichlorocyclopropanatedpolybutadiene copolymers with 5 mol% of either a benzamide [2]catenanes, or a macrocyclic or linear mimic were sonicated for a prolonged time in order to reach the limiting molecular weights (M_{lim}) below which mechanical activation does not occur [64]. All three polymers show similar limiting molecular weights (M_{lim}) and limiting lengths (L_{lim}), which confirms their similar mechanochemical reactivity (Figure 1.13).

The previous results suggest that bond scission occurs outside the topological features of these polymers (i.e. in the linear segments in between); however, the location and nature of the scissile bond were not investigated. This issue was tackled in a

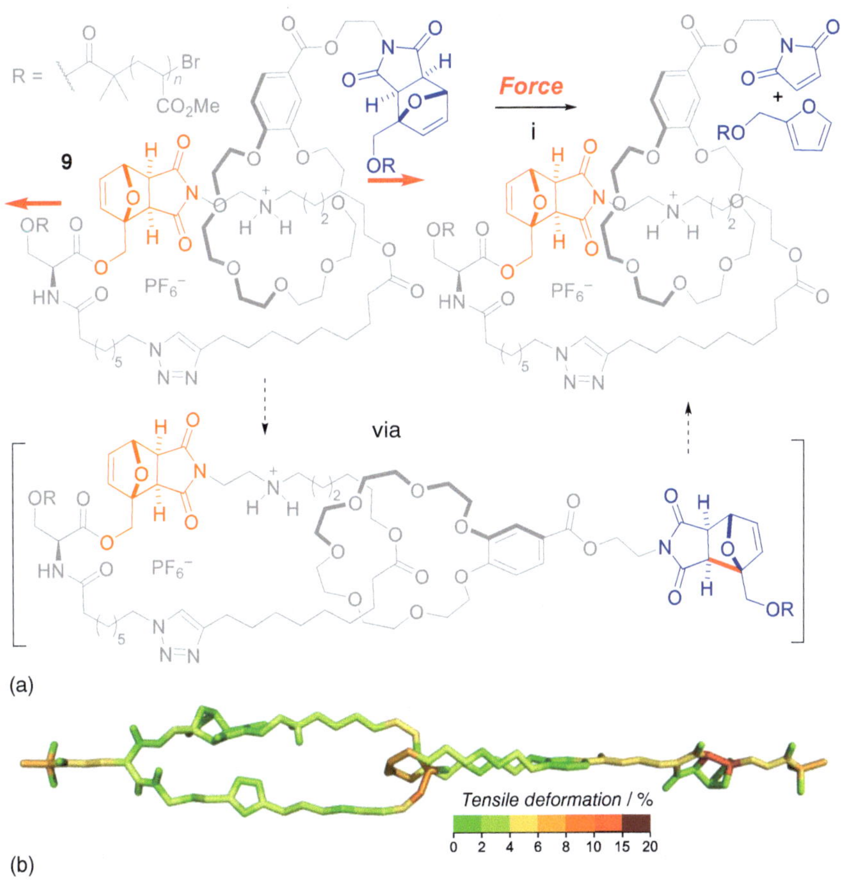

Figure 1.14 Mechanical activation of catenane mechanophore **9**. (i) US (20 kHz, 13.0 W cm^{-2}, 1 s ON/2 s OFF), CH_3CN, 5–10 °C, 240 min. Red arrows indicate the direction of force. (b) Distribution of tensile stress in catenane mechanophore obtained from CoGEF calculations (DFT B3LYP/6-31G*).

recent study by Zhang and De Bo, where they investigated the mechanical activation of a [2]catenane containing two identical mechanophores, one placed inside the main ring of the catenane and the other one in the linear segment joining the other ring to the actuating polymer (Figure 1.14a) [65]. The catenane is formed of a B21C7 macrocycle threaded onto a larger 39-membered ring macrocycle and contains two identical *proximal-exo* DA mechanophores. The *intracyclic* mechanophore is embedded in the main ring, while the *extracyclic* mechanophore separates the smaller B21C7 macrocycle from the actuating polymer. Sonication of catenane **9** leads to the sole activation of the *extracyclic* mechanophore. Indeed, as the catenane is stretched, the macrocycles will rotate along each other to a position where the tension is spread equally over the two rings. As a result, the *intracyclic* mechanophore experiences a much lower level of deformation than its *extracyclic* counterpart in the linear section of the construct (Figure 1.14b).

1.4 Changes in Optical Properties – Reversible and Irreversible Changes of Optical Properties by Movement of Macrocycle in a Rotaxane

External force can also be applied to polymers containing mechanical bonds in the solid state. This activation has previously been used in the development of irreversible force sensors as the cleavage of covalent bonds in mechanophores gives an optical response [66]. Polymers containing mechanical bonds provide access to reversible force probes. Gradually applying force in the solid state will first disrupt weak non-covalent interactions in the mechanophores, followed by bond dissociations as the force is increased or many cycles are applied. This behavior has resulted in the development of optical force sensors that act both reversibly and irreversibly and have the ability to visualize both small and large forces.

Slide ring gels contain polymer chains that are covalently cross-linked by a figure of eight cross-links. These cross-links are neither covalently linked to nor have physical interactions with the polymer chains, allowing for the links to pass over the chains, causing the dispersion of applied tension over the material. The slide ring effect causes these gels to have improved mechanical properties compared to polymers containing covalent cross-links [63].

In an effort to visualize the slide-ring effect, Takata, Otsuka, and coworkers assembled a slide-ring gel in which macrocycles are connected by a diarylbibenzofuranone (DABBF) [67] mechanochromophore [68]. As the gel is stretched, mechanical energy is dissipated via the sliding of the macrocycle (slide-ring transition). In this context, the DB24C8 macrocycle is pulled away from the secondary ammonium station over the C12 linker until it reaches the terminal 2,5-dimethylphenyl stopper. Further elongation would activate the DABBF mechanochromophore to produce a blue-colored arylbenzofuranone (ABF) stable radical, effectively revealing where a slide-ring transition has occurred (Figure 1.15a). Covalent (CCP) and rotaxane (RCP) cross-linked polymers were formed by reacting a methyl acrylate (MA)-based random copolymer modified with 3-isopropenyl-α,α-dimethylbenzyl isocyanate

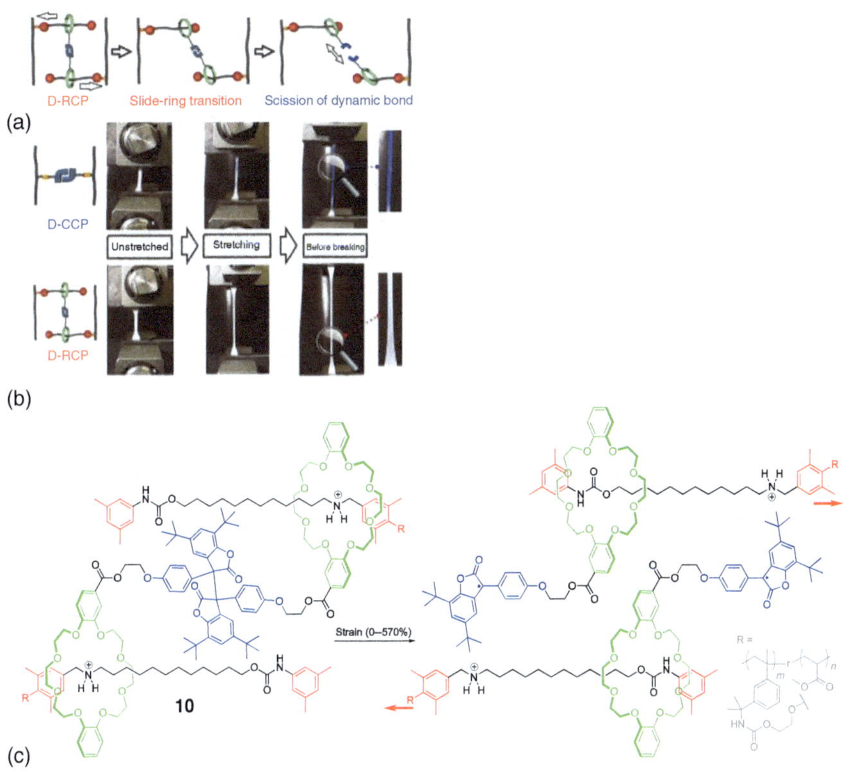

Figure 1.15 (a) Schematic illustration of the activation of a rotaxane crosslinked polymer **10**. (b) Photographs of polymers during tensile tests. (c) Activation of rotaxane crosslinked polymer at 570% elongation. Red arrows indicate the direction of force. Source: Lu et al. [68] / Royal Society Of Chemistry.

(TMI) with DABBF or DABBF-linked [3]rotaxane cross-linkers respectively. Both CP were subjected to tensile tests, and the mechanochromic behavior of DABBF was monitored visually while the formation of the ABF radical was quantified by electron paramagnetic resonance (EPR). Both CCP and RCP form white films when no strain is applied. At maximum strain (CCP: 490%, RCP: 570%), CCP color turns to a deep blue over the entire sample, unlike RCP, which only displays a faint blue domain in the neck of the sample (Figure 1.15b). This picture was confirmed by EPR, where the dissociation rate of RCP was found to be about four times lower than that of CCP. The low degree of activation in the rotaxane cross-linked system was attributed to the slide-ring transition itself (i.e. the mechanical energy is dissipated before the DABBF can be activated) and to a faster recombination of ABF radical in RCP than CCP due to the higher mobility of the rotaxane junction. However, a dissociation of the rotaxane by dethreading cannot be excluded as the 2,5-dimethylphenyl stopper can be easily crossed by the macrocycle under tension [69].

Stoddart' s earliest example of the mechanochemistry of rotaxane-centered polymers involved a change of optical properties due to exposure of a DNP unit after disruption of CT interaction (Figure 1.10) [57]. However, in this work, the change in optical properties was used only to determine the mechanical change depending on whether the polymer is still intact. More recently, groups have been able to develop mechanical bonds containing mechanochromic devices to give an optical response to a force. Mechanochromic mechanophores have been previously reported [70], but they usually rely on the rupture of a covalent bond and hence require high activation force (nN regime). The ability to sense lower force is particularly desirable to detect mechanobiological processes, which typically occur in the low pN regime [71]. This force regime is the realm of non-covalent interactions, and a rotaxane is an ideal platform to build such a sensor as its subcomponents are often held in place with non-covalent interactions. This strategy was recently used by Sagara and Weder to design a mechanochromic rotaxane in which a fluorophore-bearing macrocycle is wrapped around an electron-poor aromatic quencher [72]. The rotaxane consists of a 1,5-dinaptho[38]crown-10 macrocycle modified with a 4,7-bis(phenylethynyl)-2,1,3-benzothiadiazole (BTH) fluorophore around an axle containing a 1,4,5,8-napthalenetetracarbocyclic diimide (NpI) quencher. As tension is applied, the macrocycle moves away from the quencher and green emission is observed; the emission then immediately disappears as the film is stretched and released, and the macrocycle can re-locate to the quencher (Figure 1.16a,b).

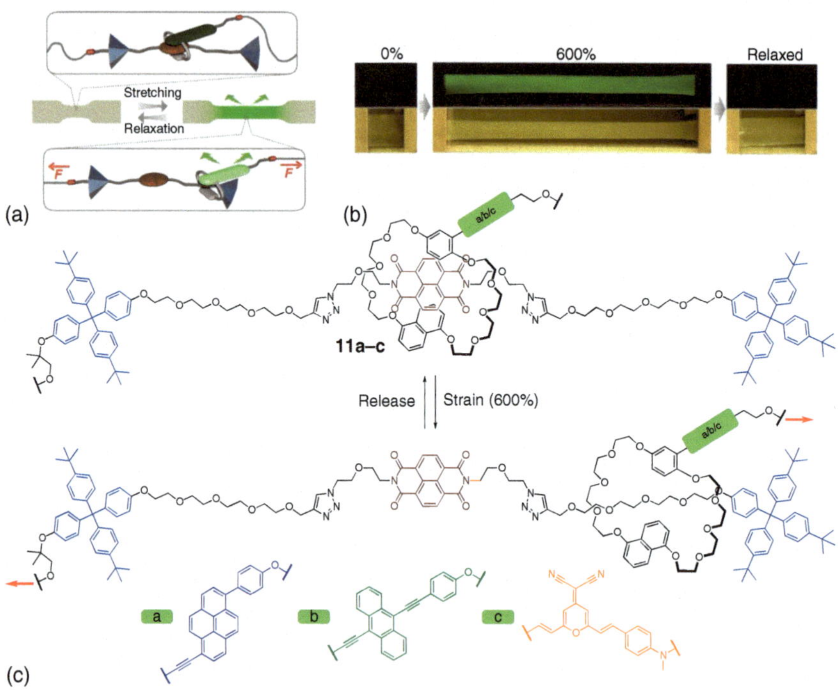

Figure 1.16 (a) Schematic illustration of reversible rotaxane mechanophores **11a–c**. (b) Photographs of mechanophore-containing thin film during elongation. (c) Activation of rotaxane mechanophores at 600% elongation. Red arrows indicate the direction of force. Source: Sagara et al. [72] / American Chemical Society.

Changing the luminophore part gives access to blue (π-extended pyrene), green (π-extended anthracene), and orange (π-extended 4-(dicyanomethylene)-2methyl-6-(4-dimethylaminostyryl)-4*H*-pyran) photoluminescence (Figure 1.16c). A white-light emission elastomer can be built by combining **11a–c** rotaxanes in the same elastomer [73].

These systems can be activated at low force in a reversible fashion, as they only require the rupture of non-covalent interactions between the macrocycle and its station. However, this design does not enable the fine-tuning of the activation force as the optical and mechanical responses are coupled. In addition, some applications would benefit from an irreversible activation. For example, when the stress history of the material is important (e.g. to indicate when a material has experienced a certain deformation). To address this challenge, Sagara, Weder, and coworkers reported a rotaxane-based mechanophore that can display both reversible and irreversible responses depending on the magnitude of the applied force (Figure 1.17a) [74]. At low tension, the macrocycle is pulled away from the quencher, but it is still retained

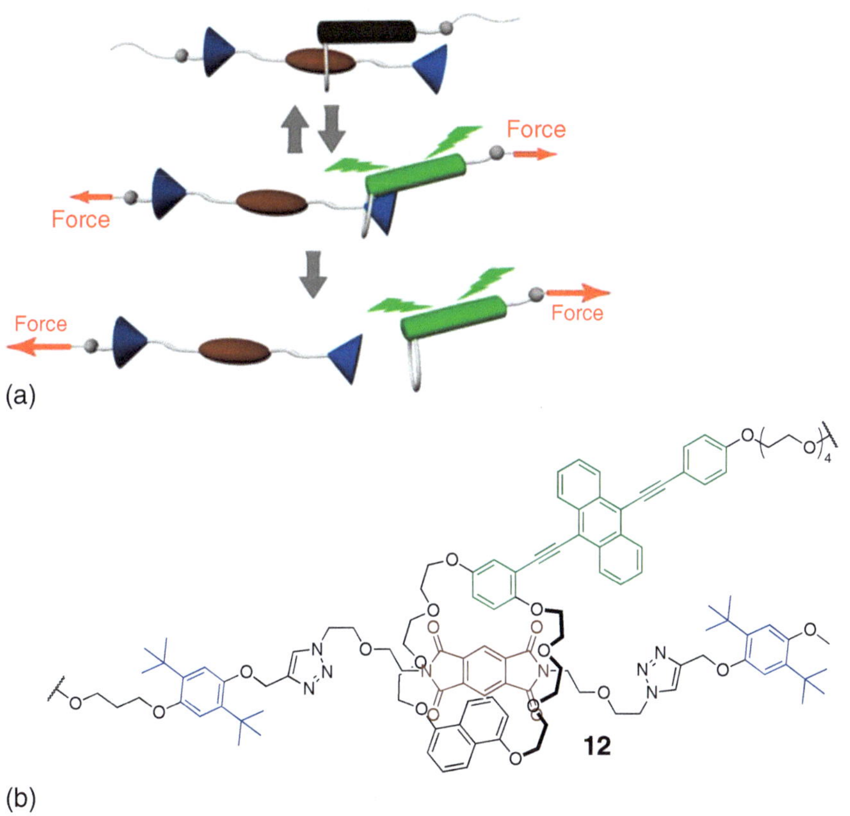

Figure 1.17 (a) Schematic illustration of reversible and irreversible rotaxane mechanophores; red arrows indicate the direction of force. (b) Example structure of dual response rotaxane mechanophore **12**. Source: Muramatsu et al. [74] / American Chemical Society.

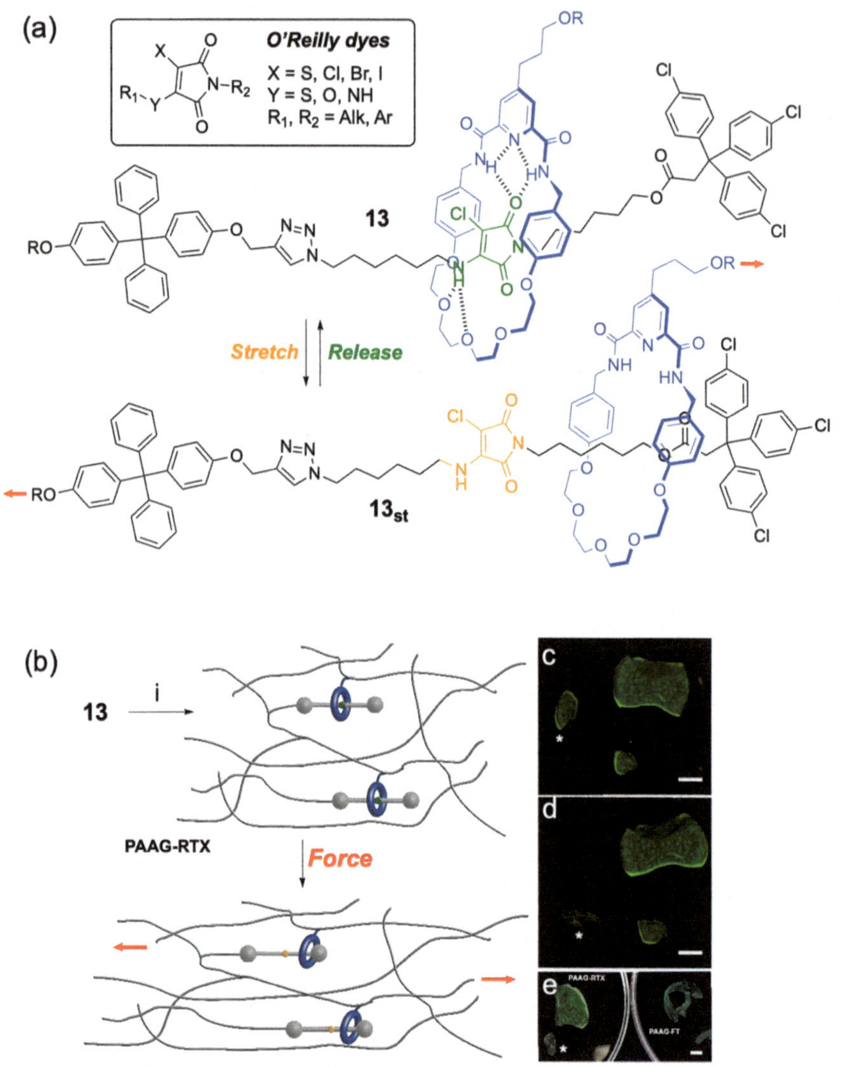

Figure 1.18 (a) Activation of a mechanochromic hydrogen-bonded rotaxane **13** containing ACM O'Reilly dye after compression. (b) Schematic illustration of force applied to polyacrylamide gel containing rotaxane mechanophore as cross-linker, red arrows indicates the direction of force. Conditions (i) acrylamide (12% w/v), bis-acrylamide (0.3% w/v), rotaxane 1 (0.3% w/v), *N,N,N′,N′*-tetramethylethylenediamine, $(NH_4)_2S_2O_8$, THF/H_2O (1 : 1). (c) Fragments of rotaxane crosslinked gel under 254 nm UV light. (d) The same fragments after compression of * with tweezers for 5 s. (e) comparison of compressed * and native PAAG-RTX fragments with PAAG-FT. Scale bar = 5 mm. Source: Sandoval-Torrientes et al. [75] / John Wiley & Sons / CC BY 4.0.

by the stopper, and releasing the tension restores the initial state. Beyond a certain force threshold, the macrocycle can be pulled over the stopper, and the activation is irreversible. This feat is achieved with a smaller stopper (*p*-di-*tert*-butylphenyl) than in the previous design (Figure 1.17b). Consequently, the size of the macrocycle has been reduced, and a smaller pyromellitic diimide quenching unit is used. Reversible activation of thin film of polyurethane elastomers incorporating the rotaxane (0.05 mol%) was achieved over 50 strain/release cycles (300% maximal strain). However, when the strain was increased to 600%, the emission intensity in the relaxed state increased gradually with the number of cycles and the applied strain. The effect was more prominent at higher strain, which suggests an increasing amount of irreversible activation by dethreading. This study shows how a small change in the architecture of a rotaxane can result in both reversible and irreversible responses to force. Notably, the rotaxanes that become dethreaded over time can allow for the recording of the mechanical history of a material undergoing repeated stain as an optical signal.

In 2021, De Bo and coworkers were able to introduce a rotaxane force sensor into a polyacrylamide gel (PAAG), which can serve as a model for living tissue [75]. The rotaxane **13** is centered around an aminochloromaleimide (ACM) derivative (O'Reilly dye) [76], which acts as both a fluorescent dye and a recognition motif for the macrocycle (Figure 1.18a). These dyes exhibit a strong emission due to their donor–acceptor architecture, which is quenched in the presence of a strong hydrogen bond donor [77]. Additionally, their optical properties can be tuned by altering the substitution pattern [78]. It was hypothesized that the macrocycle, containing both a hydrogen bond donor unit (pyridyl-2,6-dicarboxyamide) and a hydrogen bond acceptor crown ether-like region, would be a suitable host for the ACM unit and that the fluorescence of ACM would be altered due to the presence of two electron-rich aromatic flanking groups and the hydrogen bond network (Figure 1.18a). The rotaxane was incorporated as a cross-linker in a PAAG, a popular matrix model to mimic living tissues due to their similar viscoelastic properties (Figure 1.18b). The green fluorescence of the resulting PAAG-RTX gel is greatly reduced when the gel is compressed (Figure 1.18c–e); however, the activation proved irreversible, probably due to stronger hydrogen bonds between polar groups in the gel and components of the rotaxane, than between the components themselves (i.e. between the dye and the macrocycle). Though a reversible activation was not achieved, this system offers a versatile platform for the assembly of a force sensor for *in vivo* applications.

1.5 Conclusions

As the number and complexity of mechanical bonds expand, new force-responsive transformations will become available. SMFS and ultrasound activation can work hand-in-hand to uncover new and interesting dissociation pathways and the real-time events leading to bond cleavage. The different ways in which rotaxanes and catenanes direct force will have important applications in the design of future materials containing these architectures, which will be subjected to high force, either accelerating the cleavage or shielding specific sites from force. The way that component parts can be pulled either toward or away from each other can be exploited to unlock new reactivity within the mechanical bond structure. Continuous development of force-responsive materials could result in increased responses to multiple levels of force, allowing for materials to display more detailed information on the stress applied and offering a wider range of potential applications in different settings.

References

1 Bruns, C.J. and Stoddart, J.F. (2016). *The Nature of the Mechanical Bond*. Hoboken, NJ: Wiley.

2 De Bo, G. (2018). Mechanochemistry of the mechanical bond. *Chem. Sci.* 9 (1): 15–21.

3 Hinterdorfer, P. and Dufrêne, Y.F. (2006). Detection and localization of single molecular recognition events using atomic force microscopy. *Nat. Methods* 3 (5): 347–355.

4 Neuman, K.C. and Nagy, A. (2008). Single-molecule force spectroscopy: optical tweezers, magnetic tweezers and atomic force microscopy. *Nat. Methods* 5 (6): 491–505.

5 Mora, M., Stannard, A., and Garcia-Manyes, S. (2020). The nanomechanics of individual proteins. *Chem. Soc. Rev.* 49 (19): 6816–6832.

6 Bustamante, C., Chemla, Y.R., Forde, N.R., and Izhaky, D. (2004). Mechanical Processes in Biochemistry. *Annu. Rev. Biochem.* 73 (1): 705–748.

7 Bustamante, C. et al. (2008). Single-molecule theme. *Annu. Rev. Biochem.* 77 (1): 45–228.

8 Müller, D.J. and Dufrêne, Y.F. (2008). Atomic force microscopy as a multifunctional molecular toolbox in nanobiotechnology. *Nat. Nanotechnol.* 3 (5): 261–269.

9 Puchner, E.M. and Gaub, H.E. (2009). Force and function: probing proteins with AFM-based force spectroscopy. *Curr. Opin. Struct. Biol.* 19 (5): 605–614.

10 Liang, J. and Fernández, J.M. (2009). Mechanochemistry: one bond at a time. *ACS Nano* 3 (7): 1628–1645.

11 Duwez, A.-S. and Willet, N. (ed.) (2011). *Molecular Manipulation with Atomic Force Microscopy*. CRC Press.

12 Marszalek, P.E. and Dufrêne, Y.F. (2012). Stretching single polysaccharides and proteins using atomic force microscopy. *Chem. Soc. Rev.* 41 (9): 3523.

13 Žoldák, G. and Rief, M. (2013). Force as a single molecule probe of multidimensional protein energy landscapes. *Curr. Opin. Struct. Biol.* 23 (1): 48–57.

14 Hughes, M.L. and Dougan, L. (2016). The physics of pulling polyproteins: a review of single molecule force spectroscopy using the AFM to study protein unfolding. *Rep. Prog. Phys.* 79 (7): 076601.

15 Nathwani, B., Shih, W.M., and Wong, W.P. (2018). Force spectroscopy and beyond: innovations and opportunities. *Biophys. J.* 115 (12): 2279–2285.

16 Bao, Y., Luo, Z., and Cui, S. (2020). Environment-dependent single-chain mechanics of synthetic polymers and biomacromolecules by atomic force microscopy-based single-molecule force spectroscopy and the implications for advanced polymer materials. *Chem. Soc. Rev.* 49 (9): 2799–2827.

17 Brown, C.L. and Craig, S.L. (2015). Molecular engineering of mechanophore activity for stress-responsive polymeric materials. *Chem. Sci.* 6 (4): 2158–2165.

18 Ghanem, M.A., Basu, A., Behrou, R. et al. (2021). The role of polymer mechanochemistry in responsive materials and additive manufacturing. *Nat. Rev. Mater.* 6 (1): 84–98.

19 Bowser, B.H., Wang, S., Kouznetsova, T.B. et al. (2021). Single-event spectroscopy and unravelling kinetics of covalent domains based on cyclobutane mechanophores. *J. Am. Chem. Soc.* 143 (13): 5269–5276.

20 Zhang, Y., Wang, Z., Kouznetsova, T.B. et al. (2021). Distal conformational locks on ferrocene mechanophores guide reaction pathways for increased mechanochemical reactivity. *Nat. Chem.* 13 (1): 56–62.

21 Janke, M., Rudzevich, Y., Molokanova, O. et al. (2009). Mechanically interlocked calix[4]arene dimers display reversible bond breakage under force. *Nat. Nanotechnol.* 4 (4): 225–229.

22 Xing, H., Li, Z., Wang, W. et al. (2020). Mechanochemistry of an interlocked poly[2]catenane: from single molecule to bulk gel. *CCS Chem.* 2 (1): 513–523.

23 Sluysmans, D., Zhang, L., Li, X. et al. (2020). Viologen tweezers to probe the force of individual donor–acceptor π-interactions. *J. Am. Chem. Soc.* 142 (50): 21153–21159.

24 Devaux, F., Li, X., Sluysmans, D. et al. (2021). Single-molecule mechanics of synthetic aromatic amide helices: ultrafast and robust non-dissipative winding. *Chem* 7 (5): 1333–1346.

25 Lussis, P., Svaldo-Lanero, T., Bertocco, A. et al. (2011). A single synthetic small molecule that generates force against a load. *Nat. Nanotechnol.* 6 (9): 553–557.

26 Van Quaethem, A., Lussis, P., Leigh, D.A. et al. (2014). Probing the mobility of catenane rings in single molecules. *Chem. Sci.* 5 (4): 1449.

27 Naranjo, T., Lemishko, K.M., de Lorenzo, S. et al. (2018). Dynamics of individual molecular shuttles under mechanical force. *Nat. Commun.* 9 (1): 4512.

28 Sluysmans, D., Hubert, S., Bruns, C.J. et al. (2018). Synthetic oligorotaxanes exert high forces when folding under mechanical load. *Nat. Nanotechnol.* 13 (3): 209–213.

29 Sluysmans, D., Devaux, F., Bruns, C.J. et al. (2018). Dynamic force spectroscopy of synthetic oligorotaxane foldamers. *Proc. Natl. Acad. Sci.* 115 (38): 9362–9366.

30 Sluysmans, D., Lussis, P., Fustin, C.-A. et al. (2021). Real-time fluctuations in single-molecule rotaxane experiments reveal an intermediate weak binding state during shuttling. *J. Am. Chem. Soc.* 143 (5): 2348–2352.

31 Brough, B., Northrop, B.H., Schmidt, J.J. et al. (2006). Evaluation of synthetic linear motor-molecule actuation energetics. *Proc. Natl. Acad. Sci.* 103 (23): 8583–8588.

32 Altieri, A., Bottari, G., Dehez, F. et al. (2003). Remarkable positional discrimination in bistable light- and heat-switchable hydrogen-bonded molecular shuttles. *Angew. Chem. Int. Ed.* 42 (20): 2296–2300.

33 Ashton, P.R., Baxter, I., Fyfe, M.C.T. et al. (1998). Rotaxane or pseudorotaxane? That is the question! *J. Am. Chem. Soc.* 120 (10): 2297–2307.

34 Affeld, A., Hübner, G.M., Seel, C., and Schalley, C.A. (2001). Rotaxane or pseudorotaxane? Effects of small structural variations on the deslipping kinetics of rotaxanes with stopper groups of intermediate size. *Eur. J. Org. Chem.* 2001 (15): 2877.

35 McGonigal, P.R., Li, H., Cheng, C. et al. (2015). Controlling association kinetics in the formation of donor–acceptor pseudorotaxanes. *Tetrahedron Lett.* 56 (23): 3591–3594.

36 Gómez-Durán, C.F.A., Liu, W., de Lourdes Betancourt-Mendiola, M., and Smith, B.D. (2017). Structural control of kinetics for macrocycle threading by fluorescent squaraine dye in water. *J. Org. Chem.* 82 (16): 8334–8341.

37 Martinez-Cuezva, A., Rodrigues, L.V., Navarro, C. et al. (2015). Dethreading of tetraalkylsuccinamide-based [2]rotaxanes for preparing benzylic amide macrocycles. *J. Org. Chem.* 80 (20): 10049–10059.

38 Saito, S., Takahashi, E., Wakatsuki, K. et al. (2013). Synthesis of large [2]rotaxanes. The relationship between the size of the blocking group and the stability of the rotaxane. *J. Org. Chem.* 78 (8): 3553–3560.

39 Clifford, T., Abushamleh, A., and Busch, D.H. (2002). Factors affecting the threading of axle molecules through macrocycles: binding constants for semirotaxane formation. *Proc. Natl. Acad. Sci.* 99 (8): 4830–4836.

40 Heim, C., Affeld, A., Nieger, M., and Vögtle, F. (1999). Size complementarity of macrocyclic cavities and stoppers in amide-rotaxanes. *Helv. Chim. Acta* 82 (5): 746–759.

41 Raymo, F.M., Houk, K.N., and Stoddart, J.F. (1998). The mechanism of the slippage approach to rotaxanes. Origin of the "all-or-nothing" substituent effect. *J. Am. Chem. Soc.* 120 (36): 9318–9322.

42 Wenz, G., Han, B.-H., and Müller, A. (2006). Cyclodextrin rotaxanes and polyrotaxanes. *Chem. Rev.* 106 (3): 782–817.

43 Dunlop, A., Wattoom, J., Hasan, E.A. et al. (2008). Mapping the positions of beads on a string: dethreading rotaxanes by molecular force spectroscopy. *Nanotechnology* 19 (34): 345706.

44 Bowman, K.A., Aarstad, O.A., Stokke, B.T. et al. (2016). Sliding contact dynamic force spectroscopy method for interrogating slowly forming polymer cross-links. *Langmuir* 32 (48): 12814–12822.

45 Dunlop, A., Bowman, K., Aarstad, O. et al. (2017). Polymer sequencing by molecular machines: a framework for predicting the resolving power of a sliding contact force spectroscopy sequencing method. *Nanoscale* 9 (39): 15089–15097.

46 Evans, E. and Ritchie, K. (1997). Dynamic strength of molecular adhesion bonds. *Biophys. J.* 72 (4): 1541–1555.

47 Friddle, R.W., Noy, A., and De Yoreo, J.J. (2012). Interpreting the widespread nonlinear force spectra of intermolecular bonds. *Proc. Natl. Acad. Sci.* 109 (34): 13573–13578.

48 Li, J., Nagamani, C., and Moore, J.S. (2015). Polymer mechanochemistry: from destructive to productive. *Acc. Chem. Res.* 48 (8): 2181–2190.

49 Klein, I.M., Husic, C.C., Kovács, D.P. et al. (2020). Validation of the CoGEF method as a predictive tool for polymer mechanochemistry. *J. Am. Chem. Soc.* 142 (38): 16364–16381.

50 Nixon, R. and De Bo, G. (2020). Three concomitant C–C dissociation pathways during the mechanical activation of an *N*-heterocyclic carbene precursor. *Nat. Chem.* 12 (9): 826–831.

51 Liu, Y., Holm, S., Meisner, J. et al. (2021). Flyby reaction trajectories: chemical dynamics under extrinsic force. *Science* 373 (6551): 208–212.

52 Lenhardt, J.M., Ong, M.T., Choe, R. et al. (2010). Trapping a diradical transition state by mechanochemical polymer extension. *Science* 329 (5995): 1057–1060.

53 Chen, Z., Mercer, J.A.M., Zhu, X. et al. (2017). Mechanochemical unzipping of insulating polyladderene to semiconducting polyacetylene. *Science* 357 (6350): 475–479.

54 Piermattei, A., Karthikeyan, S., and Sijbesma, R.P. (2009). Activating catalysts with mechanical force. *Nat. Chem.* 1 (2): 133–137.

55 Michael, P. and Binder, W.H. (2015). A mechanochemically triggered "click" catalyst. *Angew. Chem. Int. Ed.* 54 (47): 13918–13922.

56 Küng, R., Göstl, R., and Schmidt, B.M. (2022). Release of molecular cargo from polymer systems by mechanochemistry. *Chem. Eur. J.* 28 (17): e202103860.

57 Stoll, R.S., Friedman, D.C., and Stoddart, J.F. (2011). Mechanically interlocked mechanophores by living-radical polymerization from rotaxane initiators. *Org. Lett.* 13 (10): 2706–2709.

58 May, P.A. and Moore, J.S. (2013). Polymer mechanochemistry: techniques to generate molecular force via elongational flows. *Chem. Soc. Rev.* 42 (18): 7497.

59 De Bo, G. (2022). Mechanochemical dethreading of a cyclobis(paraquat-*p*-phenylene)-derived [2]rotaxane. figshare. Figure. https://doi.org/10.6084/m9.figshare.20153708.v1.

60 Zhang, M. and De Bo, G. (2018). Impact of a mechanical bond on the activation of a mechanophore. *J. Am. Chem. Soc.* 140 (40): 12724–12727.

61 Stevenson, R. and De Bo, G. (2017). Controlling reactivity by geometry in retro-Diels–Alder reactions under tension. *J. Am. Chem. Soc.* 139 (46): 16768–16771.

62 Zhang, M. and De Bo, G. (2019). Mechanical susceptibility of a rotaxane. *J. Am. Chem. Soc.* 141 (40): 15879–15883.

63 Mayumi, K., Ito, K., and Kato, K. (2015). *Polyrotaxane and Slide-Ring Materials.* Cambridge: Royal Society of Chemistry.

64 Lee, B., Niu, Z., and Craig, S.L. (2016). The mechanical strength of a mechanical bond: sonochemical polymer mechanochemistry of poly(catenane) copolymers. *Angew. Chem. Int. Ed.* 55 (42): 13086–13089.

65 Zhang, M. and De Bo, G. (2020). A catenane as a mechanical protecting group. *J. Am. Chem. Soc.* 142 (11): 5029–5033.

66 Davis, D.A., Hamilton, A., Yang, J. et al. (2009). Force-induced activation of covalent bonds in mechanoresponsive polymeric materials. *Nature* 459 (7243): 68–72.

67 Imato, K., Irie, A., Kosuge, T. et al. (2015). Mechanophores with a reversible radical system and freezing-induced mechanochemistry in polymer solutions and gels. *Angew. Chem. Int. Ed.* 54 (21): 6168–6172.

68 Lu, Y., Aoki, D., Sawada, J. et al. (2020). Visualization of the slide-ring effect: a study on movable cross-linking points using mechanochromism. *Chem. Commun.* 56 (23): 3361–3364.

69 De Bo, G. (2022). Mechanochemical dethreading of a dibenzo-24-crown-8 (DB24C8) macrocycle connected to a diarylbibenzofuranone (DABBF) mechanochromophore. figshare. Figure. https://doi.org/10.6084/m9.figshare.20340873.v1.

70 Chen, Y., Mellot, G., van Luijk, D. et al. (2021). Mechanochemical tools for polymer materials. *Chem. Soc. Rev.* 50 (6): 4100–4140.

71 Roca-Cusachs, P., Conte, V., and Trepat, X. (2017). Quantifying forces in cell biology. *Nat. Cell Biol.* 19 (7): 742–751.

72 Sagara, Y., Karman, M., Verde-Sesto, E. et al. (2018). Rotaxanes as mechanochromic fluorescent force transducers in polymers. *J. Am. Chem. Soc.* 140 (5): 1584–1587.

73 Sagara, Y., Karman, M., Seki, A. et al. (2019). Rotaxane-based mechanophores enable polymers with mechanically switchable white photoluminescence. *ACS Cent. Sci.* 5 (5): 874–881.

74 Muramatsu, T., Okado, Y., Traeger, H. et al. (2021). Rotaxane-based dual function mechanophores exhibiting reversible and irreversible responses. *J. Am. Chem. Soc.* 143 (26): 9884–9892.

75 Sandoval-Torrientes, R., Carr, T., and De Bo, G. (2021). A mechanochromic hydrogen-bonded rotaxane. *Macromol. Rapid Commun.* 42 (1): 2000447.

76 Robin, M.P., Wilson, P., Mabire, A.B. et al. (2013). Conjugation-induced fluorescent labeling of proteins and polymers using dithiomaleimides. *J. Am. Chem. Soc.* 135 (8): 2875–2878.

77 Mabire, A.B., Robin, M.P., Quan, W.-D. et al. (2015). Aminomaleimide fluorophores: a simple functional group with bright, solvent dependent emission. *Chem. Commun.* 51 (47): 9733–9736.

78 Xie, Y., Husband, J.T., Torrent-Sucarrat, M. et al. (2018). Rational design of substituted maleimide dyes with tunable fluorescence and solvafluorochromism. *Chem. Commun.* 54 (27): 3339–3342.

2

Colloidal Nanomaterials with Mechanically Interlocked Parts

Euan R. Kay

University of St Andrews, EaStCHEM School of Chemistry, Purdie Building, North Haugh, St Andrews, KY16 9ST, UK

2.1 Introduction

Materials with nanoscale dimensions possess unique size-dependent properties, exhibit emergent collective behaviors, and present myriad possibilities for composites through integration with other components. Colloidal nanoparticles (NPs), which can be prepared, analyzed, and processed using familiar wet chemistry techniques and infrastructure, provide a particularly attractive entry point to the nanoscale world for synthetic chemists. Composed of a core material surrounded by peripheral stabilizing structures, these multicomponent species span a wide spectrum of core types (including carbonaceous, metallic, semiconductor, or (macro)molecular), surface functionality (organic or inorganic), particle sizes, and shapes. State-of-the-art synthetic methods now achieve rational control over each of these parameters, giving access to a vast array of nanoscale entities and, consequently, a diverse library of size-tunable properties.

The development of colloidal nanomaterials, in many respects, mirrors that of mechanically interlocked molecules (MIMs) over similar time periods. Transformational synthetic strategies, allied with critical developments in characterization methods, have opened up these two entirely new regions of chemical space. The early explorers encountered new properties, discovered new phenomena, and amassed new knowledge. Having now established the synthetic access routes and analytical pathways for these new territories and begun to map their conceptual and phenomenological landscapes, reconnoitering the borderland where these two cutting-edge research zones meet now represents a realistic and productive endeavor; it is from these frontiers that new technologies will ultimately emerge.

Integrating structurally and functionally sophisticated MIMs with nanomaterials poses significant synthetic and analytical challenges but presents attractive prospects for augmenting nanoscale properties with dynamic, switchable and adaptive molecule-controlled behaviors. Colloidal NPs have a further valuable characteristic in that they can be amenable to analytical tools capable of revealing

Mechanically Interlocked Materials: Polymers, Nanomaterials, MOFs, and more, First Edition.
Edited by Emilio M. Pérez.

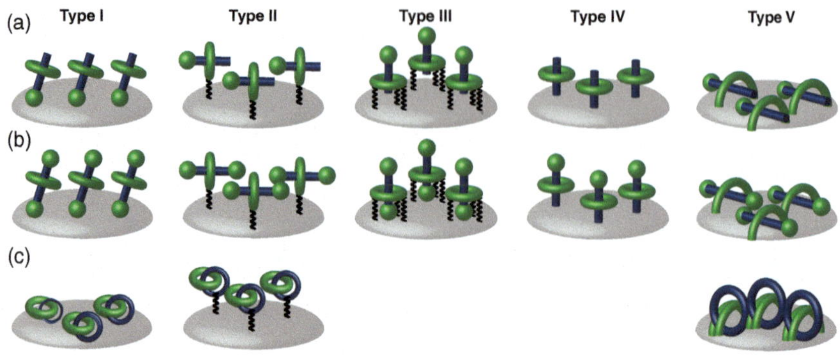

Figure 2.1 Schematic description of architectural categories for surface-immobilized mechanically interlocked molecular structures exemplified for (a) [2]pseudorotaxanes, (b) [2]rotaxanes, (c) [2]catenanes. Type *I:* Physisorption of intact interlocked architectures at arbitrary orientations. Type II*:* Rim-tethered macrocycle interlocked architectures. Type III*:* Face-tethered macrocycle interlocked architectures. Type IV*:* End-tethered linear component interlocked architectures. Type *V:* Interlocking through a surface-included cyclic component generated by chemisorption at two sites on one linear molecular unit. For Types IV and V, the surface plays an intrinsic architectural role in maintaining the interlocked nature of the structure. Note that Type V-c can be equally described as a surface-included [2]catenane or a multiply-tethered linear component [2]rotaxane.

molecular-level detail from ensemble measurements. Therefore, structure–property maps are within reach, which will ultimately guide the prospectors working to survey the new landscape defined by MIM–NP hybrids.

In this chapter, we will introduce seminal examples of mechanically interlocked molecular structures integrated with colloidal nanomaterials. A number of architectural categories can be defined (Figure 2.1). We consider both MIMs (catenanes and rotaxanes) and threaded supramolecular complexes (pseudorotaxanes). The latter are frequently a synthetic precursor to the former and, therefore, present similar challenges for preparation and analysis. Furthermore, in some scenarios, the threaded supramolecular complexes offer distinct dynamic behaviors and functional possibilities compared to their fully interlocked analogs. Early studies naturally focused on establishing synthetic and analytical capabilities; these have now led to the majority of architectural categories being realized using one or more of the well-established MIM structural types. Attention quickly turned to properties and functional capabilities. We explore the application of MIMs to tune and switch nanomaterial properties, to control the uptake and release of cargos from nanodelivery vehicles and to construct NP assemblies. Examining these significant advances at the intersection of two cutting-edge fields of twenty-first century chemical endeavor points the way to an exciting future for devices and materials that integrate functional components on several length scales.

The focus of this chapter is multicomponent NPs comprising a core material bearing surface-attached molecular species. Colloidal polymer nanostructures – which, in their simplest form, are compositionally homogeneous agglomerates of polymer chains – are not covered in detail, although it should be noted that similar opportunities exist within this nanomaterial class. Even our restricted definition of a NP encompasses vast structural diversity, with quite stark differences in synthetic strategies, analytical challenges, and functional properties. Particularly relevant for the current discussion is the nature of the microenvironment experienced by the surface-confined MIMs and the analytical challenge of assessing complex molecular-scale structure and dynamics in this setting. Monolayer-stabilized noble metal cores (e.g. Au or Ag) are often considered as the canonical NP. Here, the surface-bound molecules are critical to the stability of the nanomaterial, as well as defining solvent compatibility for forming stable colloidal dispersions. With core sizes typically <10 nm, these structures have high densities of surface-bound molecules (ca. 5 molecules nm^{-2}) corresponding to ~10–1000 molecules per particle. The crowded surface-confined environment can be expected to have a considerable influence on all sorts of molecular properties, not least of which are the characteristic structural dynamics of MIMs. Often, mixed-ligand monolayers may be employed, where the functional molecule is diluted as a minor constituent alongside a structurally simpler species chosen to confer surface stability and physicochemical property tuning while minimizing interactions between the more structurally complex species. Particles in the >10 nm size regime typically require electrostatic or polymer stabilizers to retain their nanoscale dimensions. Some core materials, such as metal(oid) oxides (e.g. SiO_2, TiO_2) possess intrinsic surface charge so that molecular surface stabilizers are not required for NP kinetic stability. In these cases, the density of surface-bound functional molecules can consequently be very low (e.g. ca. 0.01–0.1 molecules nm^{-2}) although with much larger surface areas, this can still correspond to several hundred molecules per particle (ca. 100–1000 molecules for cores in the 50–500 nm range). Crowding effects affecting MIM dynamic behaviors are correspondingly significantly less likely in these systems compared to the more densely functionalized nanomaterials. Conversely, for large particles, the vast majority of a sample volume is composed of the NP core material, presenting considerable analytical challenges for assessing the structure and function of the molecular species present at extremely low concentrations.

2.2 Installing and Actuating Mechanically Interlocked Molecular Architectures at Colloidal Nanoparticle Surfaces

Several of the early pioneers of artificial interlocked molecules also made important initial forays toward combining these novel molecular architectures with newly accessible colloidal NPs that were emerging around the same time. These efforts often adopted similar synthetic strategies to contemporaneous efforts aimed at integrating MIMs with extended planar surfaces [1–4]. These heterogeneous environments, however, present distinct analytical challenges. Thus, establishing sample purity and characterizing the compositional makeup, structural identity, and functional behavior of surface-bound species was (and to a large extent remains) at least as demanding as the synthetic hurdles.

Several architectural categories can be distinguished for installing interlocked molecular structures on a nanoscale (or bulk) surface (Figure 2.1). Each geometry entails different implications for synthetic strategy and for maintaining any switchable behavior of the surface-bound molecular component. In the simplest approach, an intact interlocked structure may be directly physisorbed on a bare solid substrate (Figure 2.1, Type I). Despite a considerable number of examples on extended surfaces [1–4], this has not been commonly implemented on colloidal nanosurfaces – relatively weak physisorption interactions are either insufficient to retain small molecules on the surface in the face of enthalpic and entropic forces favoring dissociation and dissolution, or the formation of molecular multilayers rapidly leads to NP aggregation. Specific bonding interactions are generally preferred for anchoring surface functionality on colloidal NPs, most commonly at sub-monolayer densities. Intact interlocked structures may be immobilized on a surface through one or more tethers to either of the interlocked components, corresponding to a variety of architectural categories (Figure 2.1, Types II–V). A particularly popular approach has been the surface immobilization of macrocycles through one (Figure 2.1, Type II) or multiple (Figure 2.1, Type III) linkages [5]. Although fully interlocked structures of this sort are possible (Figure 2.1, Types II-b, II-c, III-b), these geometries are more readily applicable for presenting surface-bound pockets to which linear components may associate noncovalently, generating pseudorotaxane geometries (Figure 2.1, Types II-a, III-a). Alternatively, tethering one end of a linear molecule provides a surface-bound pillar around which a macrocyclic component may be complexed or mechanically interlocked (by including a bulky stopper unit at the peripheral end of the linear component) to form pseudorotaxane or rotaxane structures, respectively (Figure 2.1, Type IV). Finally, tethering both ends of a linear unit to the same surface creates a surface-bound hoop through which a second molecular component may be complex or interlocked (Figure 2.1, Type V). In each of Types IV and V, the surface itself is intrinsic to the interlocked architecture. The fidelity of the surface-tethering interactions, therefore, becomes critical to the integrity of the mechanical bond. Furthermore, this role for the surface can blur conventional definitions of molecular topology. For example, a macrocycle interlocked with a surface-included hoop (Figure 2.1, Type V-c) can be equally described as a surface-included [2]catenane or a [2]rotaxane that is multiply tethered at either end of the linear component.

In each of the chemisorbed architectural categories (Figure 2.1, Types II–V), the choice of molecule–surface linkage is largely dictated by the chemical constitution at the surface of the NP core; a relatively small set of established molecule-anchoring chemistries are in common use for each of the major nanomaterial categories [6]. Just as for extended surfaces, self-assembled monolayers that exploit the strong gold–sulfur interaction [7–9] are one of the canonical surface–molecule interfaces used on NPs.

End-tethered [2]rotaxanes (Figure 2.1, Type IV-b) were first accessed from water-soluble citrate-stabilized AuNPs ($\langle d \rangle = 13\,\text{nm}$) by Kaifer and coworkers using a ligand-exchange route [10]. An excess of α-cyclodextrin (α-CD) had been used to solubilize various hydrophobic alkanethiols in water, delivering them to the gold surface, producing AuNPs that consequently became soluble in apolar solvents. By contrast, when using alkanethiol **1**, functionalized with a bulky ferrocene terminal unit, the NP product was only soluble in polar aprotic solvents. Even after purification, signals corresponding to both the alkanethiol and α-CD were observed at a ratio of 1 : 0.7 in the ^{1}H NMR (nuclear magnetic resonance) spectrum. Crucially, significant line broadening evidenced that all signals originated from surface-bound species, while characteristic chemical shift changes indicated encapsulation of the alkyl chain by the macrocycle to form a [2]rotaxane, stoppered at one end by the ferrocene and the other by the AuNP surface (AuNP-$\mathbf{1}_{0.3}\mathbf{2}_{0.7}$, Figure 2.2a).

Following the same synthetic strategy, Sortino and coworkers later constructed photoresponsive end-tethered [2]rotaxanes on AuNPs using an alkanethiol bearing a terminal azobenzene photoswitch [13]. The absence of signatures of π–π stacking in ultraviolet–visible (UV–Vis) spectra suggested that the steric bulk of the water-solubilizing α-CD macrocycles prevents close packing of the azobenzene

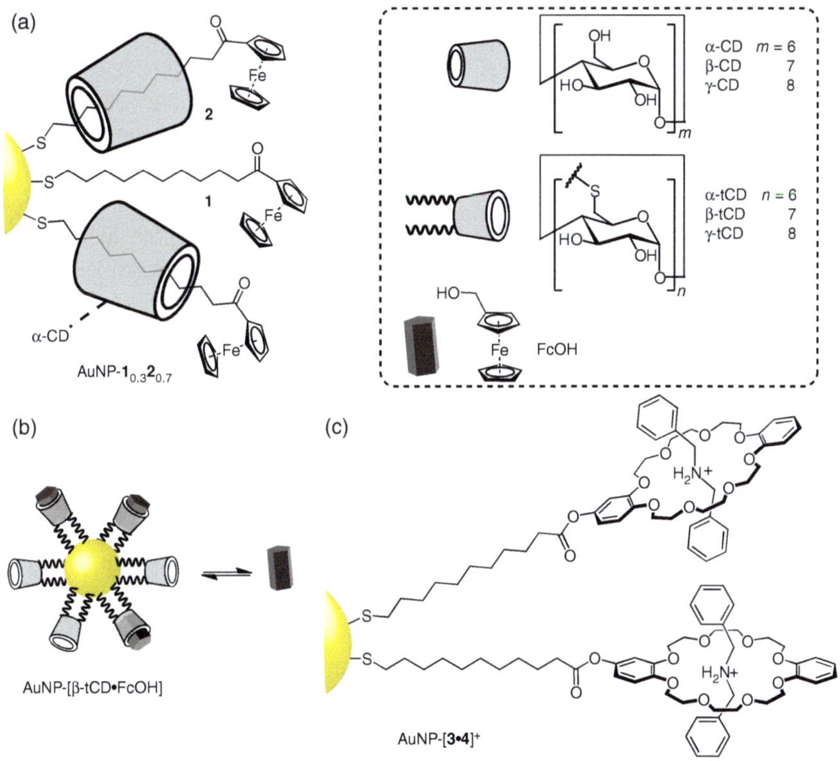

Figure 2.2 Early examples of interlocked molecular structures constructed on AuNP cores. (a) End-tethered [2]rotaxanes (**2**) as the major component in a mixed-ligand monolayer with the non-interlocked thread (**1**): AuNP-$\mathbf{1}_{0.3}\mathbf{2}_{0.7}$ [10]; (b) face-tethered pseudorotaxanes AuNP-[β-tCD•FcOH] formed by guest encapsulation in surface-bound perthiolated cyclodextrins [11]; (c) rim-tethered pseudorotaxanes AuNP-[**3•4**]$^{+}$ formed at the periphery of an alkylthiyl monolayer [12].

units on the surface. In contrast to earlier photoswitch-functionalized AuNPs, the azobenzenes underwent reversible photoisomerization at rates that were unchanged from the azobenzene molecules in bulk solution, presumably reflecting the free volume created within the low-density monolayer brought about by the steric bulk of the interlocked architecture [13].

Kaifer and coworkers went on to demonstrate two synthetic routes to face-tethered macrocycle-stabilized AuNPs (Figure 2.1, Type III-a). Perthiolated CDs could either be introduced post-synthesis via ligand exchange on citrate-stabilized AuNPs ($\langle d \rangle = 13$ nm) [14], or alternatively applied directly to stabilize the growth of AuNPs on reduction of $HAuCl_4$ with $NaBH_4$ [11]. The direct synthesis procedure – which has the advantage that the thiolated organic molecule is the only strongly surface-binding species present – produced AuNP-α/β/γ-tCD with size distributions that depended on the identity and concentration of the stabilizing macrocyclic ligand ($\langle d \rangle = 2.7$–6.4 nm). The smaller water-soluble NPs were amenable to analysis by ^{1}H NMR spectroscopy, where broad signals again provided evidence for surface immobilization of the CD ligands. The formation of surface-bound inclusion complexes was probed by electrochemical methods using ferrocenemethanol (FcOH) – a redox-active guest for β-CD. The reduction potential for FcOH shifted to more positive values in the presence of AuNP-β-tCD, consistent with the formation of AuNP-[β-tCD•FcOH] (Figure 2.2b) [11]. Similar voltammetric shifts were observed in the presence of solution-phase β-CD but, interestingly, the accompanying decrease in anodic current was more pronounced for the NP-bound hosts. A consequence of the far more significant reduction in diffusion coefficient compared to binding a small molecule host, this observation served as an early reminder that (supra)molecular-level behavior can be significantly modified by the distinct features of a nanoscale system.

Contemporaneous with these early contributions, Fitzmaurice and Stoddart exploited a similar direct synthesis strategy to create organic-soluble AuNP-**3** ($\langle d \rangle = 4.2$ nm), stabilized by a dense single-component monolayer bearing peripheral dibenzo[24]crown-8 macrocycles (Figure 2.2c) [12]. Diagnostic of surface attachment, extreme broadening rendered signals for protons close to the NP surface unobservable in the ^{1}H NMR spectrum. Meanwhile, resonances assigned to protons near the monolayer periphery were well resolved. Thus, NMR spectroscopy could be used to monitor the formation of rim-tethered [2]pseudorotaxanes (Figure 2.1, Type II-a). In the presence of dibenzylammonium cation **4**$^+$, complexation-induced chemical shifts were observed similar to those observed for [**3**•**4**]$^+$ pseudorotaxanes in solution. In contrast to solution-phase behavior, however, quantitative analysis revealed evidence of positive binding cooperativity for AuNP-**3**. Ascribed to the progressively higher affinity of the cationic guest at the increasingly polar surface, this observation represented another early indication that surface-specific effects can induce new phenomena in the interactions of NP-bound molecules.

Taking advantage of the strong affinity of phosphonic acids for metal oxide surfaces [15], Fitzmaurice and coworkers assembled end-tethered [2]pseudorotaxane and [2]rotaxane structures (Figure 2.1, Types IV-a, IV-b) on TiO_2 NPs [16, 17]. It was first established that treating bipyridinium-functionalized

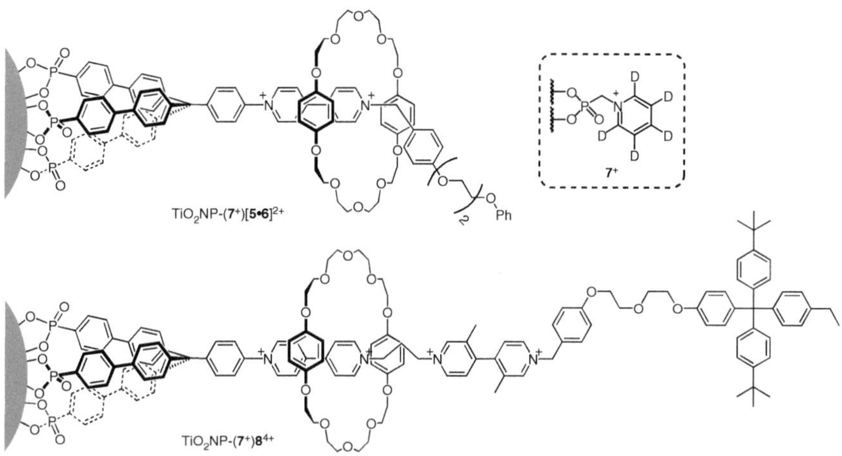

Figure 2.3 End-tethered redox-responsive [2]pseudorotaxane and [2]rotaxane architectures constructed on TiO_2NP cores [16, 17].

TiO_2NP-(7^+)5^{2+} (<*d*> = 5 nm) with bis(dioxybenzene) crown ether **6** led to formation of surface-bound [2]pseudorotaxanes TiO_2NP-(7^+)[**5•6**]$^{2+}$ (Figure 2.3). Low functionalization densities are typically achieved on metal oxide particles (see Section 2.1); in this example, each particle supported only an average of 13 copies of 5^{2+} with the remaining surface Ti^{4+} binding sites saturated by stabilizer molecule 7^+. Nevertheless, the strongly charge-stabilized particles formed stable high-concentration colloidal solutions so that structural characterization by ^{1}H NMR spectroscopy was feasible. Solution-phase evidence for the surface-bound structure could be derived from both NMR (complexation-induced chemical shift changes for resonances from the surface-bound bipyridinium units) and UV–Vis spectroscopies (charge–transfer absorption band). Cyclic voltammetry on nanostructured thin film electrodes prepared from TiO_2NPs was also consistent with the encapsulation of the bipyridinium redox center by the electron-rich macrocycle. Differential broadening of signals in the ^{1}H NMR spectrum suggested that the bulky and rigid tripodal anchoring unit projects the bipyridinium site away from the surface. Together with the low functionalization density, this should minimize surface-specific effects. Indeed, the voltammetric shifts for the surface-bound species were almost identical to those for solution-phase analogs, suggesting that, in this structure, the supramolecular interactions are not significantly affected by the surface-confined environment. The subtle challenge posed by the rational design of surface-confined systems, however, was emphasized in the observation that a related but more flexible bipyridinium ligand showed no evidence of NP-bound pseudorotaxane formation [16].

Taking advantage of the optical transparency of metal oxide nanomaterials, fluorescence spectroscopy can provide a highly sensitive probe of surface-bound molecular interactions that is applicable to low-concentration samples. For example, the optical response of a fluorescent indolocarbazole guest on SiO_2NPs (<*d*> ~ 10 nm) provided evidence for anion-templated end-tethered

[2]pseudorotaxane assembly [18]. Importantly, the anion-specific optical responses were consistent with those observed for analogous systems attached to much larger polystyrene microparticles (<*d*> ~ 100 μm) for which further structural evidence for [2]pseudorotaxane formation was available from high resolution-magic angle spinning (HR-MAS) NMR spectroscopy.

Incorporating two redox-responsive bipyridinium binding sites, TiO_2NP-(**7**$^+$)**8**$^{4+}$ (<*d*> = 6 nm) introduced the prospect of remote-controlled co-conformational switching in NP-bound [2]rotaxanes. Evidence from both ^{1}H NMR spectroscopy and cyclic voltammetry (on nanostructured TiO_2 thin film electrodes) suggested a strong preference for macrocycle binding to the surface-proximal bipyridinium site, in-line with expectations based on solution-phase analogs. Significant macrocycle interaction with this binding site appears to be maintained on reduction to the radical cation, with switching of the dominant co-conformation to the distal bipyridinium only achieved in the diradical state [17]. Unfortunately, it was not possible to characterize the finer details of the co-conformational equilibria or dynamics using these methods.

Redox-responsive switches are particularly attractive for application in heterogeneous systems because electrochemical methods are capable of simultaneously delivering the activating stimulus while also reporting on the molecular environment and interactions of the redox-active unit. A common ligand exchange strategy, which circumvents the incompatibility of sensitive molecular functionality with the highly reducing conditions required for metal NP synthesis, has been adopted to construct a variety of NP-bound bistable interlocked redox switches. Successful self-assembly of end-tethered [2]pseudorotaxanes AuNP-(**9**)$_{0.94}$[**10**Me•CBPQT]$^{4+}{}_{0.06}$ (<*d*> = 2 nm, Figure 2.4a) was established by Cooke et al. using evidence from ^{1}H NMR spectroscopy [19]. De-threading of the surface-bound complexes could be triggered and monitored using square wave voltammetry experiments. Alternatively, UV–Vis and NMR spectroscopies evidenced de-threading on the addition of tetrathiafulvalene (TTF) – a competitive guest for the macrocycle [19].

Grzybowski, Stoddart, and coworkers adopted a similar synthetic strategy to install bistable redox-switched [2]catenanes, end-tethered [2]pseudorotaxanes, and [2]rotaxanes (Figure 2.1, Types II-c, IV-a, and IV-b, respectively) in mixed monolayers on AuNP (<*d*> = 4.6 and 6.6 nm), PtNP (<*d*> = 2.4 nm), and PdNP (<*d*> = 2.4 and 2.7 nm) cores (e.g. Figure 2.4b) [20, 21]. In the absence of molecular structural characterization, zeta potential measurements evidenced the redox-switched assembly–disassembly of end-tethered [2]pseudorotaxanes AuNP-(**11**)$_{0.75}$[**12**•CBPQT]$^{4+}{}_{0.25}$. However, particle surface charge is not sensitive to the submolecular conformational dynamics of fully interlocked surface-bound structures; consequently, the researchers turned to cyclic voltammetry experiments to probe the co-conformational preference of surface-bound **13**$^{4+}$ and **14**$^{4+}$. Comparison to the electrochemical behavior of analogs in bulk solution confirmed that redox-triggered mechanical switching was retained in the NP-bound systems. Quantitative analysis revealed increasing negative cooperativity for the surface-bound redox potentials as the mole fraction of redox-active units is

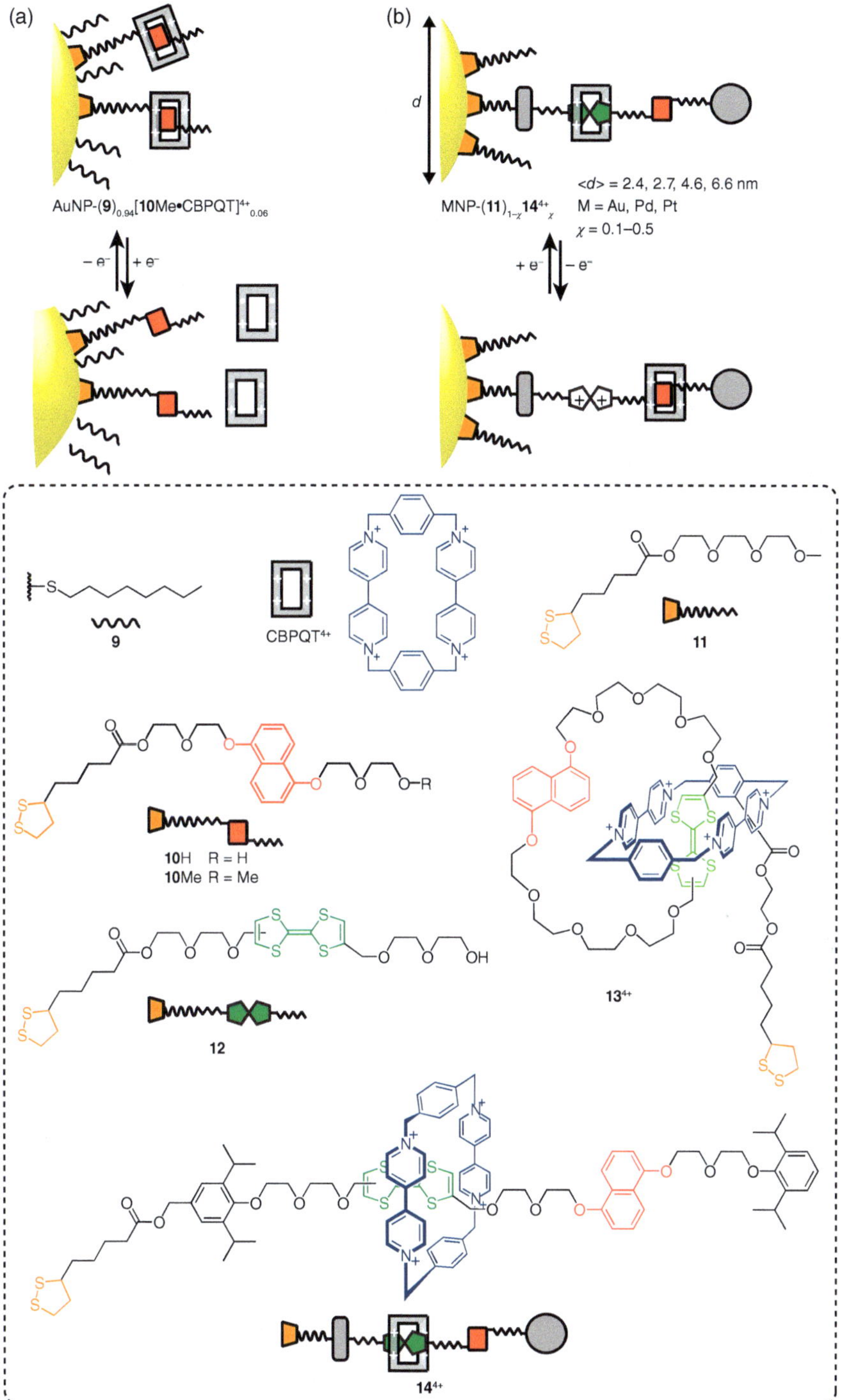

Figure 2.4 Redox-responsive MIMs on metal NP cores. (a) Redox-switchable complexation of an end-tethered [2]pseudorotaxane on AuNPs [19]. (b) End-tethered bistable [2]rotaxanes with redox potentials and co-conformational switching dynamics that are sensitive to both the core and monolayer structural parameters [20].

increased – electrostatic repulsion between positively charged units serving to render oxidation increasingly difficult. Furthermore, the rates of slow sub-molecular co-conformational motions also depended on functionalization density and NP surface curvature. Both of these surface-specific effects could be reproduced with remarkable accuracy from a theoretical model of the interfacial electrostatic potential, suggesting novel design parameters for predictably tuning the sub-molecular stimuli-responsive mechanical behavior of surface-confined MIMs [20, 21].

Further architectural variations for surface-bound interlocked molecules are possible in the form of doubly or multiply tethered structures (Figure 2.1, Type V). For example, a doubly tethered [2]rotaxane was prepared as the sole component on the surface of gold NPs AuNP-**15** (<*d*> = 3.5 nm, Figure 2.5a) in a direct synthesis approach [22]. As a consequence of the bulky tertiary amide substituents in this design, the ligand precursor **15** is itself a [2]rotaxane. The surface is, therefore, not intrinsic to the interlocked architecture in this example. However, this structure is topologically identical to a surface-hoop-based [2]catenane (Figure 2.1, Type V-c). Doubly tethered [3]rotaxanes have also been prepared in a stepwise ligand-exchange protocol involving the orthogonal assembly of [α-CD•azobenzene] complexes, followed by pyridine–platinum coordination to give [3]pseudorotaxanes, then finally sulfur–gold binding to produce AuNP-**16** (<*d*> = 15 nm, Figure 2.5b) [23]. Although the number of intact [3]rotaxanes was not incontrovertibly established, induced circular dichroism absorption of the azobenzene unit provided evidence of co-conformational movement of the α-CD ring on the application of a photoisomerizing stimulus.

These examples were important for establishing that MIMs can be installed on colloidal NP surfaces. A variety of architectural categories have been realized (Figure 2.1) on several types of NP cores, and the feasibility of translating

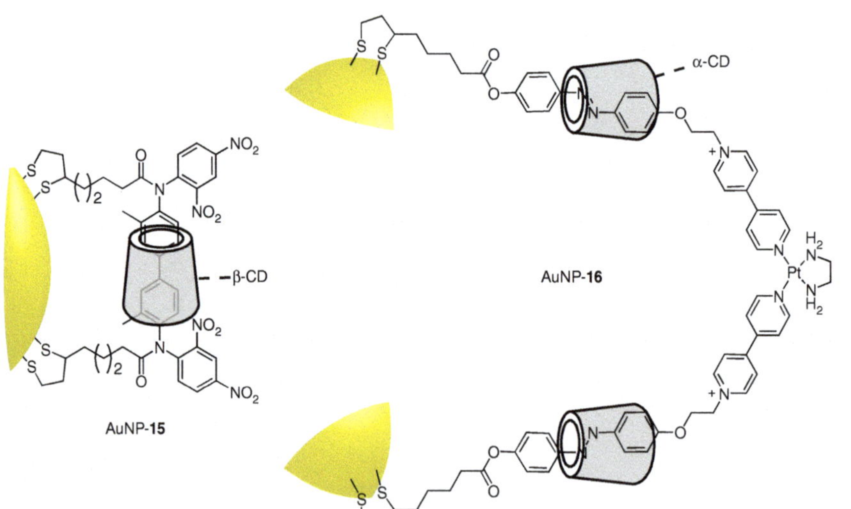

Figure 2.5 Doubly tethered [2]rotaxanes and [3]rotaxanes on AuNP cores [22, 23]. For molecular structures of α-CD and β-CD, see Figure 2.2.

stimuli-responsive control over molecular mechanical motions to this interfacial environment has been demonstrated. At the same time, however, these pioneering studies serve to highlight challenges inherent in creating and operating functional hybrid (supra)molecular–NP systems.

The nature of the core material is a critical factor in dictating viable synthetic strategies, defining compatible surface-linking chemistry and determining attainable functionalization densities. Direct synthesis approaches are capable of installing sophisticated molecular architectures at high density, possibly as the sole surface-stabilizing component. Yet, this strategy demands that the interlocked structures remain intact under the NP synthesis conditions, which, at best, places limits on the range of compatible functionality and, at worst, renders this approach completely infeasible for many nanomaterial categories. Ligand exchange or conjugation to inherently stable colloids allows sophisticated molecular structures to be introduced under benign conditions yet entails challenges in terms of controlling and characterizing the constitution of the surface-attached molecular population. Rather than installing the interlocked architecture directly, either synthetic strategy can be adopted to produce a surface-bound intermediate that is then converted to an interlocked structure in a post-synthesis association or conjugation step. Under these circumstances, establishing evidence for successful generation of the surface-bound interlocked architecture becomes particularly crucial.

In contrast to solution-phase small-molecule systems, the intrinsic population dispersity, batch-to-batch variability and heterogeneous multicomponent nature of hybrid (supra)molecular–NP constructs are all hurdles to structural and functional characterization. In many cases, mixed monolayers have been required to attain workable physicochemical characteristics such as solubility. Solubility limits, functionalization densities and mixed monolayer compositions all, in turn, have implications for the analytical methods that are practically applicable. NMR spectroscopy sets a benchmark for rich, quantitative structural and dynamical information but is limited to select systems on account of relatively low sensitivity, incompatibility with paramagnetic materials and the effects of surface attachment on chemical shift dispersity and nuclear spin relaxation [24, 25]. Electrochemical methods can provide sensitive and quantitative probes of noncovalent interactions in both homogeneous and heterogeneous settings, provided at least one component is electroactive [26, 27]. Most often, evidence from a suite of judiciously chosen but indirect methods must be combined with careful comparison to representative analogs that can be studied in bulk solution [28–30].

The noncovalent interactions that stabilize interlocked complexes and define the co-conformational preferences of fully interlocked structures can be significantly perturbed by surface-confinement. The interfacial setting defines a local microenvironment that can be quite different from a bulk solution. Hence, interactions at the NP surface experience unique solvation, dielectric and steric conditions. Forming multiple host–guest complexes in close proximity can also lead to either positive or negative cooperative effects, while arrays of closely spaced surface-tethered units present the opportunity to exploit quite significant multivalency in the formation of association complexes. Furthermore, stimuli-induced transformations typically exploited in switchable (supra)molecular systems (e.g. protonation state, photoisomerization, and redox reactions) can be equally strongly influenced by surface confinement. Notwithstanding these challenges, mechanically interlocked molecular structures have been applied in NP-bound settings to achieve a variety of innovative functions and intriguing effects.

2.3 Modulating Nanoparticle Physicochemical Properties Using Switchable Mechanically Interlocked Architectures

The recruitment or release of molecular components to/from a NP surface in the form of threaded pseudorotaxane complexes offers a route to stimuli-responsive property switching. Dense surface coverages of switchable mechanically interlocked structures can tune the global physicochemical properties of the nanoscale object.

In their pioneering preparation of [2]rotaxane stabilized AuNP-$\mathbf{1}_{0.3}\mathbf{2}_{0.7}$ (Section 2.2), Kaifer and coworkers demonstrated that the concept of macrocycle encapsulation to modify molecular properties could be extended to modulate NP solubility. The solution saturation concentration of AuNP-$\mathbf{1}_{0.3}\mathbf{2}_{0.7}$ in water (measured by UV–Vis absorbance at the AuNP surface plasmon resonance) increased in a concentration-dependent manner with the addition of excess β-CD, which is large enough to form an inclusion complex with the ferrocene rotaxane stopper [10].

Operationally simple and efficient methods for phase transfer of NPs are highly desirable, particularly for bridging incompatible solvent requirements of synthetic methods to those of applications [29, 31]. Following a similar α-CD [2]pseudorotaxane-based approach for the delivery of hydrophobic alkanethiols to AuNP ($\langle d \rangle = 3.2$ nm) surfaces in water, unidirectional water-to-organic phase transfer could be achieved simply by shaking a biphasic mixture with chloroform, resulting in de-threading of the hydrophilic interlocked complex [32]. Phase transfer in the opposite direction has been demonstrated by mixing hexane solutions of oleic acid-stabilized Fe_2O_3NPs ($\langle d \rangle = 10$ nm) or AgNPs ($\langle d \rangle = 8$ nm) with α-CD in water [33]. In this scenario, the relatively weak affinity of oleic acid for the NP surface is likely important to allow the density of the surface-stabilizing monolayer to adjust to accommodate the bulky water-soluble macrocycles during the formation of end-tethered [2]pseudorotaxane complexes with the fatty acid tails. Informed by these early reports, bidirectional phase transfer was achieved for CdSe/ZnS core/shell quantum dots (QDs) ($\langle d \rangle = 5.2$ nm) stabilized by a low-density ferrocene-terminated alkylthiyl monolayer [34]. Chloroform-to-water transfer occurred via pseudorotaxane formation with β-CD. Although not possible spontaneously, the reverse (water → chloroform) transfer was triggered by introducing a large excess of a competitive water-soluble guest for β-CD, such as adamantane carboxylate. Irreversible changes to the QD optical properties, however, indicated that the bi-directional process was not entirely reversible in all respects [34].

NPs stabilized by face-tethered perthiolated CDs (Figure 2.2b) are inherently hydrophilic on account of the hydroxylated secondary CD face at the NP–solvent interface and form inclusion complexes driven by hydrophobic effects (see Section 2.2). Amphiphilic guests (alkyldimethyl(ferrocenylmethyl)ammonium ions) have been shown to transfer hydrophilic AuNP-β-tCD (<*d*> = 3 nm) into chloroform in a concentration-dependent manner [35]. NPs stabilized by α-tCD were transferred far less efficiently, consistent with the selectivity of ferrocenes as guests for the larger host macrocycles. The intriguing observation that the inclusion complexes apparently remain intact in the apolar solvent suggests that a layer of water is retained within the monolayer, stabilizing the interlocked structure and solvating the accompanying counterions [35]. More recently, this concept was extended to achieve fully reversible remote-controlled NP phase transfer [36], exploiting the well-characterized photoswitchable host–guest interaction between α-CD and azobenzenes [37, 38]. Hydrophilic AuNP-α-tCD transferred from water to toluene in the presence of an *E*-azobenzene-terminated surfactant, consistent with the formation of face-tethered [2]pseudorotaxanes (Figure 2.1, Type III-a). Irradiation with UV light initiated return to the aqueous phase, consistent with dissociation of the inclusion complex as a result of photoisomerization to the *Z*-azobenzene isomer, which is a poor guest for α-CD. Subsequent irradiation with visible light to trigger the back isomerization returned the particles again to the toluene phase. The efficiency of the photoswitched water → toluene phase transfer was sufficient to achieve on-demand quenching of an aqueous phase AuNP-catalyzed reaction [36].

Modulating the interactions between a NP and its surrounding matrix can influence the formation of hybrid NP composite materials. Despite a relatively high estimated monolayer density, end-tethered polypseudorotaxanes were constructed by treating AuNP-**17** (<*d*> = 16 nm, Figure 2.6) with α-CD [39]. In contrast to AuNP-**17** alone, which spontaneously produces amorphous aggregates from aqueous solution, AuNP-[**17**•*n*(α-CD)] was resistant to precipitation. Yet, self-assembly of AuNP-[**17**•*n*(α-CD)] was observed on drop-casting, producing ordered layer structures, which could subsequently be fully re-dispersed in solution [39].

Surface-bound addressable pseudorotaxane complexes have more recently been applied to modulate the chemical properties of NPs in demanding biological contexts. Masking and unmasking AuNP-bound alkyldiammonium ligands through reversible encapsulation by cucurbituril (CB) macrocycles has been shown to regulate electrostatic and hydrophobic interactions between proteins and the particle surface [40]. Mass spectrometry and ^{1}H NMR spectroscopy confirmed essentially quantitative formation of end-tethered pseudorotaxanes in mixed monolayers that have sufficient free volume to accommodate the macrocycle. Selective de-threading was observed in the presence of competitive guests. Threaded AuNP-[**18**•CB[7]] (<*d*> = 2.5 nm, Figure 2.7a) are readily taken up into the endosomes of human breast cancer cells, showing no apparent cytotoxicity at concentrations ≤50 μM. However, subsequent incubation with 1-adamantylamine (ADA) – a bioorthogonal competitive guest for CB[7] – triggered endosomal escape and a cytotoxic response very similar to that elicited by AuNP-**18** alone; presumably the result of intracellular disruption of the NP-bound host–guest complex by ADA [41].

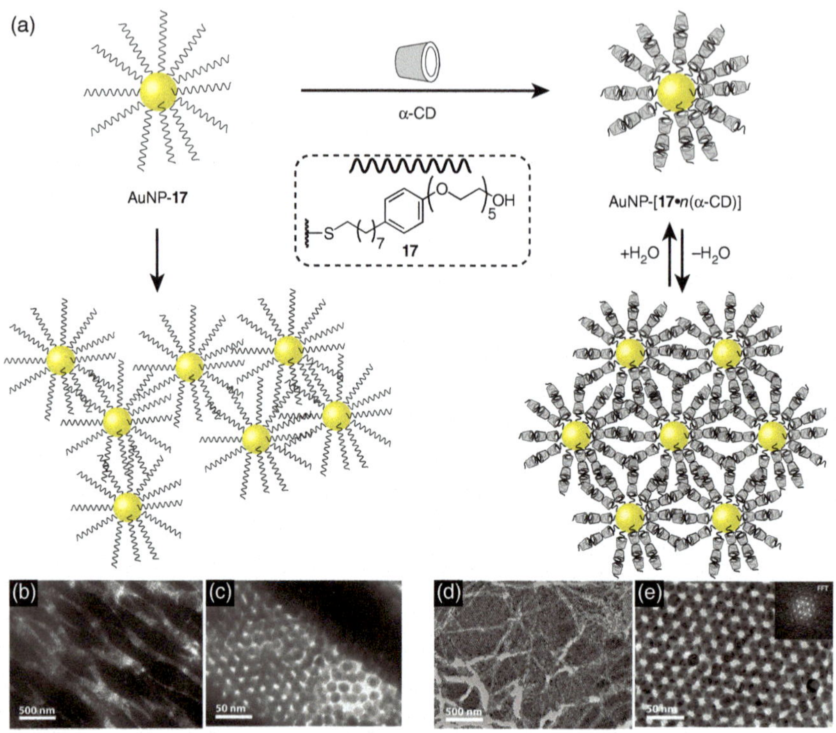

Figure 2.6 Modulation of hierarchical AuNP composite structure by mechanically interlocked surface functionality [39]. (a) AuNP-**17** spontaneously forms amorphous aggregates from an aqueous solution; AuNP-[**17**•*n*(α-CD)] reversibly assembles into ordered supercrystals. (b, c) TEM images of amorphous AuNP-**17** aggregates. (d, e) TEM images and corresponding fast Fourier transform (inset) of hexagonally packed bilayer assemblies of AuNP-[**17**•*n*(α-CD)]. For molecular structure of α-CD, see Figure 2.2. Source: (b–e) Reproduced with permission from Coelho et al. [39], copyright 2014 John Wiley and Sons.

A similar strategy was implemented to achieve intracellular control of a bioorthogonal catalytic reaction [42]. The sterically bulky end-tethered pseudorotaxanes on AuNP-[**19**•CB[7]](RuCat) ($\langle d \rangle = 2.5$ nm, Figure 2.7b) block substrate access to an organometallic catalyst intercalated within the alkyl chain region of the monolayer. The competitive guest ADA, therefore, acted as a selective switch for restoring the catalytic activity by triggering pseudorotaxane de-threading. Using this approach, deallylation reactions could be initiated with spatiotemporal control in living cells. For example, the anticancer drug 5-fluorouracil (5FU) was activated only on co-localization of a propargyl-protected 5FU prodrug with an NP-bound, CB-protected catalyst and ADA (Figure 2.7b) [42].

The strategy of interfacing mechanically switchable (supra)molecular components with colloidal NPs to achieve stimuli-responsive property control has not been restricted to mechanically interlocked architectures [43–45]. Volume phase changes of thermosensitive polymer surface coatings [46–51], surface-bound photochromic switches [52–57], and responsive oligonucleotide structures [58–60] have all been

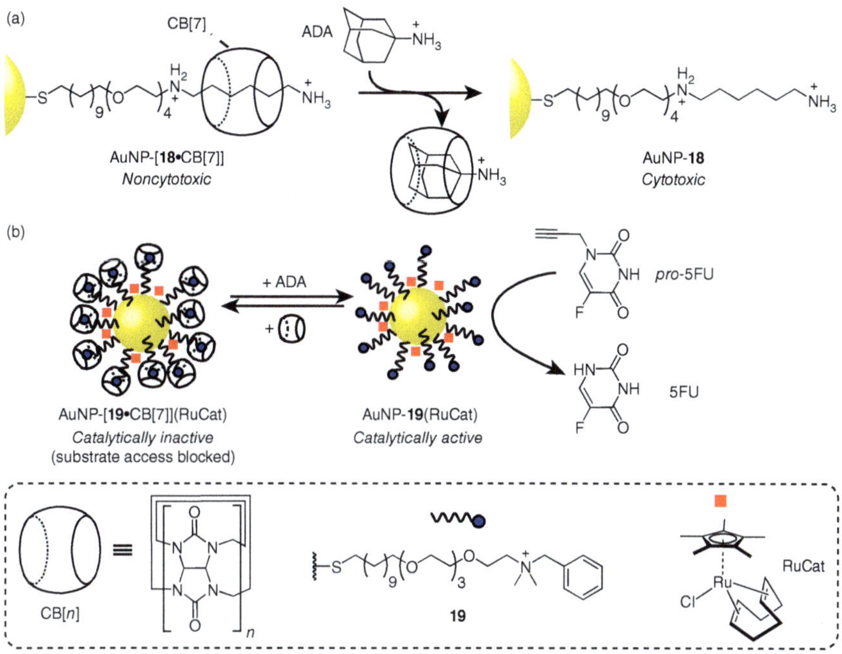

Figure 2.7 End-tethered [2]pseudorotaxanes for spatiotemporal modulation of AuNP properties in living cells. (a) Intracellular decomplexation of biocompatible AuNP-[**18**•CB[7]] reveals cytotoxic AuNP-**18** [41]. (b) Intracellular decomplexation of AuNP-[**19**•CB[7]] enables access to a monolayer-embedded Ru catalyst (RuCat), triggering prodrug activation [42].

used to modulate a variety of NP-based properties and functions [43–45]. Just as for their mechanically interlocked counterparts, the nanoconfined environment has been observed to significantly influence the behavior of the switchable molecular component; many of the design, construction, and characterization challenges are shared, irrespective of the switchable molecular architecture.

2.4 Interlocked Gates for Nanoparticle Pores: From Cargo Release to Nanoscale Communication

Some of the first works to demonstrate that structurally sophisticated interlocked molecular architectures could be integrated with colloidal nanomaterials to achieve useful functions focused on mesoporous silica nanoparticles (MSNPs). Tunable porous structures with huge internal surface areas, stable and benign chemical composition, optical transparency and well-developed surface functionalization methodologies make MSNPs attractive nanocarrier platforms, particularly for drug delivery applications [61–63]. Spatiotemporal control over cargo release, triggered by either endogenous or externally applied stimuli, would allow the selective delivery of high local concentrations of sensitive or toxic cargo. Although several strategies have been implemented [64–66] – including charge modification of polymer surface coatings, conformational switching of surface-tethered peptides, pore-mounted photochromic switches and self-immolative pore-blocking molecules – actuation of surface-attached stimuli-responsive MIM structures has become a touchstone strategy for controlling pore access [67–69]. Beyond the delivery of therapeutic agents *in vivo*, these responsive hybrid nanocarriers have more recently been applied to engineer impressive examples of nanoscale communication and logic systems [70].

Switchable steric control over pore accessibility can be achieved through a number of different configurations involving MIM structures positioned near the pore openings (Figure 2.8). Irrespective of the structural details, a common modus operandi is adopted (illustrated for one example in Figure 2.9). The functionalized mesoporous carrier in a "gate-open" state is incubated with a solution of the intended cargo, allowing diffusion into the mesoporous matrix. Conversion to the "gate-closed" state by application of an external stimulus or *in situ* assembly of the MIM structure traps some proportion of the cargo in the mesoporous network. Non-entrapped cargo can then be removed, usually by NP filtration or precipitation and washing cycles, to yield the loaded nanocarrier. Cargo release is then activated by applying one of the stimulus-induced gate-opening mechanisms (Figure 2.9). In proof-of-concept studies, small molecule dyes are the most common test cargos; release from a solid sample of the NPs into a supernatant solution can be monitored by fluorescence spectroscopy or, quantitatively, by spectrophotometry. From an applications perspective, there are several features that demand careful

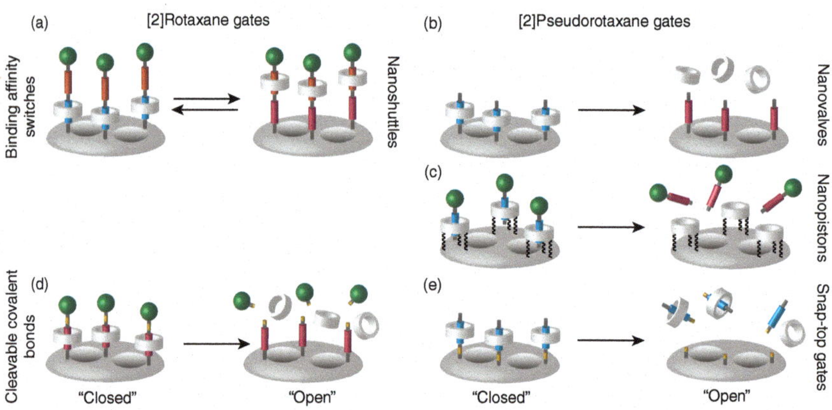

Figure 2.8 Design strategies for stimuli-responsive control over MSNP pore access using mechanically interlocked (supra)molecular structures categorized by structural class (columns) and operating mechanism (rows). (a) "Nanoshuttle" gates constructed from bistable end-tethered [2]rotaxanes. (b) "Nanovalves" are constructed from end-tethered [2]pseudorotaxanes. (c) "Nanopiston" gates constructed from face-tethered macrocycle [2]pseudorotaxanes. (d) "Snap-top" gates constructed from end-tethered [2]rotaxanes. (e) "Snap-top" gates constructed from end-tethered [2]pseudorotaxanes. Note that pore opening from architectures (b) and (c) can be alternatively brought about through a stimulus-triggered change to the macrocycle structure, the introduction of a competitive binder for either the linear or macrocyclic component or an increase in local temperature.

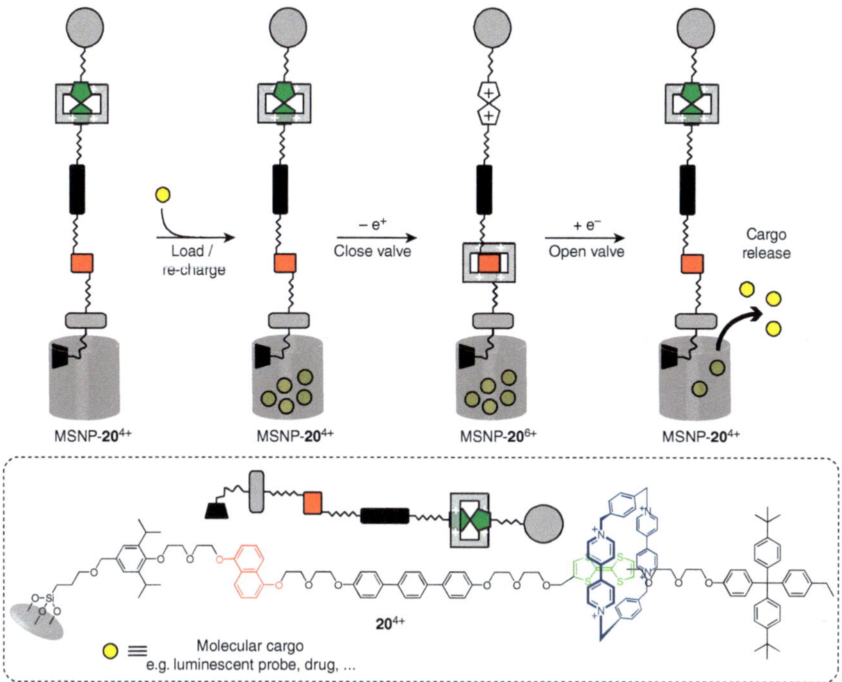

Figure 2.9 Redox-controlled cargo loading and release in MSNPs via co-conformational switching of bistable end-tethered [2]rotaxane nanoshuttles [71, 72].

consideration and optimization, including (i) loading capacity, (ii) background cargo release in the gate-closed state, (iii) yield and rate of cargo release in the gate-open state, (iv) switching fidelity and specificity; (v) stimulus compatibility with the target application (e.g. biocompatibility).

Pioneered by a collaboration between the groups of Stoddart (MIMs) and Zink (MSNPs) [73, 74], each of the structural designs shown in Figure 2.8 has now been implemented using a variety of switching stimuli [65]. The first, and in many regards still the most structurally sophisticated, design incorporated end-tethered bistable molecular shuttles on the surface of MSNPs (Figure 2.8a) [71, 72]. The gate-open and gate-closed states of MSNP-**20** correspond to surface-distal and surface-proximal co-conformations of a cyclobis(paraquat-*p*-phenylene) ($CBPQT^{4+}$) macrocycle along its rotaxane thread unit (Figure 2.9). Switching between these two states is triggered by oxidation ($TTF \rightarrow TTF^{2+}$; open → closed) and reduction (closed → open) and can be followed *in situ* through quenching of fluorescence from the dioxynaphthalene site when encapsulated by the $CBPQT^{4+}$ macrocycle. Constructed on MCM-41 mesoporous silica particles ($d \sim 400$–900 nm) with pore sizes ~ 2 nm, luminescent dyes were used as proof-of-concept cargos. The positional distribution of the switchable MIM units across the nanoscale platform can be controlled by the order of operations during the particle construction and functionalization procedure. It was found that restricting the switchable MIMs to the outside of the porous materials was optimal for achieving the largest loading capacity.

Structure–function optimization further revealed that short linkers between the rotaxane and MSNP surface are critical to minimize leakage of small cargos in the closed state; longer linkers were only efficient for trapping larger species. On the other hand, the rigid terphenylene spacer connecting the recognition sites was shown to have minimal influence on functional efficiency [72].

2.4.1 Optimizing Mechanisms for Cargo Release

Notwithstanding the technical achievement of incorporating and operating bistable [2]rotaxane molecular switches confined to NP surfaces, simpler structural designs are more likely to hold realistic promise for efficient prototyping and translation to end applications. With this in mind, both macrocycle-tethered and end-tethered [2]pseudorotaxanes (Figure 2.8, Types b, c, e) have been more extensively investigated as so-called "nanopistons" or "nanovalves," respectively, for controlling cargo release from MSNPs. One advantage of [2]pseudorotaxane gates is that the final pore-capping step is the formation of a noncovalent complex, which can be performed under very mild conditions compatible with sensitive cargos. Although redox-switched examples have been reported [75], pH-switched systems predominate, largely motivated by the potential for endogenously triggered drug delivery (see Section 2.4.2). End-tethered [2]pseudorotaxanes formed by crown ether recognition of surface-attached dibenzylammonium stalks were constructed on MCM-41 MSNPs in the presence of a coumarin guest [76, 77]. Deprotonation of the dibenzylammonium cation to disrupt the mechanically interlocked architecture triggered guest release at a rate that depended on both the pK_a and steric bulk of the organic base used. Alternatively, the addition of solution-phase guests for the crown ether (dialkylammonium ions or, more effectively, alkali/alkaline earth metal cations) induced valve opening by a competitive binding mechanism [77].

Importantly, none of these early systems were compatible with operation in aqueous environments that would be required for drug delivery applications. Combining components already proven to be biocompatible, pore-blocking end-tethered polypseudorotaxanes were constructed on MSNP cores ($<d> \sim 425$ nm) by threading α-CD or γ-CD macrocycles onto surface-attached polyethyleneimine (the larger macrocycles giving doubly threaded interlocked structures). The pore-blocking interlocked structures were assembled at pH 11, while protonation of the polyamine backbone at pH 5.5 resulted in de-threading and cargo release [78]. The integrity of these nanovalve gates was not assessed at physiological pH values, however. The ion–dipole binding between cucurbit[6]uril (CB[6]) macrocycles and bisammonium recognition sites offers an alternative potentially biocompatible molecular recognition motif. End-tethered [2]pseudorotaxane nanovalves constructed from MSNP-attached bisammonium stalks **21**($2H^+$) and CB[6] macrocycles could be opened by deprotonating the ammonium recognition units on raising the solution pH to around 10 (Figure 2.10a) [79]. For pH-triggered drug delivery applications, however, cargo release at values below physiological pH is more relevant (corresponding to acidic intracellular compartments or acidic interstitial spaces characteristic of some disease states). To this end, an end-tethered

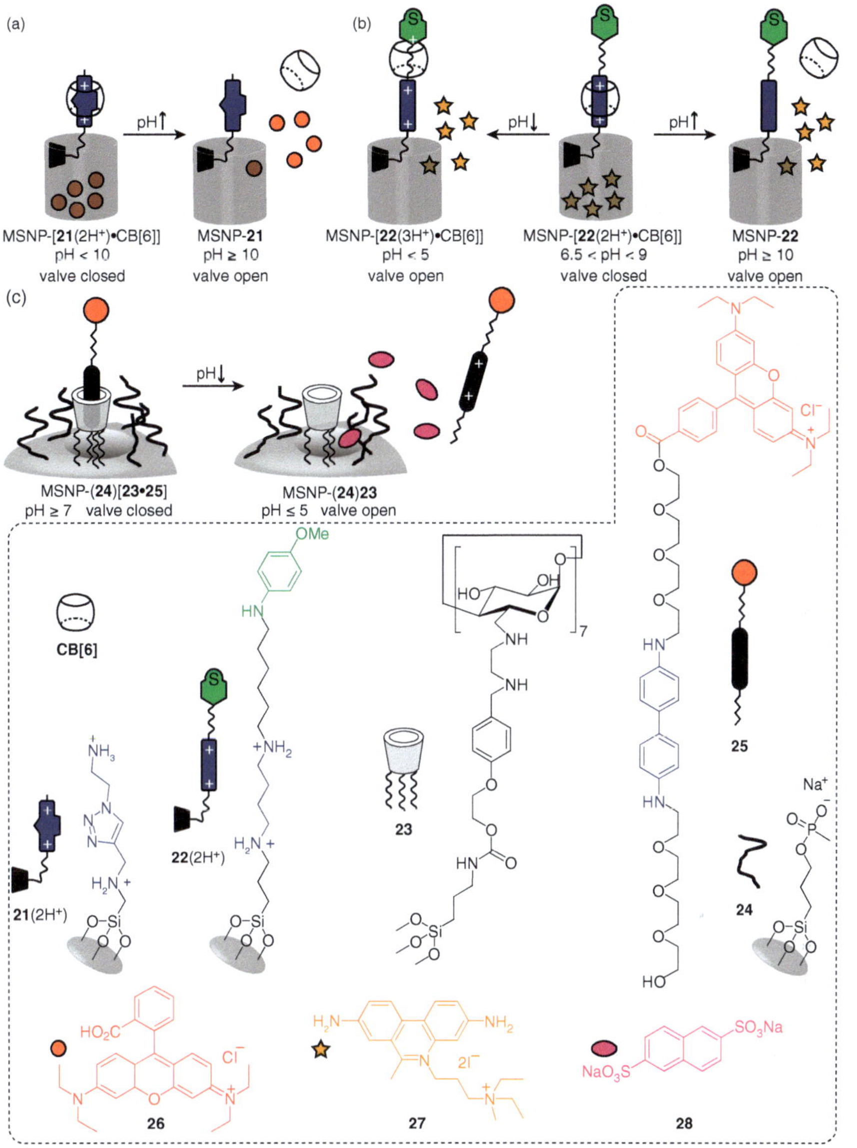

Figure 2.10 pH-Triggered cargo release mechanisms for MSNP carriers compatible with aqueous environments. (a) Nanovalve release at high pH [79]; (b) dual-mechanism nanoshuttle release at low pH (switching pH tuneable according to the electronic nature of substituent S) and nanovalve release at high pH [80]; (c) nanopiston release at low pH [81]. For the molecular structure of CB[6], see Figure 2.7.

[2]pseudorotaxane bistable shuttle [**22**(2H$^+$)•CB[6]] was anchored on MSNPs (<*d*> ~ 130 nm, Figure 2.10b) [80]. At neutral pH values, the CB[6] macrocycle resides on the proximal tetramethylenediammonium binding site, blocking cargo escape. At lower pH values, the anilinium unit is protonated, creating a distal hexamethylenediammonium binding site of higher affinity to the proximal site. Shuttling of the macrocycle to the surface-distal co-conformation opens the pores, allowing cargo to be released. The switching was tunable according to the pK_a of the anilinium unit, and the rate of cargo release was observed to depend on the acidity of the environment. Alternatively, at high pH values (~10), deprotonation of all amines disrupts the intercomponent molecular recognition interactions entirely, resulting in de-threading of the interlocked architecture [80]. De-threading at low pH values has also been achieved for an alternative end-tethered [2]pseudorotaxane design incorporating a dialkyl bispyridinium binding site alongside a terminal carboxylate unit, which, on becoming protonated, released a CB[6] macrocycle, opening access to the MSNP pores [82].

Cargo release at low pH values in aqueous environments has also been achieved using a macrocycle-tethered [2]pseudorotaxane "nanopiston" design (Figure 2.8c). To direct the attachment of gating units near the pore orifices, a phosphonate-terminated modifier **24** was used to block exterior silanol sites. Thus, only the pore regions were available for modification with hepta-functionalized-β-CD macrocycles producing MSNP-(**24**)**23** (<*d*> ~ 100 nm) [81]. The pore-localized surface-attached macrocycles provide recognition sites for benzidine "piston" **25**. Molecular structural characterization for modifications on MSNPs is considerably more challenging than smaller monolayer-stabilized NPs, with many studies relying on largely phenomenological evidence. In this example, strong support for the successful covalent attachment of the β-CD macrocycles was provided by solid-state NMR spectroscopy. Selective excitation of fluorescent probe cargo **28** and fluorophore-appended piston **25** allowed their release profiles to be monitored simultaneously. The surface-attached [2]pseudorotaxanes effectively blocked cargo release at pH 7. On lowering the pH, both piston and cargo were released at pH-dependent rates, which were faster for the positively charged piston than the small interior-trapped cargo [81].

While pH is a potentially useful endogenous stimulus for cargo release in biological contexts, photo-responsive systems offer operator-guided spatiotemporal control over cargo release in virtually any setting to which the light trigger can be delivered. Photo-release was achieved from water-soluble end-tethered [2]pseudorotaxane MSNP-[*E*-**29**•β-CD-Py] (<*d*> ~ 400 nm, Figure 2.11a) [83]. Consistent with other designs using this architecture, the intact surface-attached MIM architecture effectively blocks cargo release. Photoisomerization of the azobenzene unit to generate MSNP-*Z*-**29** disrupts the molecular recognition interaction, leading to de-threading and cargo release. The release of a fluorescent cargo (**26**) and macrocycle (β-CD-Py) could be monitored simultaneously, again showing correlated release profiles with the capping macrocycle removed more rapidly than the appearance of the cargo [83]. A similar design was realized using CB[7] macrocycles and surface-tethered photoisomerizable cinnamamide recognition sites; photoinduced cargo release was

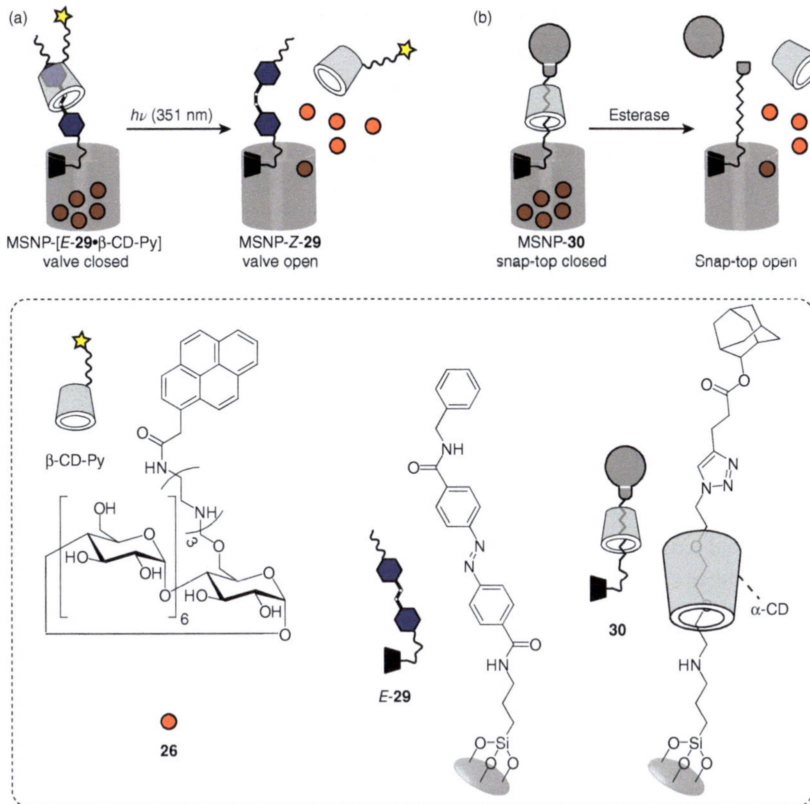

Figure 2.11 Cargo release by (a) light-triggered nanovalve opening [83], and (b) enzyme-catalyzed covalent bond cleavage of snap-top gates [86]. For molecular structure of **26**, see Figure 2.10 and of α-CD, see Figure 2.2.

achieved in both phosphate buffer solution and fetal bovine serum, more accurately modeling physiological environments [85]. Furthermore, the cargo could be released in multiple short bursts by pulsatile application of the light stimulus.

For structures in which the noncovalent interactions between interlocked components are relatively weak, cleavage of a distal covalent bond in the surface-attached component can trigger disassembly of the MIM architecture and hence cargo release (Figure 2.8d); alternatively, a surface-proximal cleavage site can release the interlocked structure in its entirety (Figure 2.8e). End-tethered [2]rotaxane **30** was constructed on MSNPs (<*d*> ~ 400 nm) through covalent attachment of a bulky stopper unit to a pseudorotaxane pre-complex in the presence of the probe cargo **26** [86] (Figure 2.11b). Using a universal irreversible stoppering reaction (azide–acetylene cycloaddition) allows a variety of stoppers incorporating different cleavable bonds to be installed. So-called "snap-top" MSNP-**30** exhibited virtually no cargo leakage until the hydrolysis of the ester-linked stopper was catalyzed by an esterase enzyme, leading to cargo release [86]. A similar design that incorporated a reductively cleavable disulfide linkage in the surface-tethered [2]rotaxane was constructed [84], pointing the way to applications that take advantage of the thiol-rich reducing environment inside cells.

A powerful consequence of the modular nature of MIM–NP hybrids is that concepts developed on one core type should be readily translatable to other core materials. Metal–organic framework nanoparticles (MOFNPs) possess several favorable characteristics for biomedical applications, including drug delivery [87], not least of which is their almost infinitely tunable material composition and exceptionally high internal surface areas. The end-tethered [2]pseudorotaxane nanovalve architecture (Figure 2.8b) has been applied on UNCM-1-NH2 MOFNPs to achieve pH and competitive binding controlled cargo release [88]. In this example, alkylpyridinium stalks were covalently conjugated at amine functional handles throughout the MOF structure. Threading by negatively charged carboxylatopillar[5]arene macroycles to assemble [2]pseudorotaxane gates trapped cargos within the mesoporous structure. Payload release was observed on either lowering pH (<5) to protonate the negatively charged macrocycle gates or on introducing a competitive guest (methyl viologen [MV^{2+}]).

With a broad pallet of mechanistic options and structural realizations (varying in both the molecular and nanomaterial components) to choose from, attention has more recently shifted toward multifunctional and multicomponent constructs that exhibit sophisticated capabilities and applications-optimized properties.

2.4.2 Autonomous Drug Delivery Triggered By Endogenous Conditions

Nanocarriers that can release high concentrations of a therapeutic agent on encountering specific local physical or chemical cues have been a long-time target for the delivery of toxic, unstable or immunogenic therapeutic agents selectively to a disease site only. As well as protecting the active agent from degradation or metabolism and modifying undesirable pharmacokinetics, endogenously triggered release mechanisms can maximize therapeutic indices and minimize side effects, all without requiring covalent modification of the drug molecule or external intervention. To realize this theoretical promise, however, successful delivery vehicles must overcome multifarious challenges posed by the *in vivo* environment, requiring stepwise optimization of multiple components.

To maximize cargo retention at normal physiological pH values and release under the acidifying conditions of the endosomal uptake pathway, a nanovalve design based on end-tethered *N*-alkyl benzimidazole pseudorotaxanes was found to have the optimal pK_a (MNSP-[**31**•β-CD](**32**)(**33**), Figure 2.12) [89]. Several features of the multicomponent nanoscale scaffolds were tuned to optimize delivery: (i) core size ~ 100 nm for optimal endosomal uptake; (ii) surface charge – positively charged ammonium ions (**32**) for rapid release of the cationic chemotherapeutic doxorubicin while retaining good loading capacity; (iii) surface functionalization with a permanently linked fluorophore (**33**) enabled nanocarrier tracking by confocal microscopy. Efficient endosomal uptake by squamous carcinoma (KB-31) cells resulted in NP localization at the perinuclear region. Release of the fluorescent drug molecules to the nucleus followed, resulting in apoptotic cell death 80 h after

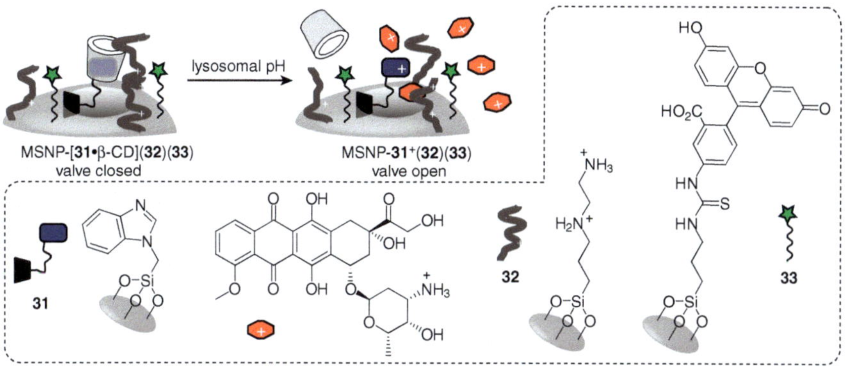

Figure 2.12 MSNP nanocarriers optimized for live cell and *in vivo* pH-triggered delivery of a cationic chemotherapeutic [89]. For molecular structure of α-CD, see Figure 2.2.

treatment [89]. The same gating mechanism has been successfully implemented on porous silica nanorods for delivery to human pancreatic carcinoma PANC-1 cells [90]. It has also been shown that similar end-tethered pseudorotaxane gates can be translated to hollow core MSNPs. This marked a significant advance, demonstrating sufficiently efficient surface functionalization to control leakage from multiple interconnected pores while retaining pH-triggered cargo release from nanodelivery vehicles that have much higher loading capacity than conventional MSNPs [91].

Building on these optimized systems, endogenous stimuli-triggered MSNPs have been translated *in vivo* for the delivery of anti-cancer agents and the treatment of diseases caused by intracellular pathogens such as tuberculosis and pneumonic tularemia. Taking advantage of the efficiency by which macrophages ingest nano-sized objects, MSNPs have been optimized to deliver moxifloxacin – a broad-spectrum antibiotic – to human mononuclear phagocytes, which are the primary host for the *Francisella tularensis* bacterial pathogen [92, 93]. Using benzimidazole-based pH-triggered pseudorotaxane MSNP gates (**31**, Figure 2.12), *in vitro* studies in *F. tularensis*-infected differentiated THP-1 macrophages demonstrated selective delivery of the antibiotic cargo on encountering low-pH endolysosomal compartments. Cargo release was most efficient for MSNPs bearing a negative surface charge modifier (**24**, Figure 2.10). Even higher release capacities were achieved using a similar design controlled by redox-triggered end-tethered pseudorotaxane snap-top gates, which take advantage of the reducing intracellular environment [93]. In a mouse model of pneumonic tularemia, both pH- and redox-triggered nanocarriers proved to be significantly (two- to fourfold) more effective than administration of the same concentration of free drug, with consequent potential to shorten treatment regimes, avoid the emergence of drug resistance, and reduce systemic toxicity [92, 93].

The inherently multicomponent nature of NP platforms presents options for constructing sophisticated delivery vehicles that integrate multiple functionalities. In cancer therapy, it is proposed that nanocarriers can take advantage of an "enhanced permeation and retention" effect [94] as a consequence of the leaky vasculature and poor lymphatic drainage that characterizes many solid tumors [95]. Targeting strategies based on molecular interactions may further improve the accuracy of cargo delivery to diseased tissue only. Molecular targeting also enables nanodelivery approaches to be extended to cancers and other diseases for which there is no inherent selectivity for carriers with nanoscale dimensions. Targeting groups have been incorporated as part of the mechanically interlocked architectures of disulfide-linked snap-top gates on hollow MSNPs, combing ligand-targeted delivery of high-capacity nanocarriers with reductive release of the anticancer drug doxorubicin. A folic acid derivative was used as the bulky peripheral stopper group of an end-tethered [2]rotaxane to target folate receptors on HeLa cells [96], or a lactobionic acid-labeled-β-CD macrocycle was used to create an end-tethered [2]pseudorotaxane targeting the asialoglycoprotein receptor on HepG2 cells [97]. In both examples, the targeted nanocarriers exhibited superior inhibition of tumor growth compared to non-targeted MSNPs and, furthermore, NP delivery was shown to avoid the side effects of treatment with the free chemotherapeutic [96, 97].

A similar folate-stoppered design has been installed on conventional MSNPs but incorporating an esterase-cleavable linkage in the end-tethered α-CD-interlocked [2]rotaxane [98]. Ligand-targeted and enzyme-activated delivery of the anti-cancer drug camptothecin triggered apoptosis in human osteosarcoma cells. Based on the estimated dose released, the efficacy of the hydrophobic anti-cancer drug was significantly increased by NP delivery, likely a result of both improved physicochemical properties and active cellular uptake by ligand-mediated endocytosis. Enzyme-activated versions of either snap-top design (Figure 2.8d,e) have the potential as highly specific endogenously triggered delivery mechanisms [86, 99].

The ability to track nanocarriers *in vivo*, probe their mechanisms of action and ultimately achieve a real-time readout of the disease response to therapy are among the fundamental objectives for nanomedicine theranostics [100]. Exploiting the two-component nature of mechanically interlocked architectures, direct visualization of gate-opening by confocal fluorescence microscopy was enabled via a Förster resonance energy transfer (FRET) response [101]. In a variation of the reductively cleaved end-tethered [2]pseudorotaxane design (Figure 2.8e), a FRET acceptor was appended to the macrocycle and a FRET donor to the retained NP-bound stalk. The change in the FRET signal could be correlated with the release of a fluorescent drug and subsequent reduction in cell viability, suggesting this approach can be applied more generally to monitor delivery of non-fluorescent cargo.

2.4.3 Cargo Release Using Tissue-Penetrating External Triggers

Nanocarriers responsive to external tissue-penetrating stimuli could release their cargos at precisely controlled times and locations. Guided by modern imaging techniques for precisely pinpointing diseased tissue, exquisite spatiotemporal control over drug delivery should be possible, with all the attendant advantages of noninvasive drug delivery systems [67, 69].

Perhaps the most immediately obvious exogenous stimulus is light, which can be applied in a focused, directional manner with wavelength, intensity, and duration all easily controlled. However, potentially harmful photodamage and poor tissue penetration of UV and visible light can present significant challenges depending on the location of the disease site. Thus, components that are stimulated by near infra-red (NIR) light hold a particular attraction for *in vivo* applications. Several distinct mechanisms have been explored for *in vivo* light-triggered delivery from MIM–MSNP hybrid carriers, many of which make clever use of nanoscale properties to enhance the photorelease mechanism.

Photo- and thermally-switchable pore gates have been created using a bistable end-tethered [2]rotaxane molecular shuttle (Figure 2.8a). A *Z*-azobenzene moiety holds an α-CD macrocycle close to the MSNP surface, while the *E*-azobenzene isomer allows shuttling away from the surface and consequent pore opening [102, 103]. *In vivo* cargo delivery was achieved in zebrafish larvae using either visible light irradiation or bulk heating to 37 °C, with the latter proving more efficient [102]. Importantly, installing the same MIM gate on silica-coated Au nanorods (AuNRs) allowed *in vivo* gate opening to be triggered photothermally on irradiation with tissue-penetrating NIR light, exciting the strong AuNR longitudinal absorption band at 830 nm [103].

In a similar fashion, dissociation of end-tethered [2]pseudorotaxanes (Figure 2.8b) can be achieved by either bulk heating or plasmonic photothermal heating for MSNPs encapsulating 20 nm AuNP cores (UV irradiation to excite localized surface plasmon resonance (LSPR) with $\lambda_{max} = 530$ nm) [104], and on silica-coated AuNRs (NIR irradiation to excite LSPR with $\lambda_{max} = 800$ nm) [105]. These multifunctional systems also have the potential to combine chemotherapeutic delivery with local hyperthermia in a two-pronged attack on deep-tissue tumors. One drawback of the pseudorotaxane designs for *in vivo* applications, however, is the potential for disruption of the mechanically interlocked architecture by competitive binders, leading to off-target cargo release.

Incorporating photocleavable covalent bonds in the MSNP-tethered MIM structure provides another route to light-triggered cargo release using snap-top mechanisms (Figure 2.8d,e), which are not susceptible to competitive binding events. End-tethered [2]pseudorotaxanes in which a terminal coumarin is encapsulated by a β-CD macrocycle could be cleaved on irradiation with either UV light or two-photon excitation of the coumarin chromophore using a femtosecond pulsed NIR laser source, giving similar cargo release profiles in each case [106]. Alternatively, a general route to rendering redox-triggered snap-top MSNPs responsive to NIR actuation was demonstrated by incorporating photosensitizer molecules as secondary surface-bound components. Two-photon excitation of the sensitizer generated a highly reducing excited state, which could transfer electrons to the neighboring snap-top gates, resulting in disulfide cleavage and gate opening [107]. Although an external sacrificial reductant was required to suppress back electron transfer in this example, such modular approaches, which take advantage of the inherently multi-component nature of nanoscale delivery vehicles, should enable rapid, independent optimization of each component (e.g. photosensitizer, gating mechanism, and blocking MIM architecture).

Alternating magnetic fields (AMF) present an alternative mechanism for non-invasive actuation of thermoresponsive gates with superior tissue penetration but poorer spatial resolution, compared to light-stimulated systems. The key to adapting MIM-gated delivery vehicles for AMF response was to integrate the MSNPs with a second nanoscale component: superparamagnetic NPs. On exposing MSNPs grown around superparamagnetic Zn-doped Fe_3O_4 cores, the local internal heat generated was sufficient to disrupt end-tethered CB[6]•dialkylammonium [2]pseudorotaxanes (Figure 2.8b) [108]. Applied *in vitro* to a human breast cancer cell line, these carriers induced a dual cytotoxic effect as a consequence of both AMF-stimulated hyperthermia and release of the anticancer drug doxorubicin. Thermosensitive snap-top gates (Figure 2.8d,e) on superparamagnetic-enabled carriers can similarly be activated by AMF [109, 110]. Although hyperthermia induced by bulk heating in the local vicinity can be attractive for some therapies, this can be deleterious in other clinical scenarios (e.g. heat can promote cancer metastasis). Heteromaterial superparamagnetic core–shell particles (Mn-doped Fe_2O_4@Co-doped Fe_2O_4, $\langle d\rangle \approx 11$ nm) were developed to optimize thermal cleavage of dialkyl azo-linked end-tethered [2]pseudorotaxane snap-top gates ([**34**•β-CD]) anchored on MSNP outer shells ($\langle d\rangle_{total} \approx 55$ nm, Figure 2.13) without significant bulk heating [110]. Only short (2–10 min) bursts of AMF were required to initiate cargo release, avoiding cytotoxicity via local hyperthermia. Several hours after stimulation, the final dose delivered and reduction in cell viability (for doxorubicin delivery to human pancreatic cells) were found to be dependent on the duration of the initial exposure [110]. Alternatively, pulsatile cargo release was achieved on repeated

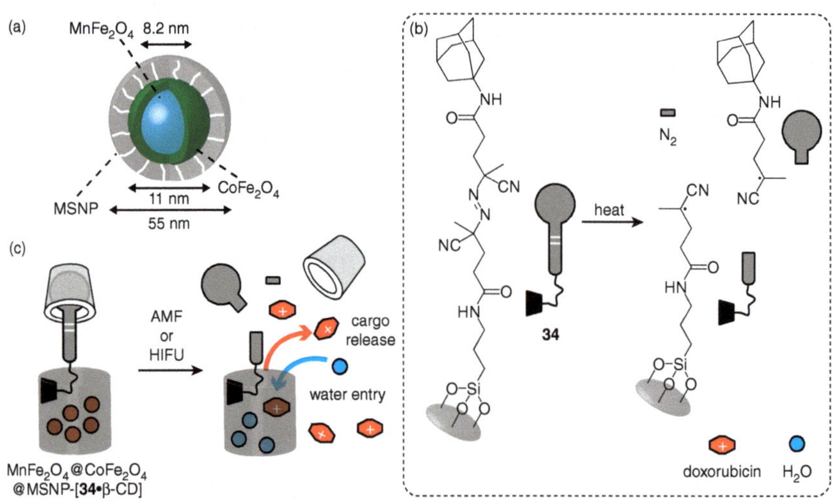

Figure 2.13 Nanoscale structure (a), molecular mechanism (b), and cartoon representation of operation (c) of multifunctional theranostic nanocarrier platforms. Remote stimulation of the superparamagnetic core by either alternating magnetic fields (AMF) or high-intensity focused ultrasound (HIFU) generates local hyperthermia, triggering the opening of the snap-top gates via a thermally cleavable azo-linkage, allowing simultaneous cargo release from and water entry into the mesoporous shell.

AMF exposure, which could be exploited to deliver a controlled therapeutic dose several times at the same site.

The same versatile MIM-gated MSNP carrier has also been activated by a third tissue-penetrating external stimulus. High-intensity focused ultrasound (HIFU) can induce both thermal and mechanical effects. Furthermore, HIFU is already in clinical use, integrated with magnetic resonance imaging (MRI) for guided therapy [69]. This inspired the development of a sophisticated NP-enabled self-reporting theranostic approach. HIFU-stimulated cargo release from $MnFe_2O_4$@$CoFe_2O_4$@MSNP-[**34**•β-CD] (Figure 2.13) is accompanied by an admission of water to the mesoporous network [111]. Increased proton relaxivity when in proximity to the encapsulated superparamagnetic cores results in enhanced MRI contrast. Thus, under MRI-guided HIFU stimulation, the dose of the drug (doxorubicin) released – tunable according to HIFU exposure duration – and consequent decrease in cell viability (human pancreatic cancer cells) could be imaged in changes to MRI contrast [111]. This ingenious exchange method, enabled by the multifunctional properties of MIM-gated nanocarriers, has exciting potential for assessing the precision of HIFU-stimulated drug release and *in situ* monitoring for real-time optimization of stimulation parameters.

2.4.4 Nanoscale Communication Between Responsive Nanoparticles

The integration of responsive MIM architectures with nanoscale platforms predicates that high levels of structural and functional sophistication can be reached by building on cutting-edge designs and capabilities for both the molecular and nanoscale components.

Combining several (supra)molecular units that are selectively responsive to specific stimuli can be exploited to achieve logic-gate cargo release responses that depend on combinations of multiple inputs. In an architecturally distinct implementation of MSNP-bound interlocked structures, bis(azobenzene)-terminated molecular struts were used to cross-link surface-tethered β-CD-side-chain polymers, thus blocking pore access [112]. Cargo release could be achieved by dissociating the [3]pseudorotaxane cross-links, either by $E \rightarrow Z$ photoisomerization of the azobenzene struts, or by introducing a competitive host for *E*-azobenzene (α-CD) to preferentially generate a solution-phase [3]pseudorotaxane. Alternatively, because the macrocycles were attached to the polymer backbone via disulfide bonds, the intact [3]pseudorotaxane cross-link could be excised on the application of a chemical reductant [112]. This multistimuli-responsive behavior corresponds to an OR logic response, with each stimulus triggering a different cargo-release mechanism and, hence, a distinct release profile. Cargo release only under a highly specific combination of stimuli can be achieved by AND logic responses. For example, pH-responsive end-tethered [2]pseudorotaxane nanovalves (Figure 2.8b) were combined with (non-interlocked) azobenzene "nanoimpellers" (photoswitchable steric blocking groups within the pores) to achieve cargo delivery only in response to the combined action of both photons and protons [113].

Selective and controlled release of multiple cargos from a single nanocarrier is particularly challenging but is highly attractive, for example, in combination drug therapies. A proof-of-concept system achieved the size-selective sequential release of two cargos, each in response to a specific trigger, by combining the snap-top and nanopiston mechanisms (Figure 2.8, Types c and e) [114]. On MSNPs with pores of diameter ~ 2.5 nm , disulfide-linked face-tethered β-CD macrocycles block access for large (>6 Å) cargos, while still allowing uptake and release of smaller (<6 Å) molecules (Figure 2.14a). Plugging the macrocycle cavities by forming a face-tethered [2]pseudorotaxane complex blocks access for both cargos. Therefore, a pH-sensitive nanovalve based on an association of methyl orange (**36**) with β-CD could control release of a small cargo (**37**) from MSNP-[**35•36**], leaving large cargos trapped (Figure 2.14a). Subsequent removal of the β-CD macrocycles under reducing conditions (addition of a monothiol) then released the larger cargo (**38**) [114]. It remains to be shown whether this type of mechanism can be extended to release clinically useful therapeutics under physiologically relevant conditions.

NP platforms present further architectural possibilities as a consequence of their nanoscale dimensions – such as spatial segregation of chemical functions – or that arise by combining the intrinsic properties of several nanomaterials in one device. Moreover, integrating artificial supramolecular and nanoscale components with biomolecules can generate yet further levels of functional sophistication. Large-pore MNSPs can accommodate biomacromolecule cargos, but effective gating of large openings is a challenge for small-molecule components. Stimuli-responsive gating of 6.5 nm diameter MSNP pores has been achieved by combining imine-linked face-tethered β-CD macrocycles with multitopic adamantane guests connected via a planar polyaromatic core (Figure 2.14b) [116]. The surface-bound [*n*]pseudorotaxane "megagates" (n = 4 or 7) selectively trap cargos according to size and are opened in a snap-top mechanism via imine hydrolysis in the presence of acid. This design was exploited to demonstrate a novel two-stage enzymatic amplification sensing strategy using MSNP-[**39•40**] [115]. The size-selective megagates were used to segregate an esterase enzyme in the large MSNP pores from a large probe substrate (ester-caged fluorophore **42**) in solution. Small molecules such as ester analyte **41**, however, were still able to access the encapsulated enzyme from the solution. Enzyme-catalyzed analyte hydrolysis, therefore, generated acetic acid in the pores, triggering hydrolytic opening of the snap-top gates and leading to enzyme release (Figure 2.14b). Subsequent hydrolysis of the probe generated **43** in solution, turning on a fluorescent signal, which is amplified by the further generation of acid such that the optical output is not stoichiometrically correlated with the initial quantity of analyte [115].

Anisotropic heteromaterial NPs provide attractive platforms for integrating biological and abiotic functional units. So-called "Janus" NPs present well-defined chemically distinct surface regions that can be used to spatially segregate chemical functions on the nanoscale. Janus Au-MSNPs functionalized with active biomolecule components on the Au face, chemically responsive (supra)molecular units on the mesoporous surface, and small-molecule cargos within the pore interiors can exhibit chemical communication between the biological and abiotic

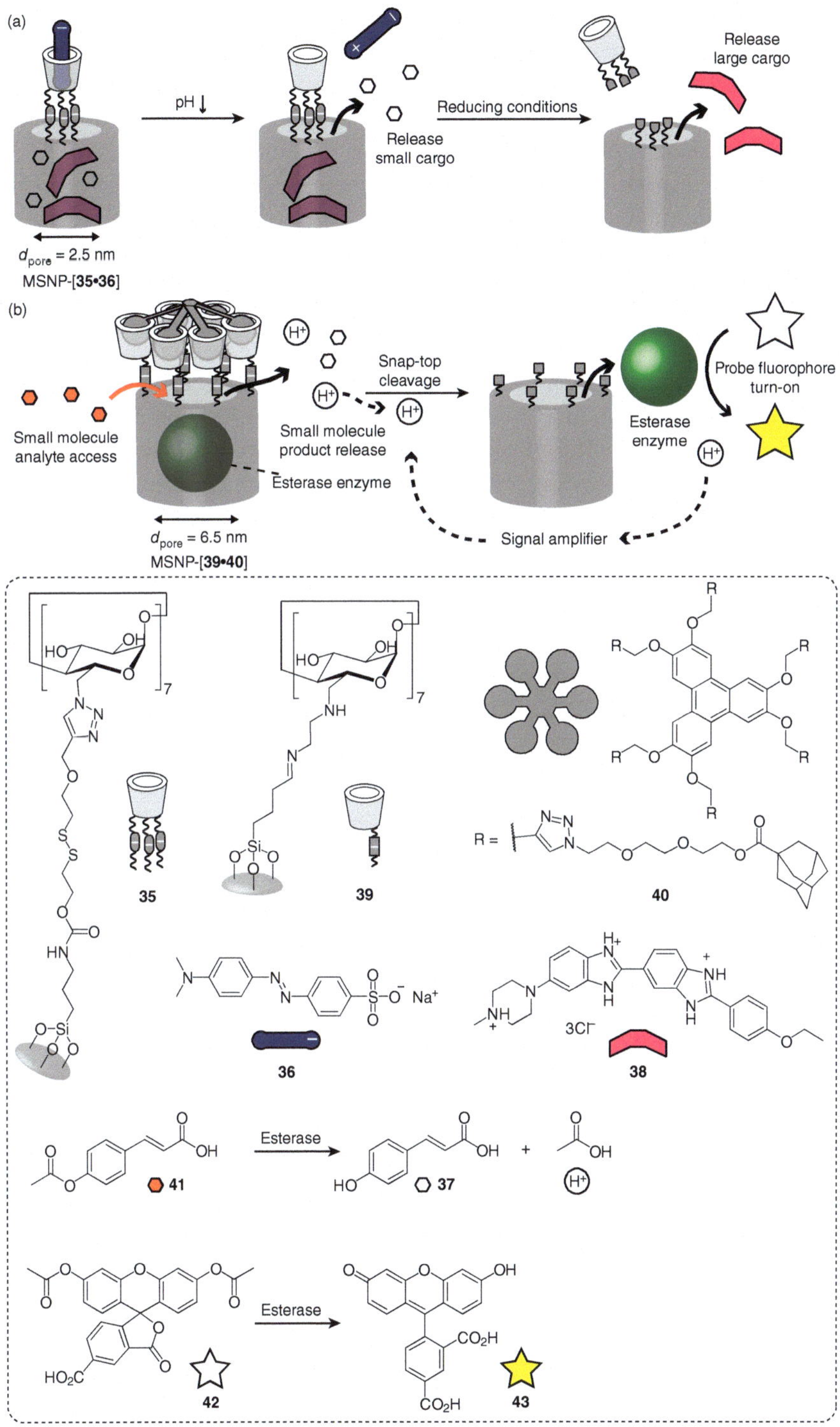

Figure 2.14 Size-selective pore access on MIM-gated MSNP nanocarriers. (a) Sequential differential stimuli-responsive dual cargo release: small cargo release under low pH nanopiston opening; large cargo release on reductive snap-top cleavage. (b) "Megagate" MIM-controlled enzyme-amplified fluorescent detection of small molecule esterase substrates.

components, either on the same particle or between different NP populations. The structural and functional sophistication of these systems has led to a variety of impressive drug delivery, logic, and communication functions [70]. Mechanically interlocked gating units are just one of the functional components that have enabled the rapid development of such platforms.

For example, internal activation of pH-responsive MSNP end-tethered [2]pseudorotaxane nanovalve gates (Figure 2.8b) has been triggered by the products of enzyme-catalyzed reactions taking place on the same Janus nanovehicle (Au-MSNP-**44**, Figure 2.15a) [117]. Esterase and glucose oxidase enzymes translated either ethyl butyrate (Input A) or D-glucose (Input B) chemical inputs into a reduction in local pH (through the generation of butyric acid or gluconic acid, respectively), thus disrupting [benzimidazole•β-CD] pseudorotaxane complexes to release cargo from the MSNP unit – an example of OR logic processing [117]. This modular design significantly expands the range of chemical signals that can be exploited for cargo release by using integrated enzymatic "encoding units" to convert any number of inputs into one of the common effectors for actuating (supra)molecular gating systems.

Furthermore, combinations of enzymatic encoders and pore-gating systems have been used to engineer communication networks mediated by the exchange of chemical messengers between distinct populations of Janus Au-MSNP carriers (e.g. Figure 2.15b,c). For example, adding lactose as an input to a mixture of Au-MSNP-**45** and Au-MSNP-**46** (Figure 2.15b) triggers the production of glucose through the action of β-galactosidase (β-Gal) on Au-MSNP-**45**. Glucose is not capable of opening the snap-top pore gates on Au-MSNP-**45** but, as a substrate for glucose oxidase, acts as an interparticle messenger to trigger gate opening on Au-MSNP-**46** via generation of internal effector gluconic acid. Consequently, a second interparticle messenger – *N*-acetyl-L-cysteine – is released from the pores of Au-MSNP-**46**, which subsequently acts to open the disulfide-linked gates on the original sender particle (Au-MSNP-**45**), which – now acting as a receiver – decodes the message to release its fluorescent payload as an output [118]. As well as exquisite two-stage selectivity only for inputs that are processed by β-Gal to produce the glucose substrate for GOx, it was found that nanoscale co-localization of GOx near the pH-sensitive gates on Au-MSNP-**46** is critical to the efficient operation of the signaling mechanism. [2]Pseudorotaxane-gated Au-MNSP-**46** has also been implemented in a similar biunivocal communication network where the other sender/receiver node is a living microorganism (budding yeast cells) [120].

The same design principles have been extended to construct a circular network of three distinct sender/received populations (Figure 2.15c) [119]. This example combines orthogonally addressable peroxide-responsive end-tethered [2]pseudorotaxane snap-top gates (Au-MSNP-**47**), pH-responsive end-tethered [2]pseudorotaxane nanovalves (Au-MSNP-**48**), and non-interlocked disulfide-linked polymer gates (Au-MSNP-**49**). The input lactose is encoded as an interparticle messenger (galactose) by Au-MSNP-**49**. Processing by galactose oxidase leads to localized production of an internal effector (hydrogen peroxide) for opening the snap-top gates on Au-MSNP-**47**. This releases a second interparticle messenger

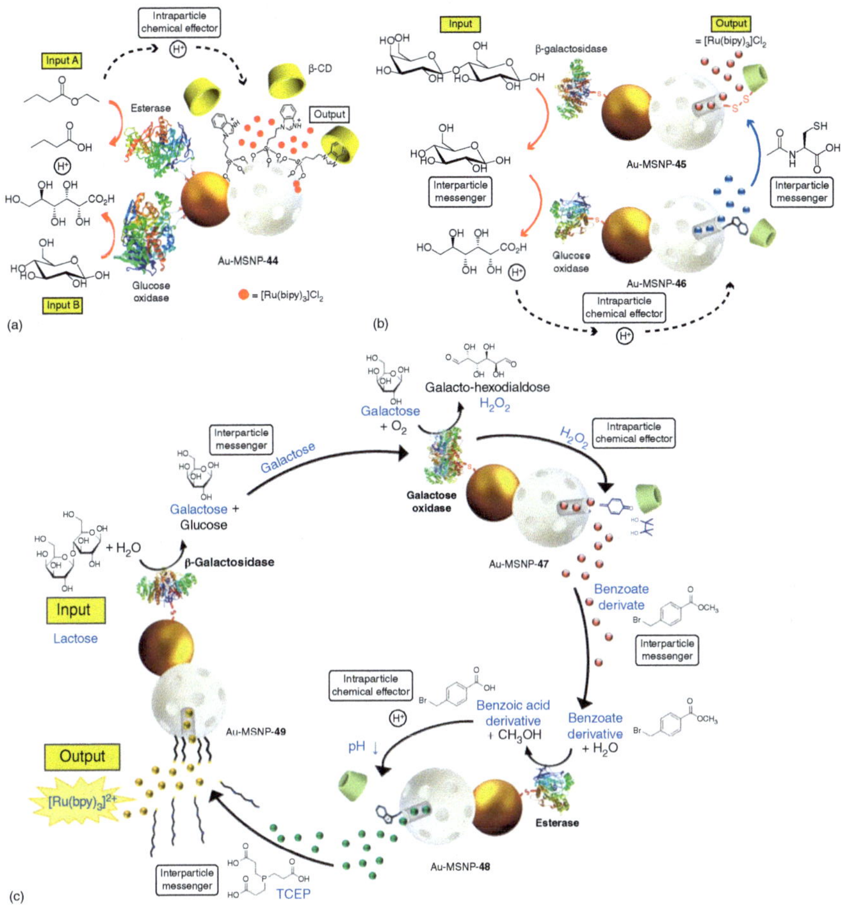

Figure 2.15 Nanoscale communication networks based on processing and exchange of chemical messengers by hybrid-responsive biomolecule–abiotic Janus Au-MSNP nanodevices. (a) Internal signal processing: fluorescent cargo release (output) is triggered by enzyme-catalyzed conversion of external input (A or B) to produce a locally generated intra-particle messenger (H^+) [117]. (b) Biunivocal communication between a pair of sender/receiver nanodevices using both enzymatically generated and cargo-release messengers. (c) Circular communication network between three sender/receiver nanodevices exploiting oxidation-sensitive end-tethered [2]pseudorotaxane snap-top gates; pH-sensitive end-tethered [2]pseudorotaxane nanovalve gates and non-MIM thiol-sensitive gates [119]. Source: Adapted with permission from Díez et al. [117] (a), copyright 2014 American Chemical Society; Llopis-Lorente et al. [118] (b), reused under CC BY license; de Luis et al. [119] (c), copyright 2021, the Royal Society of Chemistry.

(a benzoate ester), which, as a substrate for the esterase enzyme of Au-MSNP-**48**, generates an internal effector (protons) for opening pH-responsive nanovalves. A third interparticle messenger (reducing agent TCEP) is thereby released, which opens the gates on this first nanocarrier to release a fluorescent payload, signaling the output of the communication network (Figure 2.15c) [119].

2.5 Mechanically Interlocked Molecular Links for Nanoparticle Assemblies

Aggregates and assemblies can significantly modify NP properties or generate entirely new emergent behaviors. Furthermore, composite materials offer a format for augmenting NP properties with additional components and integrating nanomaterials with any number of condensed-phase technologies. The ability to direct and control NP aggregation and assembly on command is consequently a crucial capability for emerging nanotechnologies [45, 121–126]. Mechanically interlocked molecular structures present a number of opportunities for directing and controlling NP assemblies. Primarily, pseudorotaxane complexes have been applied as noncovalent interparticle linkages that are potentially stimuli-responsive [5, 127]. Fully interlocked molecular links for NP assemblies are only beginning to emerge but have exciting potential to combine robust linkages with dynamic capabilities.

2.5.1 Pseudorotaxane-linked Nanoparticle Assembly–Disassembly

Having established that on-particle host–guest inclusion complexes were formed between surface-bound β-CD macrocycles and ferrocene guests (Figure 2.2b), the pioneering early investigations of Kaifer and coworkers went on to explore pseudorotaxane-linked NP aggregates [14]. Adding a simple alkyl diferrocene molecular linker to water-soluble AuNP-β-tCD (<*d*> = 13 nm) produced a red shift in the UV–Vis spectrum, indicating surface plasmon coupling resulting from NP aggregation. Slow precipitation was then observed at a rate dependent on the concentration of linker. Monotopic FcOH could be used to partially redisperse precipitates, while no precipitation occurred in the presence of an excess of either FcOH or β-CD, providing good evidence for dynamic noncovalent interparticle linkages [14]. The easily observed optical signal generated on forming or dispersing AuNP aggregates has been used as an output signal for myriad sensing systems [128]. In an impressive example, aggregates that are linked through multivalent combination of individually weak noncovalent interactions between AuNP-β-tCD and diferrocene linkers were shown to enable an amplified sensing mechanism that is capable of detecting as few as 23 copies of an oxidizing enzyme [129]. The distinct characteristics of relatively weak noncovalent interactions with threaded guests can also be used to engineer structural control in AuNP assemblies. Xia and coworkers generated AuNPs (<*d*> = 20 nm) in phosphate buffered saline (PBS) buffer using unmodified β-CD macrocycles as both reducing and stabilizing agents [130]. Despite a very small estimated number of surface-bound β-CD, a tetra(carboxyphenyl)porphyrin linker induced NP aggregation. Electron microscopy (scanning electron microscopy (SEM) and transmission electron microscopy (TEM)) revealed 1D AuNP chains at low linker concentrations, while higher concentrations produced 2D aggregates. Interparticle distances in the 1D chains were consistent with bridges formed predominantly through 180° opposing phenylene units on the tetravalent linker, while spacing in the 2D assemblies suggested a mixture of 180° and 90° linkages. However, it was not established whether these links were exclusively mediated by formation of inclusion complexes with NP-bound β-CD, or through direct interaction between the carboxylate groups and AuNP surfaces [130].

Host–guest complexes of a number of other surface-tethered macrocycles have been investigated for several applications, including NP assembly [5, 127]. For example, decacarboxypillar[5]arene was used to stabilize AuNPs (<*d*> = 3.1 nm) produced by a direct synthesis protocol in water [131]. Introduction of a ditopic bis(alkyldipyridinium) guest induced NP aggregation. However, the AuNP cores are only weakly stabilized by the carboxylated macrocycles, so NP assembly was accompanied by Ostwald ripening, leading to significant changes in core size. In the absence of direct evidence for on-surface inclusion complex formation, it also cannot be ruled out that nonspecific electrostatic interactions with the positively charged linker leading to surface-charge neutralization play a role in driving NP aggregation in this system [131].

Notwithstanding the challenges of characterizing putative surface-bound pseudorotaxane linking structures within heterogeneous aggregates, the pioneering work of Kaifer exemplified the dynamic and adaptive nature of surface-tethered host–guest complexes [14]. Although that early study did not invoke the redox-switchable nature of the ferrocene guest, stimuli-responsive molecular recognition clearly presents an opportunity to construct remote-controlled assemblies [129]. In the context of pseudorotaxane complexes, this concept has been most widely explored using photoswitchable linkers for CD-modified NPs.

In a series of studies, Ravoo and coworkers have shown that ditopic photoisomerizable azoaromatic linkers can actuate light-switched aggregation–redispersion cycles for a variety of CD-stabilized NPs in water (Figure 2.16). The operating mechanism is independent of the surface attachment chemistry, which has been tailored to be compatible with several core materials [132–136]. In each case, the *E*,*E*-configurational isomer of the ditopic linker initiated NP assembly. Aggregation timescales (on the order of minutes to hours) were typically dependent on the linker concentration and, in the case of inherently charged NPs such as SiO_2, also tunable by adjusting solution ionic strength. Redispersion of NP aggregates was first demonstrated for SiO_2NPs (<*d*> = 58 nm) in conjunction with azobenzene linker *E*,*E*-**50** on irradiation at 356 nm [132]. Photoisomerization to the *Z*-azobenzene isomer disrupts the host–guest interaction with surface-bound β-CD, breaking the interparticle linkages. Irradiation with visible light (460 nm) triggered the reverse isomerization process, re-establishing the aggregated state. Although each photostationary state was characterized as a mixture of both configurational isomers, the changes were sufficient to generate reversible and repeatable aggregation–dispersion over at least four cycles [132].

The same strategy was applied to AuNP-β-tCD (<*d*> = 12 nm) using newly developed arylazopyrazole photoswitches that exhibit superior switching fidelity to azobenzenes and similar binding constants for β-CD hosts [133]. Optimized ditopic arylazopyrazole linker *E*,*E*-**51** generated AuNP-β-tCD aggregates with concentration-tunable sizes in the range 30–300 nm. Again, UV irradiation initiated NP redispersion and visible light-triggered aggregation [133].

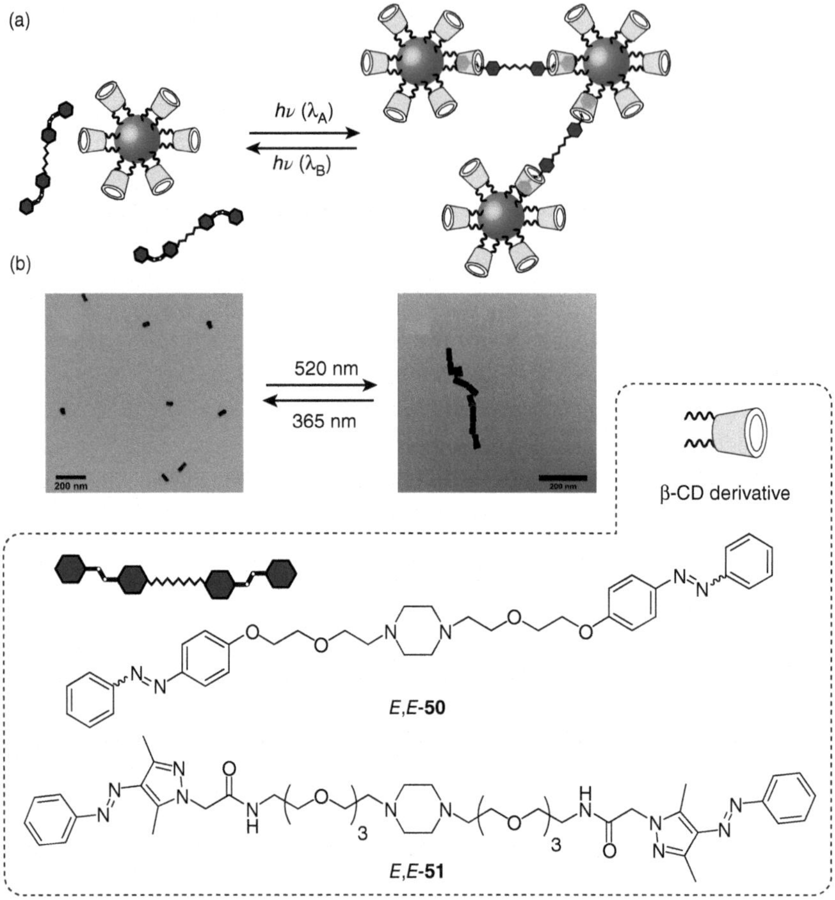

Figure 2.16 (a) Remoted-controlled NP aggregation–dispersion via photoswitchable assembly of face-tethered [3]pseudorotaxane interparticle links. β-CD derivatives were functionalized with surface-binding functionality complementary to NP core material (thiols for AuNPs; carboxylic acids for UCNPs and Fe_2O_3NPs; silyl esters for SiO_2NPs) [132–136]. (b) TEM images of dispersed and end-to-end aggregated states for β-tCD-functionalized AuNRs. (c) Representative molecular structures of bisazobenzene and bisarylazopyrazole linkers [132–136]. Source: Adapted from Niehues et al. [136], reused under CC BY license.

Translating the same systems to upconverting $LiYF_4{:}Yb^{3+},Tm^{3+},Gd^{3+}$ cores (UCNPs, $d \sim 40\,nm$) enabled the $E \rightarrow Z$ isomerization to be triggered by irradiation with a low intensity ($0.22\,W\,cm^{-2}$) NIR light source [134]. On two-photon excitation, the UCNP cores emit at 530 nm, thereby exciting photoswitches encapsulated by the surface-tethered β-CD macrocycles. Avoiding potentially damaging short-wavelength or high-intensity light is of particular interest for potential biological applications. However, in this example, there was a trade-off in the form of significantly slower switching rates and reduced compositional bias at the photostationary state [134].

Expanding this toolkit further with β-CD-stabilized superparamagnetic iron oxide NPs (<*d*> = 7.6 nm), a modular multifunctional system was demonstrated whereby three different NP cores (Fe_2O_3NP, AuNP, and UCNP) are simultaneously assembled and dispersed in response to light [135]. Crucially, energy-dispersive X-ray spectroscopy (EDX) analysis confirmed the presence of all three core materials in the aggregates visualized on a TEM grid, although a quantitative assessment to determine whether the aggregate ratio reflected the ratio of particles in the dispersed state was not carried out. An optical readout of the aggregation–dispersion process was provided by distance-dependent nature of the AuNP LSPR band in UV–Vis spectra; the magnetic NPs enabled field-directed manipulation to pattern the aggregates on a substrate; and the UCNPs again allowed disaggregation to be triggered by NIR irradiation. All three properties could be enhanced by altering the characteristics and proportion of the responsible NP component [135].

Attempting to apply the same approach to achieve directional assembly of anisotropic NPs, Ma and coworkers found that end-to-end assembly of AuNRs could be achieved by adding a pre-formed ditopic [bis(azobenzene)•2(β-CD-SH)] [3]pseudorotaxane linker [137]. Selective ligand exchange of the thiolated CD at the tips of the nanorods (where the stabilizing surfactant bilayer is loosely packed) produced short linear chains. However, disassembly by photoisomerization of the azobenzene photoswitches was only successful if accompanied by mechanical agitation (sonication) and could not be reversed. Furthermore, AuNRs pre-functionalized with β-CD-SH failed to assemble on subsequent introduction of the linker. Ravoo and coworkers solved both of these issues by using perthiolated β-tCD for selective end-functionalization of AuNRs (<*l*> = 58 nm, <*w*> = 25 nm), followed by displacement of the remaining cetyltrimethylammonium bromide (CTAB) bilayer with a monothiolated tetraehtyleneglycol stabilizer, before exploiting arylazopyrazole linker *E*,*E*-**51** for aggregation as before [136]. Likely through a combination of improved resistance to irreversible aggregation conferred by the thiolated stabilizer and the superior switching properties of arylazopyrazoles, reversible light-switched end-to-end assembly–disassembly of the nanorods was achieved over four cycles. Finally, the reverse architectural arrangement, wherein NP-bound azobenzenes are combined with a ditopic bis-β-CD molecular linker, has also been shown to effectively generate photoswitchable AuNP aggregates [138], although the focus of this study was the catalytic properties of the free bis-β-CD linker, rather than the AuNP assemblies generated.

Configurations in which the NP-bound macrocycle is appended at the periphery of a self-assembled monolayer can play a similar architectural role. In this arrangement, neither face of the macrocycle is blocked by the NP surface, affording greater flexibility for the formation of fully threaded inclusion complexes that can more readily be recognized as pseudorotaxane in nature. The expanded structural possibilities for the surface-bound monolayer also allow for incorporating additional functionality and tuning NP properties for compatibility in a wider range of media. Adopting a ligand-exchange approach using the dibenzo[24]crown-8 terminated alkanethiol **3** (Figure 2.2), Fitzmaurice, Stoddart, and coworkers prepared AgNP-$\mathbf{3}_{0.15}$(DDT)$_{0.85}$ (<*d*> = 7.3 nm) stabilized by a mixed monolayer of crown-ether-terminated **3** and unfunctionalized dodecanethiol (DDT) [139]. On adding a bis-dibenzylammonium dication to a solution of AgNP-$\mathbf{3}_{0.15}(DDT)_{0.85}$ in $CHCl_3$, dynamic light scattering (DLS) measurements indicated a gradual increase in solvodynamic size to aggregates of ca. 100 nm after 1.5 h. Despite low signal-to-noise and poor resolution, the ^1H NMR spectrum of the assembling particles was consistent with the formation of surface-bound [3]pseudorotaxanes. The dynamic and specific binding interactions responsible for particle linking were confirmed by arresting assembly on the addition of either a monotopic benzylammonium cation or unfunctionalized dibenzo[24]crown-8 macrocycles [139].

Naturally, similar strategies can be devised by appending end-tethered linear guests to NP surfaces (Figure 2.1, Type IV-a) in combination with multitopic macrocyclic linkers. Furthermore, confining macrocyclic hosts and linear guests to the surfaces of two different NPs allows for the assembly of composite NP materials directly linked by pseudorotaxane complexes. Having established the molecule-linked assembly of AgNP-$\mathbf{3}_{0.15}(DDT)_{0.85}$, Fitzmaurice went on to functionalize SiO_2NPs (<*d*> = 180 nm) with end-tethered dibenzylammonium units [140]. After combining colloidal suspensions of the two functionalized NPs, the large SiO_2NPs were clearly observed to be surrounded by small AgNPs in TEM images. The responsive nature of the specific noncovalent interparticle links was demonstrated by reversing the assembly on deprotonation of the dibenzylammonium units and further corroborated by inhibiting assembly on adding an excess of dibenzo[24]crown-8 macrocycles [140]. The combination of plasmonic, metallic, and semiconducting properties in a composite material is of interest for optoelectronic applications, which would likely require the integration of the colloidal materials with existing solid-state technologies. Using un-complexed dibenzo[24]crown-8 macrocycles on the outer surfaces of the binary NP aggregates, the same molecular recognition interaction could be used to attach the heteromaterial assemblies to silicon wafers [141]. Substrate patterning was achieved by generating silicon-bound dibenzylammonium cations via a photodeprotection reaction so that the nanoconstructs were only attached to regions that had been exposed to light.

An increasingly diverse range of macrocyclic structures have been combined with end-tethered linear guest molecules to create NP aggregates imbued with the characteristics of the pseudorotaxane linkages and properties that depend on the nature of the linking unit.

The distinctive capabilities of cucurbit[8]uril macrocycles can be exploited as a universal and stimuli-responsive supramolecular "handcuff" linkage for connecting NP building blocks without requiring the macrocycle itself to be modified or installed on the NP surface. Redox-switchable 2 : 1 complexes are formed between CB[8] and radical cation dimers in water. For example, MV^{2+} functionalized AuNP-$(\mathbf{52})_{0.92}\mathbf{53}^{2+}{}_{0.08}$ (<*d*> = 5 nm, Figure 2.17b) precipitated rapidly from solution in the presence of CB[8] on addition of a chemical reductant as a consequence of forming 2 : 1 [$(\mathbf{53}^{+\bullet})_2$•CB[8]] complexes [144]. The same host–guest complex can be generated photochemically when the MV^{2+} unit is attached to the surface of photo-excitable semiconducting NPs such as TiO_2NP-$\mathbf{54}^{2+}$ (<*d*> = 5.5 nm, Figure 2.17a) [142]. Photo-induced electron transfer from the semiconducting

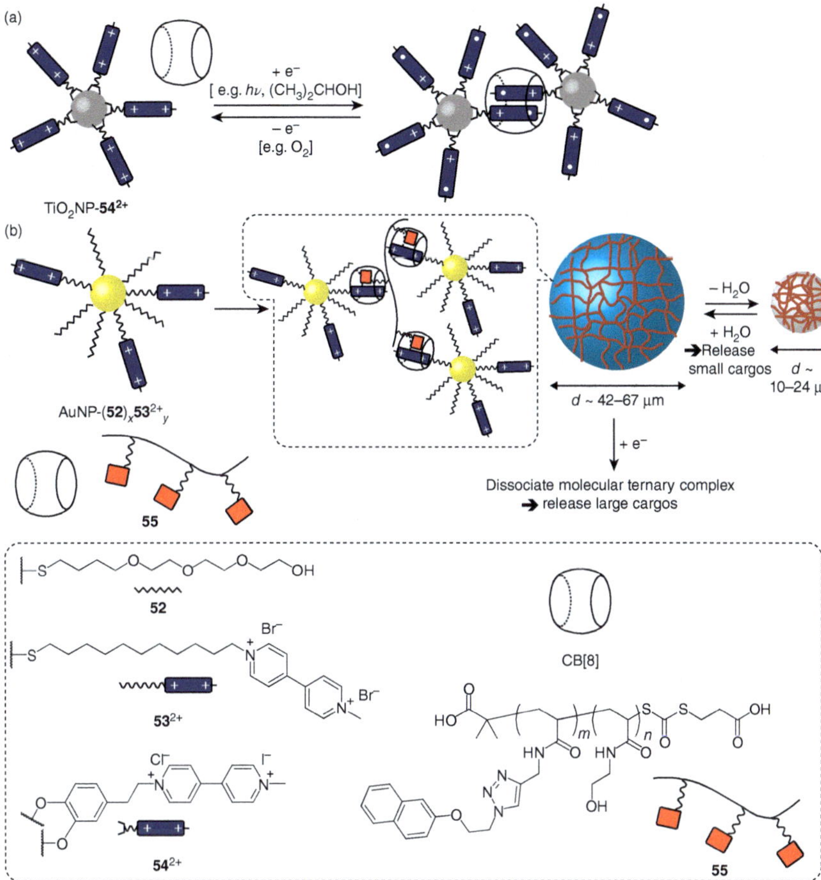

Figure 2.17 Nanoparticle assembly mediated by [3]pseudorotaxane complexes of CB[8] macrocycles. (a) Redox-switchable NP assemblies linked by 2 : 1 $[(MV^{+\bullet})_2{\bullet}CB[8]]$ handcuff complexes [142]. (b) Size-tunable water-in-oil microcapsules encapsulated by polymer–AuNP composite membranes linked by heteroternary CB[8] complexes [143]. For molecular structure of CB[8], see Figure 2.7.

NPs in the presence of a sacrificial electron donor generated surface-bound $[(MV^{+\bullet})_2{\bullet}CB[8]]$ handcuff complexes, resulting in rapid aggregation (Figure 2.17a). Redispersion was achieved on oxidation by H_2O_2 or O_2 to restore MV^{2+} dications, thus disrupting the handcuff complex. This aggregation–redispersion cycle was used to engineer catalyst separation and recycling when the TiO_2 NPs were used as photocatalysts for the reductive degradation of an organic dye [142]. A similar light- and oxygen-switched aggregation–dispersion cycle has been achieved for MV^{2+} functionalized ZnONPs [145].

Alternatively, CB[8] forms stable 1 : 1 : 1 ternary complexes with pairs of electron-rich and electron-deficient aromatic guests, allowing selective linking between two different appropriately functionalized building blocks. NP–polymer composites were assembled by combining AuNP-$(\mathbf{52})_{0.92}\mathbf{53}^{2+}{}_{0.08}$ with a multivalent water-soluble copolymer bearing naphthoxy side-chains (**55**) and CB[8] (Figure 2.17b) [144]. This ternary self-assembling system was subsequently applied in a scalable microfluidic approach for fabricating porous and responsive microcapsules [143]. Microdroplets were generated at a rate of 300 Hz by injecting an aqueous phase combining three components (AuNP-(**52**)$_{0.8}$**53**$^{2+}{}_{0.2}$ <*d*> = 20 nm, co-polymer **55**, CB[8]) at right angles to a fluorous oil carrier. Water-soluble cargos included in the aqueous phase were readily incorporated into the microcapsule interior. Evaporating the oil phase produced microcapsules with sizes in size range of 10–24 μm (tunable by varying the relative flow rates of the two phases), which were stable to both elevated temperature and reduced pressure. The cross-linked polymer–AuNP composite membranes defining the capsule walls permitted only small molecular weight cargos to escape on rehydration. However, the on-demand release of large cargo was achieved by exploiting the responsive nature of the supramolecular network that constitutes the capsule shell. Chemical reduction of **53**$^{2+}$ to **53**$^{+\bullet}$ disrupted the ternary [**53**•**55**•CB[8]] complex, releasing cargo over several hours. Furthermore, the AuNP component in the capsule shell acted as a surface-enchanced Raman spectroscopy (SERS) substrate, producing strong enhancement of signals from molecules encapsulated in the interior compartment [143]. The same high-specificity host–guest system has been used to create end-to-end AuNR assemblies from end-functionalized AuNR-CTAB-**53**$^{2+}$ (<*l*> = 40 nm, <*w*> = 10 nm) and naphthoxy-terminated ditopic linkers. Evidence for linear chains was observed in the dried state by TEM imaging, and the specificity of the molecular interaction was demonstrated by the absence of ordered assemblies in the presence of a competitive guest for CB[8] (adamantaneamine) or the smaller CB[7] macrocycle [146].

Redox-switchable [2]pseudorotaxane links have been used to reversibly construct NP assemblies with some remarkable aspects of structural and compositional selectivity. Side-chain CBPQT^{4+} functionalized poly(methyl methacrylate) **56**$^{4n+}$ acted as a multivalent cross-linker for the assembly of end-tethered TTF-functionalized particles AuNP-(**11**)$_{0.95}$**12**$_{0.05}$ through the formation of NP–polymer [2]pseudorotaxane links (Figure 2.18) [147]. Using either component in large excess fully saturated the binding sites of the other building block, producing small colloidally stable clusters. On the other hand, for stoichiometries close to 1 : 1 TTF:CBPQT^{4+}, linear chain-like assemblies were observed at short time periods, evolving into extended cross-linked aggregates that precipitated from solution (Figure 2.18a). Assembly was reversible on chemical or electrochemical oxidation to give AuNP-(**11**)$_{0.95}$**12**$^{2+}{}_{0.05}$, which disrupted the [2]pseudorotaxane NP–polymer links on account of Coulombic repulsion. The aggregated state was fully restored on reduction back to AuNP-(**11**)$_{0.95}$**12**$_{0.05}$. Redox-inactive AuNP-(**11**)$_{0.95}$**10**H$_{0.05}$ formed similar assemblies that were correspondingly unresponsive to electrochemical stimuli. The redox-switched differential affinities of host–guest complexes [**12**•CBPQT^{4+}] > [**10**H•CBPQT^{4+}] > [**12**$^{2+}$•CBPQT^{4+}] enabled the selective capture and release of either NP component from a binary mixture (Figure 2.18b). Treating an equimolar mixture of AgNP-(**11**)$_{0.95}$**12**$_{0.05}$ and AuNP-(**11**)$_{0.95}$**10**H$_{0.05}$ with **56**$^{4n+}$, selectively aggregated the silver particles, leaving a characteristically

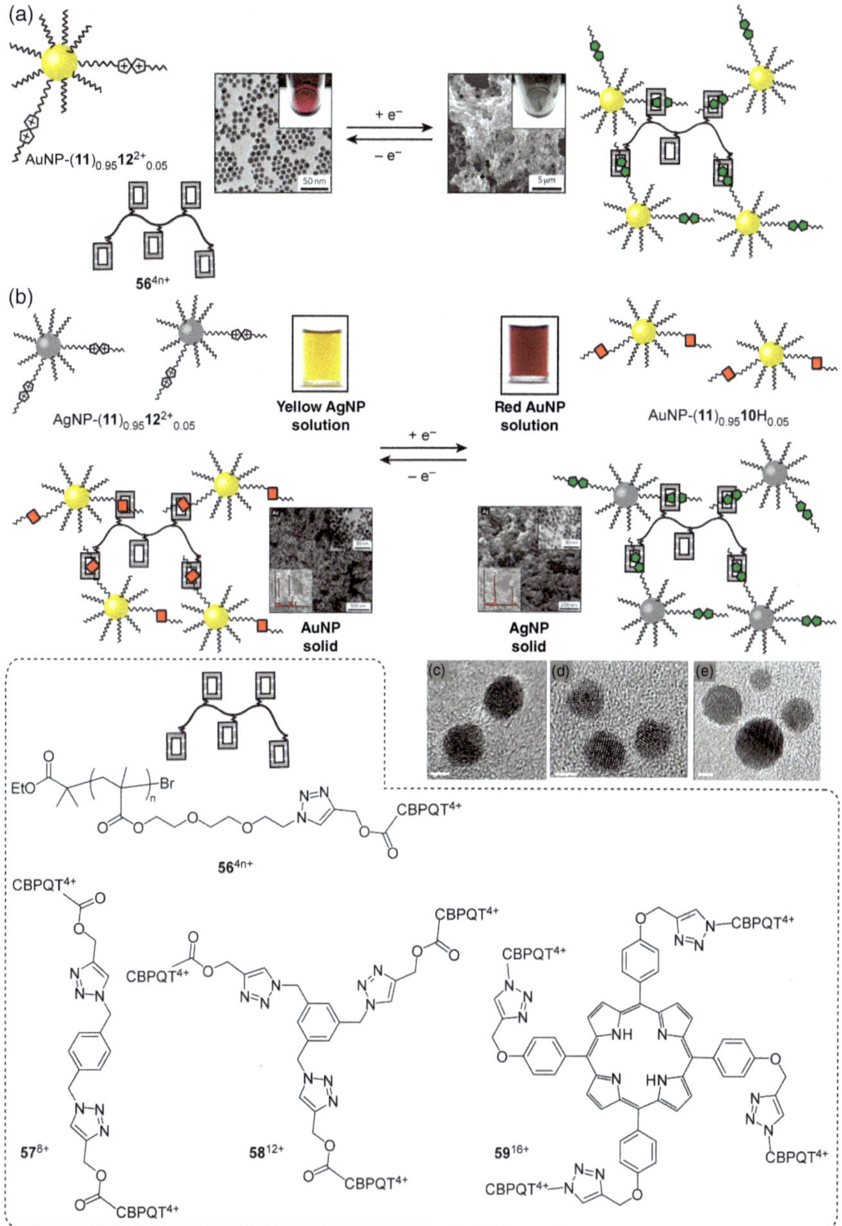

Figure 2.18 Nanoparticle assemblies linked by redox-switched association of end-tethered [2]pseudorotaxanes with multivalent macrocyclic molecular mediators [147, 148]. (a) Reversible assembly–disassembly of AuNP aggregates in the presence of polymeric $CBPQT^{4+}$ mediator $\mathbf{56}^{4n+}$ ($n \sim 160$). (b) Selective NP capture and release from a heteromaterial mixture by polymeric $CBPQT^{4+}$ mediator $\mathbf{56}^{4n+}$. (c–e) Representative TEM images of an AuNP dimer (c), trimer (d), and tetramer (e) formed by capturing AuNP-$(\mathbf{11})_{0.9935}\mathbf{12}_{0.0065}$ with divalent, trivalent or tetravalent mediators $\mathbf{57}^{8+}$, $\mathbf{58}^{12+}$, or $\mathbf{59}^{16+}$, respectively. For molecular structures of **10**H–**12** and $CBPQT^{4+}$, see Figure 2.4. Source: Reproduced with permission from Klajn et al. [147] (a, b), copyright 2009 Springer Nature and Olson et al. [148] (c–e), copyright American Chemical Society.

red solution of AuNPs. Oxidation of the NP-bound TTF units resulted in almost complete re-suspension of the silver particles as AgNP-$(\mathbf{11})_{0.95}\mathbf{12}^{2+}{}_{0.05}$, coincident with precipitation of the gold particles instead, leaving a pale yellow colloidal solution [147]. This work represents a remarkable achievement of highly selective and programmable assembly from a mixture of NP building blocks mediated by noncovalent mechanically interlocked molecular links.

Constructing discrete NP assemblies with controlled valency and morphology is a significantly more demanding challenge than producing extended amorphous aggregates. Replacing the multivalent polymer $\mathbf{56}^{4n+}$ with di-, tri-, or tetravalent $CBPQT^{4+}$ linkers $\mathbf{57}^{8+}$, $\mathbf{58}^{12+}$, or $\mathbf{59}^{16+}$, AuNP-$(\mathbf{11})_{0.9935}\mathbf{12}_{0.0065}$ ($\langle d \rangle = 4.6$ nm) could be assembled to produce dimers, trimers, or tetramers, respectively (Figure 2.18c–e) [148]. It was important to maximize the population of NPs bearing only one guest ligand **12** so as to minimize the likelihood of cross-linking. This was only possible by producing samples in which the majority of NP-bound monolayers contained no guest ligand at all. Consequently, assembly yields were very low and accompanied by poorly defined aggregates. Nevertheless, high-resolution TEM imaging revealed that the relative particle positions were consistent with the computed dimensions of each molecular template, and the assemblies were again endowed with the redox-responsive characteristics of the [2]pseudorotaxane links [148]. This remarkable result indicates the potential to use small molecule templates to direct the assembly of nanoscale building blocks.

2.5.2 Fully Interlocked Molecular Links for Nanoparticle Assemblies

One of the most exciting prospects of connecting NPs through mechanically interlocked linkages is the potential to exploit the large amplitude co-conformational motions characteristic of interlocked molecules to manipulate structural parameters on the nanoscale and, hence, modulate emergent properties. In a pioneering step toward this goal, Qu, Tian, and coworkers have achieved stimuli-responsive control over interparticle separation in AuNP dimers linked by inter-penetrated molecular muscles [149]. In response to acid and base stimuli, the preferred binding site for the dibenzo[24]crown-8 macrocycles in [c2]daisy chain rotaxane **60** is switched between a benzylalkylammonium unit and a *N*-methyltriazolium site, respectively, corresponding to a co-conformational extension/contraction of approximately 2.8 nm (Figure 2.19). Using the bis(trithiol) functionality on **60** to bridge between AuNP surfaces, small assemblies could be created in solution but were unstable to the acid/base switching stimuli. Using a stepwise approach, dimers of large ($\langle d \rangle = 60$ nm) and smaller ($\langle d \rangle = 15$ nm) AuNPs linked by $\mathbf{60}(2H^+)$ were instead constructed on transparent indium tin oxide (ITO) surfaces. AuNPs of these sizes experience strong distance-dependent coupling of their localized plasmon resonances, providing a sensitive probe of interparticle distance. Plasmon resonance scattering spectra (PRSS) with single particle resolution could be recorded on the ITO-immobilized structures using dark field microscopy, revealing that dimers were formed at approximately 40% of the large AuNPs. The inherently dynamic nature of the interparticle linkages was manifested in broad PRSS, requiring long

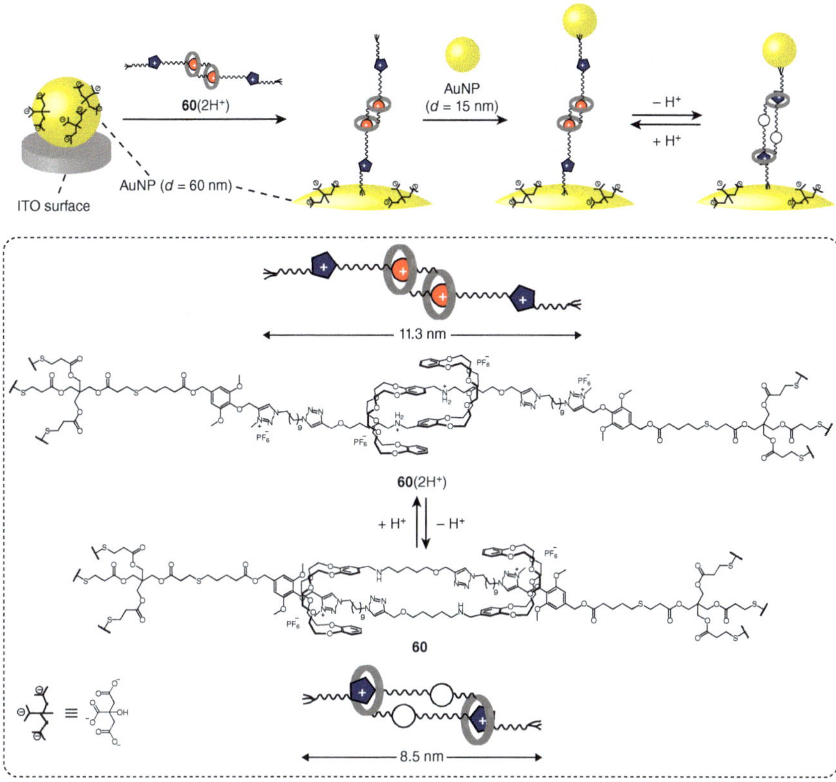

Figure 2.19 Stepwise construction of AuNP dimers linked by [c2]daisy chain MIM structures, leading to pH-actuated modulation of interparticle separation.

integration times to allow for averaging of the conformationally accessible states at the single-particle level. Monitoring individual dimers in real-time, a reversible shift in the PRSS maximum wavelength was observed over at least four cycles of base and acid stimuli. Counter-intuitively, a blue shift was recorded under contraction conditions (base) and a red shift when an extension was expected (acid). Finite element method simulations were able to explain this observation as evidence of very short interparticle distances (~ 1.2 and ~ 3.3 nm, respectively), which are likely the consequence of the bridging molecular muscles adopting folded conformations. A crucial control experiment demonstrated that AuNP dimers linked by a pH-inert daisy chain analog exhibited no change in PRSS on the application of the switching stimuli [149].

In the context of efforts to mimic the remarkable molecular mechanism and hierarchical function of biological muscles through the coordinated action of individual artificial molecular machines [150], this last example makes an important advance in transducing molecular-level motion to manipulate nanoscale structural parameters. More generally, it marks a transition from the relatively simple concept of responsive NP assembly–disassembly schemes to hierarchical amplification of large-amplitude molecular-level motions achievable in MIMs to manipulate nanoscale components and properties. This surely points toward an important theme for elaboration in the coming years.

Mechanically interlocked linkages offer unparalleled opportunities for NP assembly mediated by molecular recognition. The high specificity of recognition events involving mechanically interlocked architectures contrasts with many alternative approaches to NP assembly using other abiotic mediators. The resulting NP assemblies are imbued with the dynamic and addressable properties of the (supra)molecular scale linkers so that aggregates and composite materials can be engineered to display responsive and adaptive features. The considerable challenge of characterizing molecular structures within heterogeneous assemblies and solid-state materials means that advances in this direction must be underpinned by the ongoing accumulation of knowledge from related systems studied in other contexts, including MIM–NP hybrids in colloidal solution (Sections 2.2–2.4), and the broader efforts toward mechanically interlocked materials discussed in other chapters of this book.

2.6 From Switches to Motors and Beyond: The Future of Colloidal Nanomaterials with Mechanically Interlocked Parts

Integrating mechanically interlocked molecular structures with nanomaterials already marks a notable synthetic and analytical achievement. Actuating stimuli-responsive mechanically interlocked switches at NP surfaces to bring about property changes, effect cargo release, or direct the assembly of hierarchical composite materials has begun to demonstrate the rich potential for the functions that are enabled by building on the foundations laid by the pioneering studies discussed above.

Mechanically interlocked molecular architectures have become particularly synonymous with the development of molecular machines and motors [151–156]. A number of non-interlocked molecular switches, machines, and motors have also been successfully interfaced with, and operated on, NP surfaces [1, 4, 43, 52, 126, 157]. It is likely that integrating yet more advanced molecular machines with NPs will unlock even higher levels of sophistication in responsive multicomponent materials and devices. Beyond the additive combination of properties and functions, installing molecular machines and motors on nanoscale scaffolds has particularly exciting potential for helping to solve the imperative problems of coordinating, correlating, and amplifying molecular machine action to achieve higher-order responses [158–161].

In the first example of an interlocked molecular motor operating on a nanoscale surface, Stoddart and coworkers reported in 2021 the operation of end-tethered pseudorotaxane molecular pumps [162]. Adapting a redox-driven mechanism previously optimized for operation in solution [163], the surface-attached pumps sequentially transport a molecular adsorbate ($CBPQT^{4+}$ macrocycles) onto a nanoscale adsorbent by forming mechanical bonds to assemble end-tethered interlocked structures up to [6]pseudorotaxanes [162]. Colloidal Zr-MOF nanosheets and UiO-66 MOFNPs ($\langle d \rangle \approx 10$ nm) served as nanoscale platforms onto which the molecular pumping units could be securely installed at a density where their operation was not impeded by crowding with neighbors. Contemporaneously, Leigh and coworkers realized a similar concept using a chemically fueled molecular mechanism for pseudorotaxane pumping on micron-sized ($\langle d \rangle = 10$ μm) polystyrene beads [164]. The pulsed-fuel nature of the operating mechanism in this case allowed the sequential threading of distinguishable rings, so that information can be encoded in the surface-bound MIM structures.

The active transport of a molecular species from one compartment (bulk solution) to another (a surface) creates a nonequilibrium local concentration of the adsorbate. In doing so, chemical energy in the form of reagents is converted into a chemical potential concentration gradient and, potentially, sequence information, which can subsequently be released or read out on demand. These active nanoscale devices consequently represent exciting prototypes of energy and information storage and release systems for future nanotechnologies.

The structural and functional complexity of these final examples is illustrative of the maturity now reached in each of the fields of MIMs and colloidal NPs individually, and is a marker of the remarkable advances made in integrating these two areas of contemporary chemical research. At this intersection, where chemical components spanning multiple length scales are combined, there is huge potential for innovation in structure, properties, and functionality. Nevertheless, the design, preparation, and analytical challenges remain considerable.

Although, in many regards, MIMs have led the way, the opportunities, principles, and challenges inherent in interfacing dynamic molecular systems with colloidal NPs are broadly universal and independent of the nature of the (supra)molecular architecture in question. Significant advances made, for example, in combining NPs with non-interlocked molecular switches [43–45, 52–57], and with responsive polymer materials [46–51], suggest there is considerable opportunity for cross-fertilization of ideas, concepts, and solutions. The remarkable diversity of systems that seek to translate dynamic artificial chemical systems into the nanoworld offers a rich and intellectually stimulating resource on which future efforts will undoubtedly draw to achieve new levels of structural and functional sophistication.

The vast majority of MIM–NP hybrids have been implemented on inorganic cores – either densely functionalized metals, metal oxides and semiconductors, or sparsely functionalized silica and titania particles. Other core materials remain to be explored entirely. For example, carbon-based cores have the potential to interact with (mainly carbon-based) functional molecular components in ways that are just not possible for inorganic cores. The reader will find considerable overlap between the systems discussed in this chapter and those in other chapters where MIMs have been integrated with similar materials but not in NP format. There is undoubtedly much inspiration to be taken from considering these efforts in the round as well.

Each of the examples discussed above highlights the considerable challenges in constructing and characterizing multicomponent assemblies comprising both polydisperse nanomaterials and inherently dynamic molecular components in interfacial environments. They demonstrate the multi-disciplinary philosophy that must be adopted, often combining information from a variety of experimental and computational methods. Understanding the influence of interfacial environments on the reactions, interactions, and dynamics of confined molecules is critical to any number of molecular approaches to nanotechnology and materials design [165]. Several examples discussed here confirm that this will be equally important in regards to exploiting the dynamic features of MIMs when integrated with nanomaterials. Consequently, this will certainly be an important focus for both experimental and computational efforts in the near future. It can be expected that an increasing number of innovations will take advantage of properties that are unique to the nanomaterial, in addition to its role as a nanoscale scaffold – examples involving NP photosensitizers and delivery vehicles discussed above have only scratched the surface of what is possible. In terms of the molecular components, nonequilibrium systems [166–169] will undoubtedly play an increasingly significant role that will lead to new functional capabilities.

All things considered, it is clear that the early pioneers of MIM–NP hybrids have done a remarkable job of opening up a vast new region of chemical space. The combination of fundamental challenges, architectural novelty, structural beauty, and technological possibilities will attract new explorers in increasing number. Correspondingly, we can anticipate increasingly rapid developments in synthetic and analytical approaches to MIM–NP hybrids and their development for myriad yet-unimagined applications.

References

1 Balzani, V., Credi, A., and Venturi, M. (2008). Molecular machines working on surfaces and at interfaces. *ChemPhysChem* 9: 202–220.

2 Coronado, E., Gavina, P., and Tatay, S. (2009). Catenanes and threaded systems: from solution to surfaces. *Chem. Soc. Rev.* 38: 1674–1689.

3 Davis, J.J., Orlowski, G.A., Rahman, H., and Beer, P.D. (2010). Mechanically interlocked and switchable molecules at surfaces. *Chem. Commun.* 46: 54–63.

4 Fahrenbach, A.C., Warren, S.C., Incorvati, J.T. et al. (2013). Organic switches for surfaces and devices. *Adv. Mater.* 25: 331–348.

5 Montes-García, V., Pérez-Juste, J., Pastoriza-Santos, I., and Liz-Marzán, L.M. (2014). Metal nanoparticles and supramolecular macrocycles: a tale of synergy. *Chem. Eur. J.* 20: 10874–10883.

6 Heuer-Jungemann, A., Feliu, N., Bakaimi, I. et al. (2019). The role of ligands in the chemical synthesis and applications of inorganic nanoparticles. *Chem. Rev.* 119: 4819–4880.

7 Love, J.C., Estroff, L.A., Kriebel, J.K. et al. (2005). Self-assembled monolayers of thiolates on metals as a form of nanotechnology. *Chem. Rev.* 105: 1103–1169.

8 Häkkinen, H. (2012). The gold-sulfur interface at the nanoscale. *Nat. Chem.* 4: 443–455.

9 Reimers, J.R., Ford, M.J., Marcuccio, S.M. et al. (2017). Competition of van der Waals and chemical forces on gold-sulfur surfaces and nanoparticles. *Nat. Rev. Chem.* 1: 0017.

10 Liu, J., Xu, R.L., and Kaifer, A.E. (1998). In situ modification of the surface of gold colloidal particles. Preparation of cyclodextrin-based rotaxanes supported on gold nanospheres. *Langmuir* 14: 7337–7339.

11 Liu, J., Ong, W., Roman, E. et al. (2000). Cyclodextrin-modified gold nanospheres. *Langmuir* 16: 3000–3002.

12 Fitzmaurice, D., Rao, S.N., Preece, J.A. et al. (1999). Heterosupramolecular chemistry: programmed pseudorotaxane assembly at the surface of a nanocrystal. *Angew. Chem. Int. Ed.* 38: 1147–1150.

13 Callari, F., Petralia, S., and Sortino, S. (2006). Highly photoresponsive monolayer-protected gold clusters by self- assembly of a cyclodextrin-azobenzene-derived supramolecular complex. *Chem. Commun.* 1009–1011.

14 Liu, J., Mendoza, S., Román, E. et al. (1999). Cyclodextrin-modified gold nanospheres. Host-guest interactions at work to control colloidal properties. *J. Am. Chem. Soc.* 121: 4304–4305.

15 Pujari, S.P., Scheres, L., Marcelis, A.T.M., and Zuilhof, H. (2014). Covalent surface modification of oxide surfaces. *Angew. Chem. Int. Ed.* 53: 6322–6356.

16 Long, B., Nikitin, K., and Fitzmaurice, D. (2003). Self-assembly of a tripodal pseudorotaxane on the surface of a titanium dioxide nanoparticle. *J. Am. Chem. Soc.* 125: 5152–5160.

17 Long, B., Nikitin, K., and Fitzmaurice, D. (2003). Assembly of an electronically switchable rotaxane on the surface of a titanium dioxide nanoparticle. *J. Am. Chem. Soc.* 125: 15490–15498.

18 Zhao, L.Y., Mullen, K.M., Chmielewski, M.J. et al. (2009). Anion templated assembly of an indolocarbazole containing pseudorotaxane on beads and silica nanoparticles. *New J. Chem.* 33: 760–768.

19 Cooke, G., Garety, J.F., Hewage, S.G. et al. (2006). The tuneable complexation of gold nanoparticles. *Chem. Commun.* 4119–4121.

20 Coskun, A., Wesson, P.J., Klajn, R. et al. (2010). Molecular-mechanical switching at the nanoparticle-solvent interface: practice and theory. *J. Am. Chem. Soc.* 132: 4310–4320.

21 Klajn, R., Fang, L., Coskun, A. et al. (2009). Metal nanoparticles functionalized with molecular and supramolecular switches. *J. Am. Chem. Soc.* 131: 4233–4235.

22 Zhao, Y.-L., Chen, Y., Wang, M., and Liu, Y. (2006). Multi[2]rotaxanes with gold nanoparticles as centers. *Org. Lett.* 8: 1267–1270.

23 Zhu, L., Yan, H., Nguyen, K.T. et al. (2012). Sequential self-assembly for construction of Pt(II)-bridged [3]rotaxanes on gold nanoparticles. *Chem. Commun.* 48: 4290–4292.

24 Hens, Z. and Martins, J.C. (2013). A solution NMR toolbox for characterizing the surface chemistry of colloidal nanocrystals. *Chem. Mater.* 25: 1211–1221.

25 Marbella, L.E. and Millstone, J.E. (2015). NMR techniques for noble metal nanoparticles. *Chem. Mater.* 27: 2721–2739.

26 Boulas, P.L., Gomez-Kaifer, M., and Echegoyen, L. (1998). Electrochemistry of supramolecular systems. *Angew. Chem. Int. Ed.* 37: 216–247.

27 Nijhuis, C.A., Ravoo, B.J., Huskens, J., and Reinhoudt, D.N. (2007). Electrochemically controlled supramolecular systems. *Coord. Chem. Rev.* 251: 1761–1780.

28 Colangelo, E., Comenge, J., Paramelle, D. et al. (2017). Characterizing self-assembled monolayers on gold nanoparticles. *Bioconjug. Chem.* 28: 11–22.

29 Hühn, J., Carrillo-Carrion, C., Soliman, M.G. et al. (2017). Selected standard protocols for the synthesis, phase transfer, and characterization of inorganic colloidal nanoparticles. *Chem. Mater.* 29: 399–461.

30 Mourdikoudis, S., Pallares, R.M., and Thanh, N.T.K. (2018). Characterization techniques for nanoparticles: comparison and complementarity upon studying nanoparticle properties. *Nanoscale* 10: 12871–12934.

31 Yang, J., Lee, J.Y., and Ying, J.Y. (2011). Phase transfer and its applications in nanotechnology. *Chem. Soc. Rev.* 40: 1672–1696.

32 Lala, N., Lalbegi, S.P., Adyanthaya, S.D., and Sastry, M. (2001). Phase transfer of aqueous gold colloidal particles capped with inclusion complexes of cyclodextrin and alkanethiol molecules into chloroform. *Langmuir* 17: 3766–3768.

33 Wang, Y., Wong, J.F., Teng, X. et al. (2003). "Pulling" nanoparticles into water: phase transfer of oleic acid stabilized monodisperse nanoparticles into aqueous solutions of α-cyclodextrin. *Nano Lett.* 3: 1555–1559.

34 Dorokhin, D., Tomczak, N., Han, M. et al. (2009). Reversible phase transfer of (CdSe/ZnS) quantum dots between organic and aqueous solutions. *ACS Nano* 3: 661–667.

35 Liu, J., Alvarez, J., Ong, W. et al. (2001). Phase transfer of hydrophilic, cyclodextrin-modified gold nanoparticles to chloroform solutions. *J. Am. Chem. Soc.* 123: 11148–11154.

36 Peng, L., You, M.X., Wu, C.C. et al. (2014). Reversible phase transfer of nanoparticles based on photoswitchable host-guest chemistry. *ACS Nano* 8: 2555–2561.

37 Chen, G. and Jiang, M. (2011). Cyclodextrin-based inclusion complexation bridging supramolecular chemistry and macromolecular self-assembly. *Chem. Soc. Rev.* 40: 2254–2266.

38 Qu, D.H., Wang, Q.C., Zhang, Q.W. et al. (2015). Photoresponsive host-guest functional systems. *Chem. Rev.* 115: 7543–7588.

39 Coelho, J.P., Gonzalez-Rubio, G., Delices, A. et al. (2014). Polyrotaxane-mediated self-assembly of gold nanospheres into fully reversible supercrystals. *Angew. Chem. Int. Ed.* 53: 12751–12755.

40 Yeh, Y.-C., Rana, S., Mout, R. et al. (2014). Supramolecular tailoring of protein-nanoparticle interactions using cucurbituril mediators. *Chem. Commun.* 50: 5565–5568.

41 Kim, C., Agasti, S.S., Zhu, Z. et al. (2010). Recognition-mediated activation of therapeutic gold nanoparticles inside living cells. *Nat. Chem.* 2: 962–966.

42 Tonga, G.Y., Jeong, Y.D., Duncan, B. et al. (2015). Supramolecular regulation of bioorthogonal catalysis in cells using nanoparticle-embedded transition metal catalysts. *Nat. Chem.* 7: 597–603.

43 Klajn, R., Stoddart, J.F., and Grzybowski, B.A. (2010). Nanoparticles functionalised with reversible molecular and supramolecular switches. *Chem. Soc. Rev.* 39: 2203–2237.

44 Edwards, W. and Kay, E.R. (2016). Manipulating the monolayer: responsive and reversible control of colloidal inorganic nanoparticle properties. *ChemNanoMat* 2: 87–98.

45 Qian, Z.X. and Ginger, D.S. (2017). Reversibly reconfigurable colloidal plasmonic nanomaterials. *J. Am. Chem. Soc.* 139: 5266–5276.

46 Lu, Y. and Ballauff, M. (2011). Thermosensitive core-shell microgels: from colloidal model systems to nanoreactors. *Prog. Polym. Sci.* 36: 767–792.

47 Cayre, O.J., Chagneux, N., and Biggs, S. (2011). Stimulus responsive core-shell nanoparticles: synthesis and applications of polymer based aqueous systems. *Soft Matter* 7: 2211–2234.

48 Oberdisse, J. and Hellweg, T. (2020). Recent advances in stimuli-responsive core-shell microgel particles: synthesis, characterisation, and applications. *Colloid Polym. Sci.* 298: 921–935.

49 Kanduc, M., Kim, W.K., Roa, R., and Dzubiella, J. (2020). Modeling of stimuli-responsive nanoreactors: rational rate control towards the design of colloidal enzymes. *Mol. Syst. Design Eng.* 5: 602–619.

50 Ding, T. and Baumberg, J.J. (2020). Thermo-responsive plasmonic systems: old materials with new applications. *Nanoscale Adv.* 2: 1410–1416.

51 Talianov, P., Fatkhutdinova, L.I., Timin, A.S. et al. (2021). Adaptive nanoparticle-polymer complexes as optical elements: design and application in nanophotonics and nanomedicine. *Laser Photonics Rev.* 15: 2000421.

52 Yildiz, I., Deniz, E., and Raymo, F.M. (2009). Fluorescence modulation with photochromic switches in nanostructured constructs. *Chem. Soc. Rev.* 38: 1859–1867.

53 Tian, Z.Y., Wu, W.W., and Li, A.D.Q. (2009). Photoswitchable fluorescent nanoparticles: preparation, properties and applications. *ChemPhysChem* 10: 2577–2591.

54 Klajn, R. (2010). Immobilized azobenzenes for the construction of photoresponsive materials. *Pure Appl. Chem.* 82: 2247–2279.

55 Klajn, R. (2014). Spiropyran-based dynamic materials. *Chem. Soc. Rev.* 43: 148–184.

56 Zhang, Y.L., Zhang, K.Q., Wang, J. et al. (2015). Photoswitchable fluorescent nanoparticles and their emerging applications. *Nanoscale* 7: 19342–19357.

57 Ghavidast, A. and Mahmoodi, N.O. (2016). A comparative study of the photochromic compounds incorporated on the surface of nanoparticles. *J. Mol. Liq.* 216: 552–564.

58 Kay, E.R., Lee, J., Nocera, D.G., and Bawendi, M.G. (2013). Conformational control of energy transfer: a mechanism for biocompatible nanocrystal-based sensors. *Angew. Chem. Int. Ed.* 52: 1165–1169.

59 Zhou, P.P., Jia, S.S., Pan, D. et al. (2015). Reversible regulation of catalytic activity of gold nanoparticles with DNA nanomachines. *Sci. Rep.* 5: 14402.

60 Yang, P., Zhou, R.X., Kong, C.P. et al. (2021). Stimuli-responsive three-dimensional DNA nanomachines engineered by controlling dynamic interactions at biomolecule-nanoparticle interfaces. *ACS Nano* 15: 16870–16877.

61 Tang, F.Q., Li, L.L., and Chen, D. (2012). Mesoporous silica nanoparticles: synthesis, biocompatibility and drug delivery. *Adv. Mater.* 24: 1504–1534.

62 Li, Z.X., Barnes, J.C., Bosoy, A. et al. (2012). Mesoporous silica nanoparticles in biomedical applications. *Chem. Soc. Rev.* 41: 2590–2605.

63 Tarn, D., Ashley, C.E., Xue, M. et al. (2013). Mesoporous silica nanoparticle nanocarriers: biofunctionality and biocompatibility. *Acc. Chem. Res.* 46: 792–801.

64 Hecht, M., Climent, E., Biyikal, M. et al. (2013). Gated hybrid delivery systems: en route to sensory materials with inherent signal amplification. *Coord. Chem. Rev.* 257: 2589–2606.

65 Song, N. and Yang, Y.W. (2015). Molecular and supramolecular switches on mesoporous silica nanoparticles. *Chem. Soc. Rev.* 44: 3474–3504.

66 Aznar, E., Oroval, M., Pascual, L. et al. (2016). Gated materials for on-command release of guest molecules. *Chem. Rev.* 116: 561–718.

67 Rühle, B., Saint-Cricq, P., and Zink, J.I. (2016). Externally controlled nanomachines on mesoporous silica nanoparticles for biomedical applications. *ChemPhysChem* 17: 1769–1779.

68 Chen, W., Glackin, C.A., Horwitz, M.A., and Zink, J.I. (2019). Nanomachines and other caps on mesoporous silica nanoparticles for drug delivery. *Acc. Chem. Res.* 52: 1531–1542.

69 Lin, F.C., Xie, Y.J., Deng, T., and Zink, J.I. (2021). Magnetism, ultrasound, and light-stimulated mesoporous silica nanocarriers for theranostics and beyond. *J. Am. Chem. Soc.* 143: 6025–6036.

70 de Luis, B., Llopis-Lorente, A., Sancenón, F., and Martínez-Máñez, R. (2021). Engineering chemical communication between micro/nanosystems. *Chem. Soc. Rev.* 50: 8829–8856.

71 Nguyen, T.D., Tseng, H.-R., Celestre, P.C. et al. (2005). A reversible molecular valve. *Proc. Natl. Acad. Sci. U. S. A.* 102: 10029–10034.

72 Nguyen, T.D., Liu, Y., Saha, S. et al. (2007). Design and optimization of molecular nanovalves based on redox-switchable bistable rotaxanes. *J. Am. Chem. Soc.* 129: 626–634.

73 Saha, S., Leung, K.C.F., Nguyen, T.D. et al. (2007). Nanovalves. *Adv. Funct. Mater.* 17: 685–693.

74 Ambrogio, M.W., Thomas, C.R., Zhao, Y.L. et al. (2011). Mechanized silica nanoparticles: a new frontier in theranostic nanomedicine. *Acc. Chem. Res.* 44: 903–913.

75 Khashab, N.M., Trabolsi, A., Lau, Y.A. et al. (2009). Redox- and pH-controlled mechanized nanoparticles. *Eur. J. Org. Chem.* 1669–1673.

76 Nguyen, T.D., Leung, K.C.F., Liong, M. et al. (2006). Construction of a pH-driven supramolecular nanovalve. *Org. Lett.* 8: 3363–3366.

77 Leung, K.C.F., Nguyen, T.D., Stoddart, J.F., and Zink, J.I. (2006). Supramolecular nanovalves controlled by proton abstraction and competitive binding. *Chem. Mater.* 18: 5919–5928.

78 Park, C., Oh, K., Lee, S.C., and Kim, C. (2007). Controlled release of guest molecules from mesoporous silica particles based on a pH-responsive polypseudorotaxane motif. *Angew. Chem. Int. Ed.* 46: 1455–1457.

79 Angelos, S., Yang, Y.W., Patel, K. et al. (2008). pH-Responsive supramolecular nanovalves based on cucurbit[6]uril pseudorotaxanes. *Angew. Chem. Int. Ed.* 47: 2222–2226.

80 Angelos, S., Khashab, N.M., Yang, Y.-W. et al. (2009). pH clock-operated mechanized nanoparticles. *J. Am. Chem. Soc.* 131: 12912–12914.

81 Zhao, Y.L., Li, Z.X., Kabehie, S. et al. (2010). pH-Operated nanopistons on the surfaces of mesoporous silica nanoparticles. *J. Am. Chem. Soc.* 132: 13016–13025.

82 Khashab, N.M., Belowich, M.E., Trabolsi, A. et al. (2009). pH-Responsive mechanised nanoparticles gated by semirotaxanes. *Chem. Commun.* 5371–5373.

83 Ferris, D.P., Zhao, Y.-L., Khashab, N.M. et al. (2009). Light-operated mechanized nanoparticles. *J. Am. Chem. Soc.* 131: 1686–1688.

84 Ambrogio, M.W., Pecorelli, T.A., Patel, K. et al. (2010). Snap-top nanocarriers. *Org. Lett.* 12: 3304–3307.

85 Sun, Y.L., Yang, B.J., Zhang, S.X.A., and Yang, Y.W. (2012). Cucurbit[7]uril pseudorotaxane-based photoresponsive supramolecular nanovalve. *Chem. Eur. J.* 18: 9212–9216.

86 Patel, K., Angelos, S., Dichtel, W.R. et al. (2008). Enzyme-responsive snap-top covered silica nanocontainers. *J. Am. Chem. Soc.* 130: 2382–2383.

87 Horcajada, P., Gref, R., Baati, T. et al. (2012). Metal-organic frameworks in biomedicine. *Chem. Rev.* 112: 1232–1268.

88 Tan, L.L., Li, H.W., Qiu, Y.C. et al. (2015). Stimuli-responsive metal-organic frameworks gated by pillar[5]arene supramolecular switches. *Chem. Sci.* 6: 1640–1644.

89 Meng, H.A., Xue, M., Xia, T.A. et al. (2010). Autonomous in vitro anticancer drug release from mesoporous silica nanoparticles by pH-sensitive nanovalves. *J. Am. Chem. Soc.* 132: 12690–12697.

90 Xue, M., Zhong, X., Shaposhnik, Z. et al. (2011). pH-Operated mechanized porous silicon nanoparticles. *J. Am. Chem. Soc.* 133: 8798–8801.

91 Du, L., Liao, S.J., Khatib, H.A. et al. (2009). Controlled-access hollow mechanized silica nanocontainers. *J. Am. Chem. Soc.* 131: 15136–15142.

92 Li, Z.L., Clemens, D.L., Lee, B.Y. et al. (2015). Mesoporous silica nanoparticles with pH-sensitive nanovalves for delivery of moxifloxacin provide improved treatment of lethal pneumonic tularemia. *ACS Nano* 9: 10778–10789.

93 Lee, B.Y., Li, Z.L., Clemens, D.L. et al. (2016). Redox-triggered release of moxifloxacin from mesoporous silica nanoparticles functionalized with disulfide snap-tops enhances efficacy against pneumonic tularemia in mice. *Small* 12: 3690–3702.

94 Min, Y.Z., Caster, J.M., Eblan, M.J., and Wang, A.Z. (2015). Clinical translation of nanomedicine. *Chem. Rev.* 115: 11147–11190.

95 Matsumura, Y. and Maeda, H. (1986). A new concept for macromolecular therapeutics in cancer-chemotherapy - mechanism of tumoritropic accumulation of proteins and the antitumor agent smancs. *Cancer Res.* 46: 6387–6392.

96 Luo, Z., Ding, X.W., Hu, Y. et al. (2013). Engineering a hollow nanocontainer platform with multifunctional molecular machines for tumor-targeted therapy in vitro and in vivo. *ACS Nano* 7: 10271–10284.

97 Luo, Z., Hu, Y., Cai, K.Y. et al. (2014). Intracellular redox-activated anticancer drug delivery by functionalized hollow mesoporous silica nanoreservoirs with tumor specificity. *Biomaterials* 35: 7951–7962.

98 Porta, F., Lamers, G.E.M., Morrhayim, J. et al. (2013). Folic acid-modified mesoporous silica nanoparticles for cellular and nuclear targeted drug delivery. *Adv. Healthcare Mater.* 2: 281–286.

99 Sun, Y.L., Zhou, Y., Li, Q.-L., and Yang, Y.-W. (2013). Enzyme-responsive supramolecular nanovalves crafted by mesoporous silica nanoparticles and choline-sulfonatocalix[4]arene [2]pseudorotaxanes for controlled cargo release. *Chem. Commun.* 49: 9033–9035.

100 Lim, E.-K., Kim, T., Paik, S. et al. (2015). Nanomaterials for theranostics: recent advances and future challenges. *Chem. Rev.* 115: 327–394.

101 Lai, J.P., Shah, B.P., Garfunkel, E., and Lee, K.B. (2013). Versatile fluorescence resonance energy transfer-based mesoporous silica nanoparticles for real-time monitoring of drug release. *ACS Nano* 7: 2741–2750.

102 Yan, H., Teh, C., Sreejith, S. et al. (2012). Functional mesoporous silica nanoparticles for photothermal-controlled drug delivery in vivo. *Angew. Chem. Int. Ed.* 51: 8373–8377.

103 Li, M.H., Yan, Y., Teh, C. et al. (2014). NIR-triggered drug release from switchable rotaxane-functionalized silica-covered Au nanorods. *Chem. Commun.* 50: 9745–9748.

104 Croissant, J. and Zink, J.I. (2012). Nanovalve-controlled cargo release activated by plasmonic heating. *J. Am. Chem. Soc.* 134: 7628–7631.

105 Li, H., Tan, L.L., Jia, P. et al. (2014). Near-infrared light-responsive supramolecular nanovalve based on mesoporous silica-coated gold nanorods. *Chem. Sci.* 5: 2804–2808.

106 Guardado-Alvarez, T.M., Devi, L.S., Russell, M.M. et al. (2013). Activation of snap-top capped mesoporous silica nanocontainers using two near-infrared photons. *J. Am. Chem. Soc.* 135: 14000–14003.

107 Guardado-Alvarez, T.M., Devi, L.S., Vabre, J.M. et al. (2014). Photo-redox activated drug delivery systems operating under two photon excitation in the near-IR. *Nanoscale* 6: 4652–4658.

108 Thomas, C.R., Ferris, D.P., Lee, J.H. et al. (2010). Noninvasive remote-controlled release of drug molecules in vitro using magnetic actuation of mechanized nanoparticles. *J. Am. Chem. Soc.* 132: 10623–10625.

109 Rühle, B., Datz, S., Argyo, C. et al. (2016). A molecular nanocap activated by superparamagnetic heating for externally stimulated cargo release. *Chem. Commun.* 52: 1843–1846.

110 Chen, W., Cheng, C.A., and Zink, J.I. (2019). Spatial, temporal, and dose control of drug delivery using noninvasive magnetic stimulation. *ACS Nano* 13: 1292–1308.

111 Cheng, C.A., Chen, W., Zhang, L. et al. (2020). Magnetic resonance imaging of high-intensity focused ultrasound-stimulated drug release from a self-reporting core@shell nanoparticle platform. *Chem. Commun.* 56: 10297–10300.

112 Liu, R., Zhang, Y., and Feng, P.Y. (2009). Multiresponsive supramolecular nanogated ensembles. *J. Am. Chem. Soc.* 131: 15128–15129.

113 Angelos, S., Yang, Y.W., Khashab, N.M. et al. (2009). Dual-controlled nanoparticles exhibiting AND logic. *J. Am. Chem. Soc.* 131: 11344–11346.

114 Wang, C., Li, Z.X., Cao, D. et al. (2012). Stimulated release of size-selected cargos in succession from mesoporous silica nanoparticles. *Angew. Chem. Int. Ed.* 51: 5460–5465.

115 Xue, M. and Zink, J.I. (2013). An enzymatic chemical amplifier based on mechanized nanoparticles. *J. Am. Chem. Soc.* 135: 17659–17662.

116 Xue, M., Cao, D., Stoddart, J.F., and Zink, J.I. (2012). Size-selective pH-operated megagates on mesoporous silica materials. *Nanoscale* 4: 7569–7574.

117 Díez, P., Sánchez, A., Gamella, M. et al. (2014). Toward the design of smart delivery systems controlled by integrated enzyme-based biocomputing ensembles. *J. Am. Chem. Soc.* 136: 9116–9123.

118 Llopis-Lorente, A., Díez, P., Sánchez, A. et al. (2017). Interactive models of communication at the nanoscale using nanoparticles that talk to one another. *Nat. Commun.* 8: 15511.

119 de Luis, B., Morellá-Aucejo, A., Llopis-Lorente, A. et al. (2021). A chemical circular communication network at the nanoscale. *Chem. Sci.* 12: 1551–1559.

120 de Luis, B., Llopis-Lorente, A., Rincón, P. et al. (2019). An interactive model of communication between abiotic nanodevices and microorganisms. *Angew. Chem. Int. Ed.* 58: 14986–14990.

121 Nie, Z.H., Petukhova, A., and Kumacheva, E. (2010). Properties and emerging applications of self-assembled structures made from inorganic nanoparticles. *Nat. Nanotechnol.* 5: 15–25.

122 Grzelczak, M., Vermant, J., Furst, E.M., and Liz-Marzán, L.M. (2010). Directed self-assembly of nanoparticles. *ACS Nano* 4: 3591–3605.

123 Romo-Herrera, J.M., Álvarez-Puebla, R.A., and Liz-Marzán, L.M. (2011). Controlled assembly of plasmonic colloidal nanoparticle clusters. *Nanoscale* 3: 1304–1315.

124 Grzybowski, B.A., Fitzner, K., Paczesny, J., and Granick, S. (2017). From dynamic self-assembly to networked chemical systems. *Chem. Soc. Rev.* 46: 5647–5678.

125 Grzelczak, M., Liz-Marzán, L.M., and Klajn, R. (2019). Stimuli-responsive self-assembly of nanoparticles. *Chem. Soc. Rev.* 48: 1342–1361.

126 Bian, T., Chu, Z.L., and Klajn, R. (2020). The many ways to assemble nanoparticles using light. *Adv. Mater.* 32: 1905866.

127 Li, D. and Qi, L.M. (2018). Self-assembly of inorganic nanoparticles mediated by host-guest interactions. *Curr. Opin. Colloid Interface Sci.* 35: 59–67.

128 Saha, K., Agasti, S.S., Kim, C. et al. (2012). Gold nanoparticles in chemical and biological sensing. *Chem. Rev.* 112: 2739–2779.

129 de la Rica, R., Fratila, R.M., Szarpak, A. et al. (2011). Multivalent nanoparticle networks as ultrasensitive enzyme sensors. *Angew. Chem. Int. Ed.* 50: 5704–5707.

130 Zhao, Y., Huang, Y.C., Zhu, H. et al. (2016). Three-in-one: sensing, self-assembly, and cascade catalysis of cyclodextrin modified gold nanoparticles. *J. Am. Chem. Soc.* 138: 16645–16654.

131 Li, H., Chen, D.X., Sun, Y.L. et al. (2013). Viologen-mediated assembly of and sensing with carboxylatopillar[5]arene-modified gold nanoparticles. *J. Am. Chem. Soc.* 135: 1570–1576.

132 Krings, J.A., Vonhören, B., Tegeder, P. et al. (2014). Light-responsive aggregation of β-cyclodextrin covered silica nanoparticles. *J. Mater. Chem. A* 2: 9587–9593.

133 Stricker, L., Fritz, E.C., Peterlechner, M. et al. (2016). Arylazopyrazoles as light-responsive molecular switches in cyclodextrin-based supramolecular systems. *J. Am. Chem. Soc.* 138: 4547–4554.

134 Möller, N., Hellwig, T., Stricker, L. et al. (2017). Near-infrared photoswitching of cyclodextrin-guest complexes using lanthanide-doped $LiYF_4$ upconversion nanoparticles. *Chem. Commun.* 53: 240–243.

135 Engel, S., Möller, N., Stricker, L. et al. (2018). A modular system for the design of stimuli-responsive multifunctional nanoparticle aggregates by use of host-guest chemistry. *Small* 14: 1704287.

136 Niehues, M., Tegeder, P., and Ravoo, B.J. (2019). Reversible end-to-end assembly of selectively functionalized gold nanorods by light-responsive arylazopyrazole-cyclodextrin interaction. *Beilstein J. Org. Chem.* 15: 1407–1415.

137 Wu, J., Xu, Y., Li, D.F. et al. (2017). End-to-end assembly and disassembly of gold nanorods based on photo-responsive host-guest interaction. *Chem. Commun.* 53: 4577–4580.

138 Zhu, L.L., Yan, H., Ang, C.Y. et al. (2012). Photoswitchable supramolecular catalysis by interparticle host-guest competitive binding. *Chem. Eur. J.* 18: 13979–13983.

139 Ryan, D., Rao, S.N., Rensmo, H. et al. (2000). Heterosupramolecular chemistry: recognition initiated and inhibited silver nanocrystal aggregation by pseudorotaxane assembly. *J. Am. Chem. Soc.* 122: 6252–6257.

140 Ryan, D., Nagle, L., Rensmo, H., and Fitzmaurice, D. (2002). Programmed assembly of binary nanostructures in solution. *J. Phys. Chem. B* 106: 5371–5377.

141 Ryan, D., Nagle, L., and Fitzmaurice, D. (2004). Light-patterned and recognition-directed adsorption of nanoparticles at a silicon wafer substrate. *Nano Lett.* 4: 573–575.

142 Zhang, Q., Qu, D.H., Wang, Q.C., and Tian, H. (2015). Dual-mode controlled self-assembly of TiO_2 nanoparticles through a cucurbit[8]uril-enhanced radical cation dimerization interaction. *Angew. Chem. Int. Ed.* 54: 15789–15793.

143 Zhang, J., Coulston, R.J., Jones, S.T. et al. (2012). One-step fabrication of supramolecular microcapsules from microfluidic droplets. *Science* 335: 690–694.

144 Coulston, R.J., Jones, S.T., Lee, T.-C. et al. (2011). Supramolecular gold nanoparticle-polymer composites formed in water with cucurbit[8]uril. *Chem. Commun.* 47: 164–166.

145 Cieslak, A.M., Janecek, E.R., Sokolowski, K. et al. (2017). Photo-induced interfacial electron transfer of ZnO nanocrystals to control supramolecular assembly in water. *Nanoscale* 9: 16128–16132.

146 Jones, S.T., Zayed, J.M., and Scherman, O.A. (2013). Supramolecular alignment of gold nanorods via cucurbit[8]uril ternary complex formation. *Nanoscale* 5: 5299–5302.

147 Klajn, R., Olson, M.A., Wesson, P.J. et al. (2009). Dynamic hook-and-eye nanoparticle sponges. *Nat. Chem.* 1: 733–738.

148 Olson, M.A., Coskun, A., Klajn, R. et al. (2009). Assembly of polygonal nanoparticle clusters directed by reversible noncovalent bonding interactions. *Nano Lett.* 9: 3185–3190.

149 Zhang, Q., Rao, S.J., Xie, T. et al. (2018). Muscle-like artificial molecular actuators for nanoparticles. *Chem* 4: 2670–2684.

150 Moulin, E., Faour, L., Carmona-Vargas, C.C., and Giuseppone, N. (2020). From molecular machines to stimuli-responsive materials. *Adv. Mater.* 32: 1906036.

151 Kay, E.R., Leigh, D.A., and Zerbetto, F. (2007). Synthetic molecular motors and mechanical machines. *Angew. Chem. Int. Ed.* 46: 72–191.

152 Erbas-Cakmak, S., Leigh, D.A., McTernan, C.T., and Nussbaumer, A.L. (2015). Artificial molecular machines. *Chem. Rev.* 115: 10081–10206.

153 Sauvage, J.P. (2017). From chemical topology to molecular machines (Nobel lecture). *Angew. Chem. Int. Ed.* 56: 11080–11093.

154 Stoddart, J.F. (2017). Mechanically interlocked molecules (MIMs)-molecular shuttles, switches, and machines (Nobel lecture). *Angew. Chem. Int. Ed.* 56: 11094–11125.

155 Baroncini, M., Silvi, S., and Credi, A. (2020). Photo- and redox-driven artificial molecular motors. *Chem. Rev.* 120: 200–268.

156 Qiu, Y.Y., Feng, Y.N., Guo, Q.H. et al. (2020). Pumps through the ages. *Chem* 6: 1952–1977.

157 van Delden, R.A., ter Wiel, M.K.J., Pollard, M.M. et al. (2005). Unidirectional molecular motor on a gold surface. *Nature* 437: 1337–1340.

158 Coskun, A., Banaszak, M., Astumian, R.D. et al. (2012). Great expectations: can artificial molecular machines deliver on their promise? *Chem. Soc. Rev.* 41: 19–30.

159 Dattler, D., Fuks, G., Heiser, J. et al. (2020). Design of collective motions from synthetic molecular switches, rotors, and motors. *Chem. Rev.* 120: 310–433.

160 Krause, S. and Feringa, B. (2020). Towards artificial molecular factories from framework-embedded molecular machines. *Nat. Rev. Chem.* 4: 550–562.

161 Aprahamian, I. (2020). The future of molecular machines. *ACS Cent. Sci.* 6: 347–358.

162 Feng, L., Qiu, Y.Y., Guo, Q.-H. et al. (2021). Active mechanisorption driven by pumping cassettes. *Science* 374: 1215–1221.

163 Qiu, Y.Y., Song, B., Pezzato, C. et al. (2020). A precise polyrotaxane synthesizer. *Science* 368: 1247–1253.

164 Thomas, D., Tetlow, D.J., Ren, Y.S. et al. (2022). Pumping between phases with a pulsed-fuel molecular ratchet. *Nat. Nanotechnol.* 17: 701–707.

165 Grommet, A.B., Feller, M., and Klajn, R. (2020). Chemical reactivity under nanoconfinement. *Nat. Nanotechnol.* 15: 256–271.

166 van Esch, J.H., Klajn, R., and Otto, S. (2017). Chemical systems out of equilibrium. *Chem. Soc. Rev.* 46: 5474–5475.

167 Ragazzon, G. and Prins, L.J. (2018). Energy consumption in chemical fuel-driven self-assembly. *Nat. Nanotechnol.* 13: 882–889.

168 Das, K., Gabrielli, L., and Prins, L.J. (2021). Chemically fueled self-assembly in biology and chemistry. *Angew. Chem. Int. Ed.* 60: 20120–20143.

169 Amano, S., Borsley, S., Leigh, D.A., and Sun, Z.H. (2021). Chemical engines: driving systems away from equilibrium through catalyst reaction cycles. *Nat. Nanotechnol.* 16: 1057–1067.

3

Mechanically Interlocked Nanotubes

Alejandro López-Moreno and Emilio M. Pérez

Ciudad Universitaria de Cantoblanco, IMDEA Nanociencia, Faraday 9, Madrid 28049, Spain

3.1 Introduction

Mechanically interlocked molecules (MIMs) consist of two or more separate components that are not connected by chemical (i.e. covalent) bonds. These structures are molecular entities, as each component is intrinsically linked to the other through a mechanical bond, which prevents dissociation without cleavage of one or more covalent bonds. Catenanes, molecules in which two or more macrocycles are interlocked, and rotaxanes, which are molecular species in which one or more macrocycles are threaded onto a linear component and bulky stoppers prevent the de-threading, are examples of MIMs. This unique type of chemistry to connect submolecular fragments leads to new sets of potentially valuable properties. It is worth highlighting the possibility of controlling the large-amplitude submolecular motion of the components with respect to one another, enabling the synthesis of some of the most advanced examples of molecular machines [1, 2].

The first synthetic routes to this class of molecules were based on statistical approaches with extremely low yields. In 1960, Wasserman reported the first "preparation of interlocking rings," which was achieved in less than 1% yield [3]. The first synthesis of a [2]rotaxane was described seven years later in 6% yield after 70 reiterations using a solid-support protocol [4]. In the early 1980s, Sauvage reported the first "templated" synthesis of a [2]catenane using Cu(I) to preorganize two phenanthroline ligands perpendicular to each other and then forming the macrocycles by alkylation of the terminal phenols with iodoethyleneglycol ethers of suitable length [5, 6]. Sauvage obtained catenane in a synthetically viable 27% yield following this strategy. More importantly, this templated synthesis was the first one to use metal–ligand bonding (noncovalent interactions) to template the synthesis of a MIM.

The development of templated methods improved yields significantly and allowed the synthesis of sufficient amounts of rotaxanes and catenanes, which have allowed the study of their dynamic properties. In 2001, Leigh and coworkers reported a nearly quantitative yield of 97% for the synthesis of a [2]rotaxane [7]. Since then, the templated methods for synthesizing MIMs have allowed the construction of much more complex MIMs, such as Borromean rings [8] and ever more complex knots [9].

Mechanically Interlocked Materials: Polymers, Nanomaterials, MOFs, and more, First Edition.
Edited by Emilio M. Pérez.

The mechanical bond is a unique tool that can be used to connect two covalent entities very intimately, with an equivalent stability to their weakest covalent bond, but without modifying their structure. This exceptional characteristic makes the mechanical bond an ideal candidate for tasks that cannot be carried out by either covalent or supramolecular chemistry, and it allows one component to move with respect to the other along (or around) distances significantly larger than atomic vibrations. However, the power of templated methods to synthesize MIMs is seriously challenged when there are no directional supramolecular interactions (H-bonds, metal–ligand) between the submolecular components. This is the case of carbon nanotubes, which lack functional groups so their supramolecular chemistry is necessarily based on dispersion interactions [10]. In this chapter, we will overview the synthesis of mechanically interlocked species featuring carbon nanotubes as an integral part of their structure.

3.2 Carbon Nanotubes

The properties of single-walled carbon nanotubes (SWNTs) are some of the most interesting ones among nanomaterials, such as extremely high mass-normalized strength and thermal conductivity, metallic carbon nanotubes transport electric current ballistically, while semiconducting nanotubes feature a well-defined bandgap depending on their chirality. These properties make SWNTs apromising material for many applications, from stronger materials to advanced electronic devices [11–15].

Despite these properties, some drawbacks, such as the limitations of production of monodisperse SWNT samples with chiral selectivity or isolating them from mixtures, necessary for the applications in (opto) electronics and the tendency of SWNTs to form aggregates or bundles through strong van der Waals interactions, limited the practical application of SWNTs. To resolve these problems, SWNTs can be chemically functionalized. The main approaches for the functionalization of SWNTs can be grouped into (i) the covalent attachment of chemical groups and (ii) the noncovalent supramolecular adsorption or wrapping of molecules onto the tubes.

The covalent functionalization of SWNTs allows functional groups to be attached to tube ends or sidewalls. This kind of modification produces kinetically stable products, but the structure of SWNTs is affected, and therefore its physical properties are often not desirable [16–19]. On the other hand, in noncovalent functionalization, the tube surface can be modified via van der Waals forces and π–π interactions by adsorption or wrapping molecules or polymers, which produce much more labile products than the covalent approach. The main advantage of this type of functionalization is that functionality can be introduced without affecting the structure and electronic network of the SWNTs [20, 21]. Within this type of functionalization, some of the most interesting works include the use of polymer wrapping to sort or solubilize the SWNTs [22–28] or sorting and/or modification of the SWNTs properties using DNA [29–36].

As an alternative method, the application of the knowledge acquired in the synthesis of small-molecule MIMs to the modification of SWNTs seems very attractive to exploit the intrinsic properties of SWNTs, and/or modulate them, while preserving their chemical structure. The earliest combinations of MIMs and nanotubes were described by the groups of Stoddart and Heath based on the noncovalent sidewall modification of SWNTs with catenane [20] or pseudorotaxane architectures [37]. In 2005, Dieckmann and coworkers reported the diameter-selective solubilization of SWNTs using peptide macrocycles reversibly formed using disulfide chemistry by encapsulating SWNTs with the macrocycles [38]. Stoddart and coworkers described the formation of "acyclic and/or cyclic adducts on or around the sidewalls of SWNTs" using Pd coordination of porphyrin fragments [39]. More recently, Campidelli and coworkers reported the formation of stable covalent porphyrinic networks around multi-wall carbon nanotubes [40].

3.3 MINTs: Clipping Strategy

The unambiguous mechanical interlocking of carbon nanotubes was reported in 2014 [41]. The synthesis of these rotaxanes in which the SWNTs act as threads was based on a "clipping" strategy, using U-shaped molecules containing two recognition motifs with high affinity for SWNTs connected through an aromatic spacer and finally decorated with alkene-terminated alkyl chains. Based on previous experience with fullerenes, the U-shaped molecules were based on two π-extended tetrathiofulvalene [9,10-di(1,3-dithiol-2-ylidene)-9,10-dihydroanthracene, exTTF] which previously had been used a recognition motifs for SWNTs [42]. First, the U-shape is associated with individual SWNTs and in a second step, the rotaxane-type species was formed through a ring-closing metathesis reaction (Figure 3.1a). In the first step, the supramolecular equilibrium is governed by an association constant (K_a) that can be measured using a relatively simple titration method based on thermogravimetric analysis (TGA) [44, 45]. In the solvents in which mechanically interlocked nanotubes (MINTs) are formed [1,1′,2,2′-tetrachloroethane (TCE), N,N-dimethylformamide (DMF)] and at room temperature, the values for the binding constant are $Ka = 10^3–10^4\ M^{-1}$, which implies that at the concentrations that MINTs are prepared more than 50% of U-shapes are associated with the SWNTs at the beginning of the reaction.

After clipping reaction, the sample was profusely washed to remove impurities or subproducts, obtaining the mechanically interlocked nanotubes, which were extensively characterized by analytical, spectroscopic, and microscopic techniques. TGA of the products showed weight losses ranging from 23% to 37% depending on the size of the nanotubes and the length of the macrocycle alkyl chains. Also, the functionalization can be modulated as demonstrated by varying the relative concentration of the U-shaped molecules with respect to SWNTs, with functionalization ranging from 12% to 37% when minimum and maximum concentrations were respectively used. These data showed that the degree of functionalization reached a maximum of about 40%, corresponding to one macrocycle for every 140 nanotube carbon atoms.

This new type of functionalized tube presented the advantages of covalent and noncovalent functionalization, showing no changes in the nanotube structure (as proven by Raman and UV–vis–NIR spectroscopies) and high stability, as proven

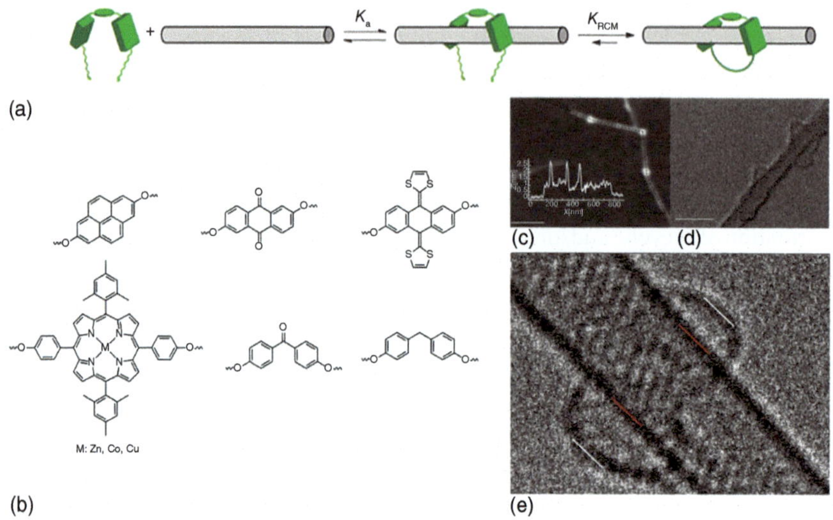

(a) (b) (c) (d) (e)

Figure 3.1 (a) Clipping strategy, based on the association of SWNTs by alkene-terminated U-shapes featuring recognition units for SWNTs; (b) chemical structure of recognition motifs used for the preparation of U-shaped molecules; Adapted from de Juan et al. [41] and Blanco et al. [43] (c) AFM topographic image of a spin-cast suspension of exTTF-MINT in TCE; the inset shows the profile along the dashed black line; (d) HR STEM bright-field image of a single SWNT surrounded by two macrocycles in the exTTF-MINT sample. Scale bar is 2 nm; and (e) AC-HRTEM image of Anthraquinone-MINT (macrocycle size of 1.72 nm and SWNTs diameter of 0.87 nm). Source: Adapted from de Juan et al. [41] and Blanco et al. [43].

by the washes of the sample with hot TCE. Raman spectroscopy showed that the changes to the electron–phonon structure of the SWNTs after MINT reaction species are small and similar to the noncovalent approach. Finally, atomic force microscopy (AFM), Figure 3.1c and transmission electron microscopy (TEM) showed the presence of rings of 2.2 nm in diameter around SWNTs (0.8 nm), confirming the existence of the mechanically interlocked nanotubes. The aberration-corrected high-resolution electron microscopy micrographs showed MINTs with close to atomic resolution (Figure 3.1d,e), and energy-dispersive X-ray (EDX) spectroscopy of the MINT samples confirmed the presence of a significant amount of sulfur (c. 1%).

A major concern when using Grubb's catalysts with bisalkene molecules is the competitive oligomerization reaction between two or more U-shapes instead of ring-closing metathesis. The investigation of the effect of concentration of the U-shaped molecule, reaction time, and catalyst concentration on the MINT-forming reaction was carried out, finding the ideal conditions to maximize the functionalization of MINTs without significant amounts of oligomeric by-products [46]. The analysis of conversion–time data proved that the macrocyclization of the U-shaped molecule in the presence of SWNTs to form MINT, or in their absence to form the macrocycle, follow the typical trends of ring-closing metathesis (RCM), but with significantly different kinetic parameters. Following a pseudo-first-order approximation, the result supported a mechanism in which the cyclization of U-shape in the MINT-forming reaction is the rate-determining step. The same work demonstrated that an excess of catalyst led to the formation of by-products (supramolecularly associated oligomers).

This clipping strategy for the synthesis of MINTs, using RCM as the final step, has proven remarkably flexible with respect to the recognition motifs that can be incorporated in the U-shaped precursor. Pyrene [47] templated the RCM of macrocycles around SWNTs to form MINTs, demonstrating that any molecule with good affinity for carbon nanotubes can be used as a recognition motif in U-shaped molecules. Later, naphthalene diimides (NDI), [48] anthraquinone [43], porphyrins, [49, 50] or benzophenones and diphenylmethane [51] were used, in some cases modifying the original U-shape design to increase the solubility and/or the flexibility of the precursor (Figure 3.1b).

Further details on the role of oligomerization in this type of functionalization were studied using NDI-based U-shaped precursors for the ring-closing around the SWNTs to form rotaxane-like MINTs or oligomerize through acyclic diene metathesis polymerization (ADMP) and then wrap the SWNTs. This can be achieved by changing the solvent. In TCE, the NDI precursors showed a lower affinity for the SWNTs, and the non-templated metathesis reaction is very slow, favoring the formation of the MINTs, and the diameter of nanotube and size of the activity must match. On the other hand, using DMF as a solvent, the non-templated reaction is significantly faster, and the NDI precursors showed much larger binding constants toward SWNTs, which results in the formation of a substantial amount of oligomer-wrapped SWNTs.

The degree of functionalization in DMF is diameter-independent and consistently leads to larger NDI loading. Each type of functionalization affected the electronic properties of the SWNTs differently.

Porphyrin-CNT hybrids have been successfully applied to the fabrication of sensors [52–54], catalysts [55–57], or optoelectronic devices [58, 59], and the preparation of mechanically interlocked SWNTs using porphyrins allowed the exploitation of their interesting properties. The first porphyrin-based MINT was reported in 2018 [49]. In this case, a hexamethylene spacer provided flexibility to template the macrocyclization around SWNTs through RCM of the terminal alkenes to form the corresponding MINTs. The reaction proceeded with chiral selectivity for the smaller diameter SWNTs. The electronic interactions between the porphyrins and the nanotubes in the ground state, as well as the charge separation and recombination processes and their underlying dynamics upon photoexcitation, were also studied. This work represented an important achievement in connecting porphyrins and SWNTs to explore different applications such as sensing or magnetism, as demonstrated with the recent study of cobalt and copper-based porphyrin MINTs [50].

Once the MINTs demonstrated versatility as a method for the chemical functionalization of SWNTs, the submolecular motion of different organic macrocycles along and around SWNTs by molecular dynamics simulations was studied [60]. These movements do not depend on the interaction energy between the macrocycle and SWNT. It was demonstrated that macrocycles in MINTs rotate and translate freely around and along SWNTs at room temperature, with an energetic cost lower than that for the rotation around the C—C bond in ethane. The slight differences between MINTs were explained in terms of macrocycle-SWNT commensurability.

A different clipping strategy to synthesize MINTs was reported in 2020 by Von Delius and coworkers [61]. The authors designed and synthesized two macrocycles that contain a curved π-extended tetrathiafulvalene (TTF) moiety with two disulfide bonds for the preparation of reversible interlocking of nanotubes. The authors carried out two control experiments to compare them with the reversible functionalization: an irreversible functionalization and the supramolecular control with the preformed macrocycle. These experiments led to different ratios of functionalization after washes.

The reversible opening and closing of these dynamic covalent bonds under mild basic conditions led to the functionalization of SWNTs with mechanically interlocked rings, as demonstrated by extensive characterization with different spectroscopic and microscopic techniques (Figure 3.2). The defunctionalization studies demonstrated that this reversible MINT synthesis approach allowed the mild and quantitative recovery of both the dithiol macrocyclic precursors and the SWNT starting materials.

This approach presented some advantages with respect to the previous one: (i) the dithiol building blocks allowed the comparative evaluation of reversible, irreversible, and supramolecular functionalization; and (ii) the disulfide bonds allow the quantitative reductive cleavage, which offers the possibility of applying this reversible functionalization to purify SWNT mixtures by achieving diameter-selective solubilization in organic solvents. The method provided selective functionalization of smaller diameter (6,4)- and (6,5)-SWNTs as opposed to the larger (7,5)- and (8,3)-SWNTs. This chirality selectivity differed for the studied macrocycles due to a combination of size match and ground-state charge transfer.

3.4 Other Strategies for the Preparation of MINTs

Miki et al. studied the direct threading of SWNTs through rigid cycloparaphenyleneacetylene (CPPA) macrocycles by mixing CPPAs and SWNTs in *o*-dichlorobenzene to obtain the complexes [62]. The authors used three different macrocycles, [8]-, [9]-, and [10]-CPPAs, which interacted with carbon nanotubes. By comparing the fit of the CPPAs cavity and the SWNTs diameter and the amount of complex obtained, the authors concluded both tube-in-ring and tube-on-ring complexes resulted from this approach. CPPAs having both larger and smaller ring sizes mainly formed "ring-on-tube" complexes with CNTs, whereas when the fit is particularly tight, like that between the [9] CPPA (1.89 nm cavity) and SWNTs of 1.1–1.4 nm diameter, the tube-in-ring predominates (Figure 3.3).

In 2018, the templated formation of self-assembled hydrogen-bonded macrocycles around SWNTs was described by González-Rodríguez and coworkers [63]. By mixing (6,5)-SWNTs with a dinucleoside monomer that is able to self-assemble in nanorings via Watson–Crick base-pairing, the authors encapsulated carbon nanotubes (Figure 3.4). This approach allowed the efficient SWNT debundling

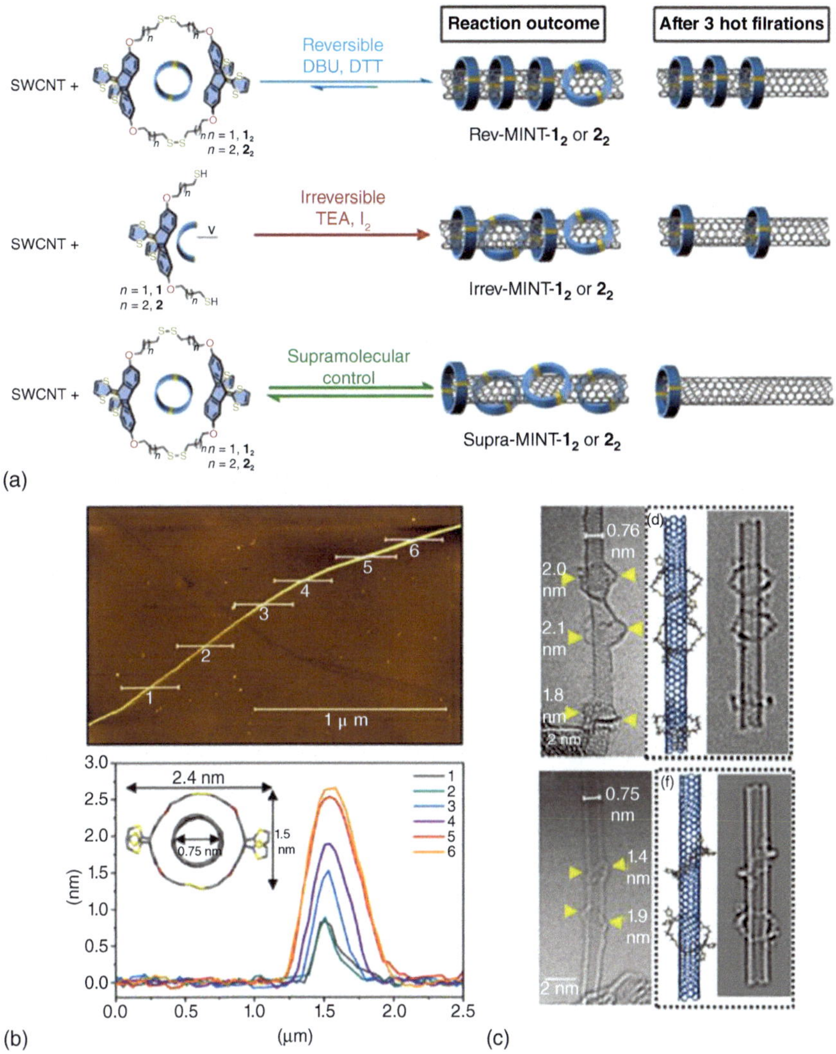

Figure 3.2 DCC strategy by von Delius. (a) Schematic representation of different functionalization approaches of SWNTs with macrocycles $\mathbf{1}_2$ and $\mathbf{2}_2$. (b) AFM topographic image of Rev-MINT-$\mathbf{2}_2$ and corresponding height profile indicated in the AFM image; inset: calculated structure, which reveals the differences in diameters upon rotation; and (c) HRTEM images of Rev-MINT-$\mathbf{2}_2$ sample and the corresponding HRTEM simulations. Source: Reproduced with permission from Balakrishna et al. [61] / John Wiley & Sons.

in apolar solvents due to the H-bonded rings, which associated intimately with SWNTs and supplied the reversibility to recover the pristine material in a simple and effective manner by using a polar solvent or heating the apolar suspension to induce the dissociation. The nanoring stability is enhanced in the presence of the SWNTs due to the interaction of more than one monomer with the carbon

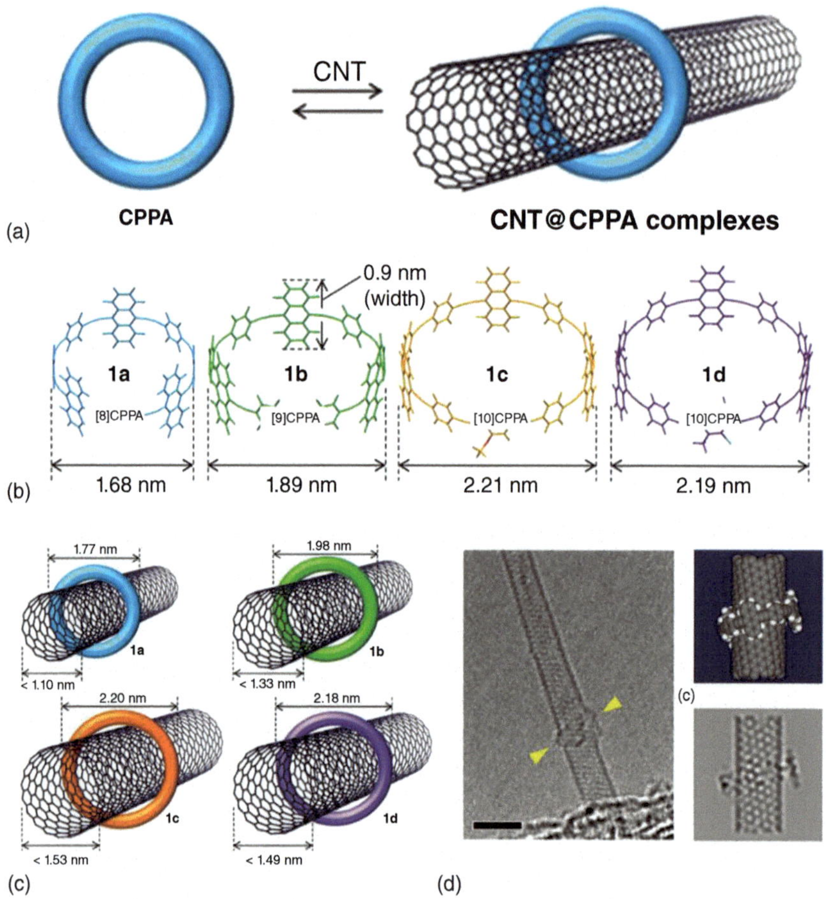

Figure 3.3 (a) Scheme of "Ring-tossing" strategy, as described by Miki et al. (b) Structure of CPPAs **1a–d** optimized at the B3LYP/6-31G(d) level of theory. (c) Schematic illustration of CNT@CPPA complexes showing the theoretical diameters of rings and SWNTs. (d) HRTEM imaging of a MINT (left, scale bar: 2 nm), a model structure of one of the complexes, and the corresponding TEM simulation for comparison. Source: Reproduced with permission from Miki et al. [62] / John Wiley & Sons.

nanotube sidewall. Furthermore, using this method of functionalization led to very concentrated, stable, and dilution-resistant SWNT dispersions. On the other hand, when noncyclable monomers were used, the dispersions were drastically worse.

Kruss, Thomas, and coworkers explored the use of α-Helical coiled-coil peptide barrels (αHBs) to solubilize and functionalize SWNTs [64]. Comprehensive spectroscopy and microscopy data demonstrated that barrels are formed around the SWNTs while optical properties are retained in SWNTs@αHBs (Figure 3.5). The outer surface of αHBs is therefore promising as functionalization platform while the SWNTs are colloidally stabilized. This approach can also be used for the design of diameter-selective αHBs and enables the attachment of additional

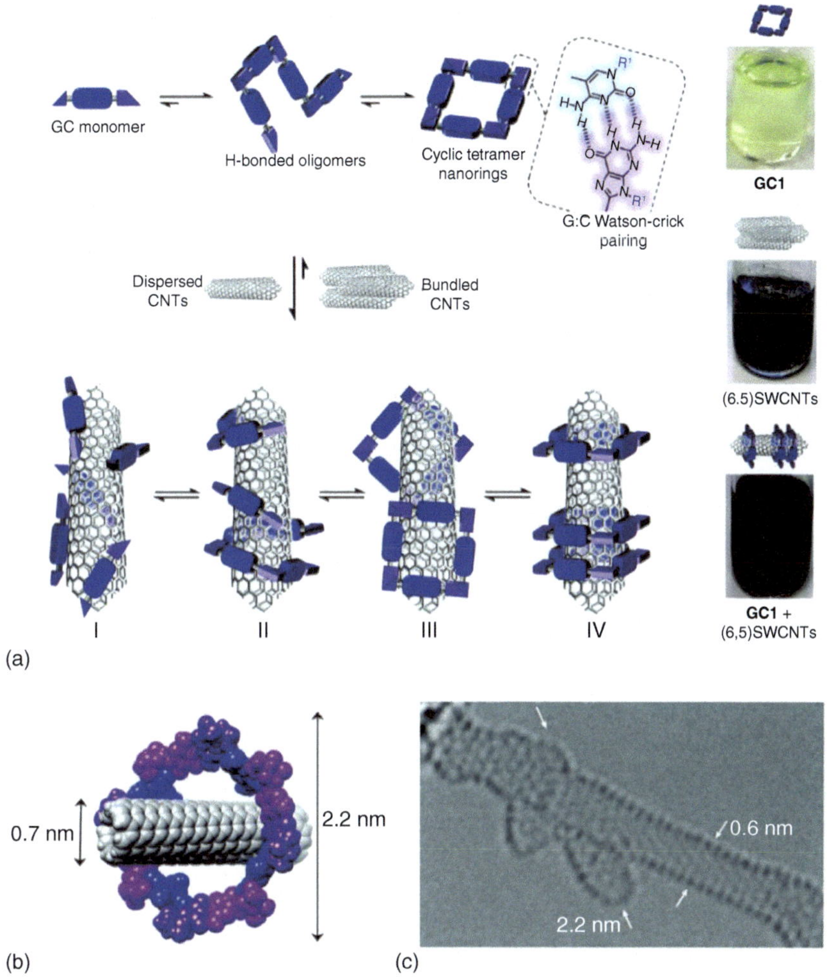

Figure 3.4 H-bonding strategy toward the encapsulation of SWNTs within self-assembled macrocycles, as described by Pérez, González-Rodríguez, and coworkers. (a) self-assembly of the cyclic tetramers using G–C base pairing and different plausible modes of association of the monomers, oligomers, and cyclic tetramers with SWNTs. Right: the image of the solutions/dispersions containing GC1, (6,5)SWNTs and their mixture in $CHCl_3$. (b) Optimized structure of the (6,5)SWNT–c**GC1**; (c) HRTEM image of a cyclic species around a SWNT. Source: Reproduced with permission from Chamorro et al. [63] / Royal Society Of Chemistry.

molecules, such as fluorophores and pharmacophores, or recognition units of the functionalized SWNT.

Finally, the last contribution to the mechanical interlocking of carbon nanotubes was reported in 2022. Despite the significant development of this type of nanotube functionalization, the threading of SWNTs through the cavity of supramolecular coordination complexes (SCCs) remained unexplored. Pérez and Peris groups reported for the use of SCCs to obtain a new type of MINT [65]. Ni- and Pd-containing

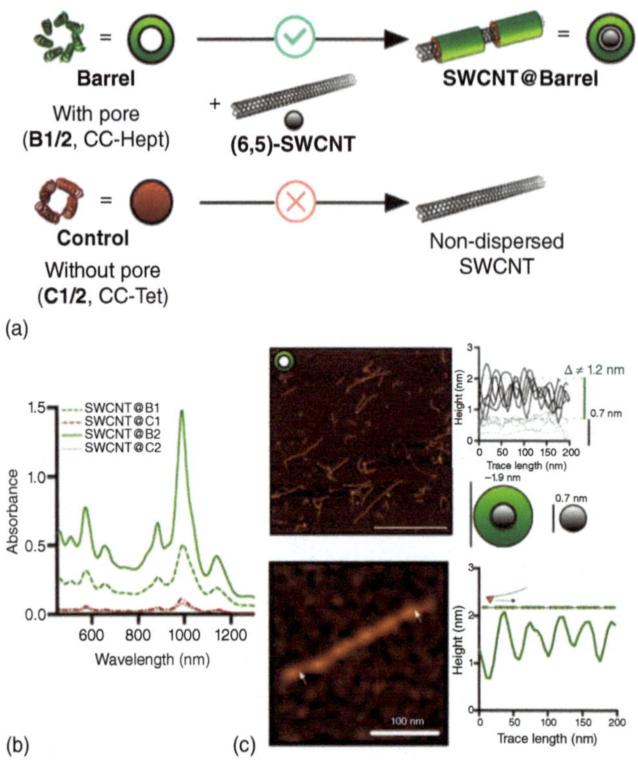

Figure 3.5 (a) α-Helical coiled-coil peptide barrels (αHBs) encapsulate SWNTs and form stable SWNT@αHBs hybrids. Coiled-coil peptides that form heptameric barrels with pores encapsulate SWNTs. Control peptides that form tetrameric barrels with no (too small) pores do not solubilize SWNTs. (b) Absorbance spectra of the peptide-dispersed SWNTs show a clear tendency in dispersion yield with the right diameter. (c) Top: AFM images of dispersed (6,5)-chirality enriched SWNTs. Height traces along the axis of SWNT@B2 (1.9 nm), which fits well to the value of free (6,5)-SWNT (≈0.7 nm) and the height additionally introduced by the heptameric α-helical barrel (≈1.2 nm). Bottom: Magnification of a single SWNT@B2 showing repeating changes in height. Source: Reproduced with permission from Mann et al. [64] / John Wiley & Sons.

metallosupramolecular assemblies based on a pyrene-bis-imidazolylidene ligand (Figure 3.6a) were previously reported, demonstrating their ability to encapsulate fullerenes (C_{60} or C_{70}). The high binding affinities of these metallo-assemblies for fullerenes and their great shape adaptability make these hosts perfect for the preparation of a new class of MINTs.

The preparation of this new metallo-MINTs can be performed either by the template formation of the metallacycles around the SWNT or by encapsulating the SWNT into the cavity of the preformed metallacycle, following a similar procedure to the one reported by Miki et al. The templated preparation of MINTs was performed using a suspension of (6,5)-SWNTs in anhydrous acetonitrile, to which the pyrene-bis(imidazolium) salt, $[Pd(allyl)Cl]_2$, $AgBF_4$ and Cs_2CO_3 were

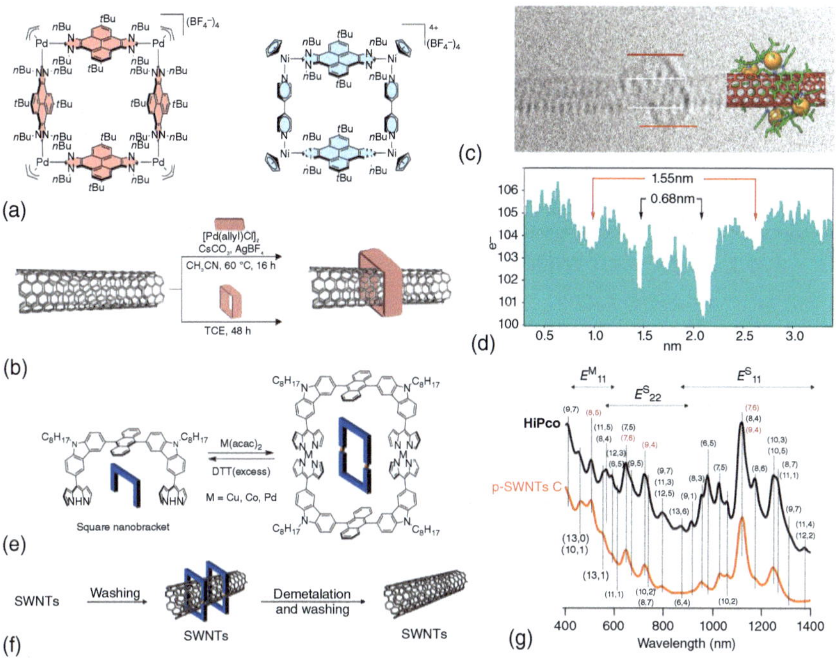

Figure 3.6 (a) Chemical structure of metallosquares used to try mechanical interlocking of SWNTs. (b) Schematic representation of MINT-**1** through two different routes. (c) ac-HRTEM image of MINT-**1**. An energy-minimized molecular model (molecular mechanics, UFF) of MINT-**1** is superimposed on the right side to serve as a visual reference. (d) Profile graph of (c), showing the dimensions of the SWNT and the macrocycle in perfect accordance with expectations. (e) Synthesis of tetragonal M-Nanobrackets from square Nanobracket. (f) Procedure for the preparation of encapsulated SWNTs after washing and extracted SWNTs after demetalation. (g) Vis–NIR of D_2O/SDBS dispersions of HiPco SWNTs and extracted SWNTs. Source: Adapted from Lopez-Moreno et al. [65] and Cheng et al. [66].

added. After 16 hours at 60 °C, the resulting MINTs were recovered by filtration. The encapsulation of SWNTs with the preformed metallosquare was achieved by adding the metallosquare to a dispersion of (6,5)-SWNTs in TCE. After 48 hours at room temperature, the MINTs were recovered by filtration and washed profusely. The formation of MINTs using both methods was confirmed using TGA, Raman and UV–vis NIR spectroscopies, PLE, AFM, and TEM, as exemplified in Figure 3.6. This work demonstrated that MINTs could be constructed by taking advantage of the metal–ligand binding and the directional control of coordination geometries. Also, the incorporation of metal atoms into the MINT has an important effect on the electronic nature of nanotubes.

In parallel, Komatsu and coworkers developed a Cu-tethered tetragonal nanobrackets [66]. Based on previous experience with nanotweezers and nanocalipers used in the selective extraction of SWNTs [67–70], Cu nanobrackets were applied in the selective separation of SWNTs with 0.90–0.92 nm in diameter due to the rigid structure of the host molecule (Figure 3.6e–g). The interlocking of Cu nanobrackets with SWNTs was confirmed by Raman spectroscopy, TEM, and AFM. In addition, these Cu nanobrackets exhibited better selectivity than previously reported "nanotweezers" and "nanocalipers." The features presented in this recent work were the strict discrimination of SWNT diameters, and the use of Cu-nanobrackets for the interlocking of SWNTs, and [3] a facile and thorough removal of the host molecule.

3.5 Application of MINTs

Since MINTs were reported, the properties of this new type of functionalized nanotubes have been explored in several works. A comprehensive experimental and theoretical investigation of the influence of linking exTTF electron donors to SWNTs through mechanical links, in comparison with non-interlocked supramolecular model systems, was reported after the first MINT preparation [71]. This study showed that, in the ground state, the degree of charge transfer between exTTF and SWNTs is negligible, whereas, upon photoexcitation, rapid electron transfer from exTTF to the SWNTs takes place. The rates of transfer and recombination are significantly different in the interlocked and non-interlocked compounds, showing the unique effect of the MINT-type modification. Later, similar conclusions were found for porphyrin-based MINTs [49].

In 2018, the catalytic properties of MINTs were studied in different systems. The first study investigated the catalytic role of MINTs in the reduction of nitroarenes with hydrazine [43]. Different types of MINTs were prepared following the RCM clipping approach: with electron donor (MINT-exTTF), acceptor (MINT-anthraquinone), or electronically neutral (MINT-pyrene) macrocycles. The comparison of these MINTs with pristine SWNTs was carried out in this work. This established a relationship between the expected electronic density at the SWNT surface and the catalytic activity. Despite the functionalization degree or available binding sites playing an important role, the catalytic activity increases with electron density on the SWNT, which is the most important parameter. The electron-rich(er) MINT-exTTF performs significantly better as a catalyst than the pristine SWNTs (black in Figure 3.7), which showed similar activity as MINT-pyrene and much better than the electron-poor(er) MINT-anthraquinone (yellow in Figure 3.7). In this strategy, the link between the effectors (the macrocycles) and the catalyst (the nanotubes) is noncovalent but stable, and its effect is remote but not allosteric since it does not change the structure of the active site (the surface of the nanotube). The MINT approach presents interesting advantages that can solve the problem of dissolution instability of physically adsorbed organic molecules, as demonstrated in this work, by testing recyclability.

More recently, the electrocatalytic application of MINTs toward oxygen reduction was explored [72]. First, anthraquinone-based MINTs showed an enhancement of stability while retaining the electrochemical properties of anthraquinone-macrocycle. A proof of concept for the electrocatalytic application of anthraquinone-MINT toward the oxygen reduction reaction (ORR) was explored,

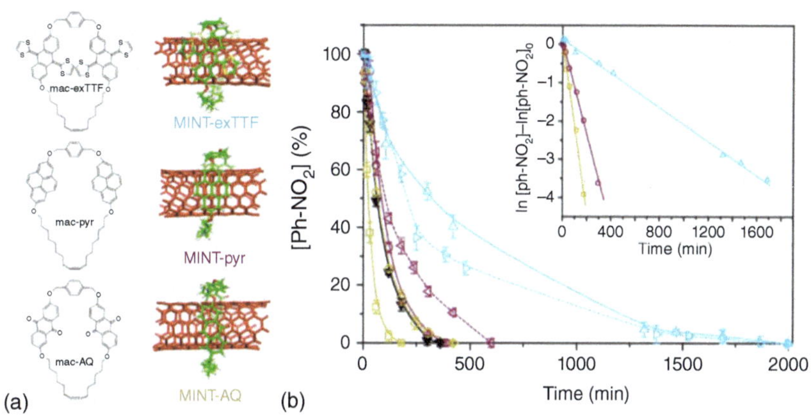

Figure 3.7 (a) Chemical structures of the electron-donating (exTTF), electronically neutral (pyr), and electron-accepting (AQ) macrocycles and energy-minimized (DFT) molecular models of their respective MINTs with (6,5)-SWNTs. (b) Kinetics of the reduction of nitrobenzene using MINT-exTTF (yellow solid line), MINT-pyr (purple solid line), MINT-AQ (cyan solid line), (6,5)-SWNT (black solid line), SWNT-exTTF (yellow dashed line), SWNT-pyr (purple dashed line), and SWNT-AQ (cyan dashed line) as catalysts. Source: Blanco et al. [43] / Springer Nature / CC BY 4.0.

showing an improvement in the H_2O_2 production rate maintaining high stability and proving the interesting properties of MINTs as immobilization strategy (Figure 3.8a,b). These promising results using MINTs open up a wide range of applications for catalysis by modulating and changing the properties of MINTs by designing different macrocycles or using different types of carbon nanotubes.

Later, the formation of MINTs using N-rich macrocycles as a novel strategy for the preparation of SWNT derivatives displaying N atoms on their surface demonstrated their ability as electrocatalysts in the ORR [51]. The N-containing MINTs, through the introduction of triphenylamine (TPA) groups in two different types of macrocycles based on benzophenones and diphenylmethane cores, demonstrated an improved ORR performance. The N-containing MINTs showed a performance improvement in comparison with pristine SWNTs and the structurally related MINTs without TPA groups, ascribed to the introduction of those nitrogen heteroatoms that act as effective active sites for the ORR. Also, the presence of CO groups in the benzophenone-based MINTs led to an improvement of the activity compared with the diphenylmethane-based MINTs (Figure 3.8c,d).

These two works indicated that the design of a more active and stable catalyst can be achieved through the modification of SWNTs by mechanical bond.

The extraordinary mechanical properties of SWNTs make them perfect candidates as fillers for the reinforcement of polymers. Due to the tendency of SWNTs to interact with themselves and self-aggregate, their chemical modification is necessary to enhance individualization and interaction with the polymer matrix. MINTs were explored as electrospun polymer fiber fillers [73]. Different types of MINTs with different macrocycles and SWNTs were explored as fillers in polystyrene fibers. Also,

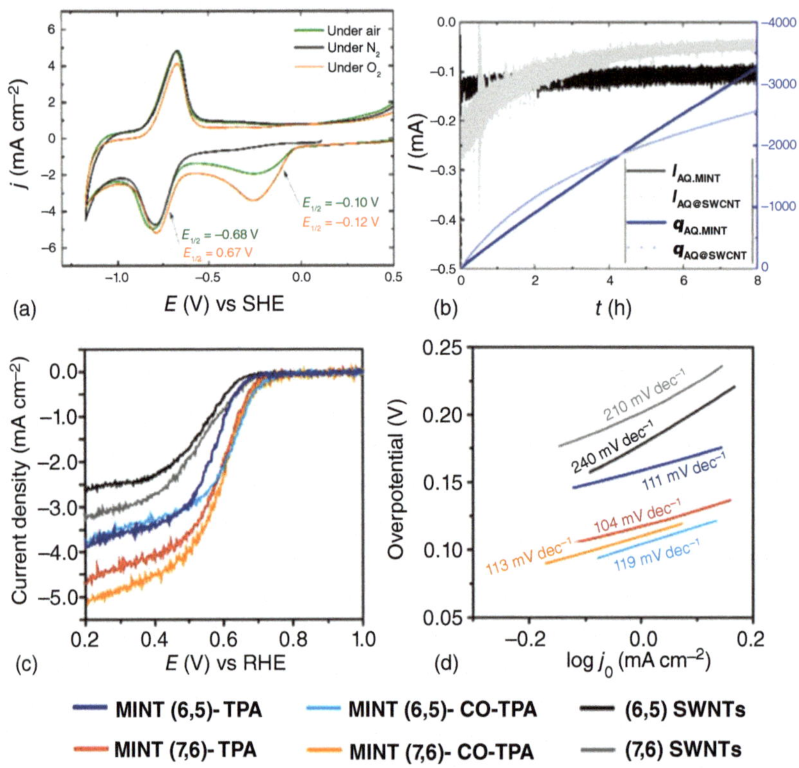

Figure 3.8 (a) Cyclic voltammograms of AQ-MINT in 0.1 M NaOH under air (green line), N_2 (black line), and O_2 (orange line) conditions. (b) Chronoamperograms and accumulated charges of AQ-MINT and AQ@SWNTs at a constant potential of −0.33 V vs SHE in 0.1 M NaOH. Electrochemical characterization of the N-rich MINT electrocatalysts. (c) Comparison of the ohmic drop corrected ORR polarization curves and (d) Tafel plots for the MINTs electrocatalysts with their respective pristine SWNTs counterpart. Source: Adapted with permission from Zhang et al. [51] and Wielend et al. [72] / Royal Society of Chemistry.

the equivalent supramolecular samples, with identical chemical composition but different topologies, and the pristine nanotubes were used as control experiments (Figure 3.9a).

Encapsulation of the SWNTs within organic macrocycles, in the forms of MINTs, results in an enhanced interaction between the polymer and the SWNTs. Electrospun polystyrene fibers doped with only 0.01% of MINTs showed a much higher both Young's modulus and tensile strength, with improvements of over 200% in some cases, while the supramolecular fillers, the non-interlocked samples showed improvement or even negligible effects on the mechanical properties compared with the electrospun fibers reinforced with pristine nanotubes or the neat polystyrene fibers without fillers (Figure 3.9b,c). This work demonstrated that mechanical interlocking is a useful strategy to optimize the performance of SWNT fillers and can be extended to many polymers, optimizing the polymer–nanotube interaction by

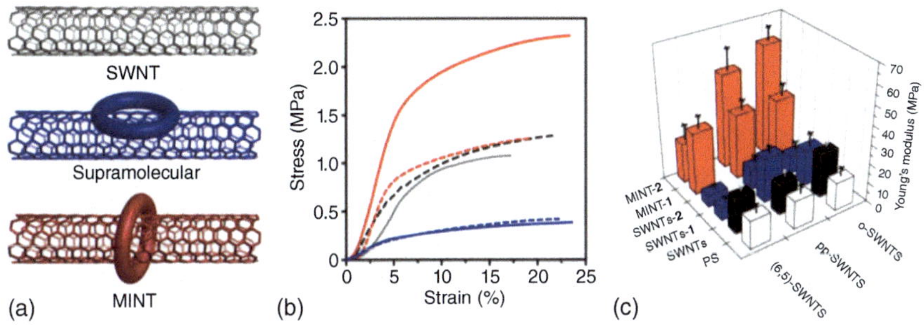

Figure 3.9 (a) Schematic representation of the three types of fillers investigated. (b) Representative stress–strain curves of electrospun fibers of polystyrene (grey), and its composites with SWNTs (dashed black), MINT-pyr (red), supramolecular mixture of SWNTs and mac-pyr (blue), MINT-exTTF (dashed red) and supramolecular mixture of SWNTs and mac exTTF (dashed blue). (c) Young's modulus for MINT-reinforced polystyrene fibers, pristine nanotubes, and supramolecular complexes. Source: Adapted with permission from López-Moreno et al. [73].

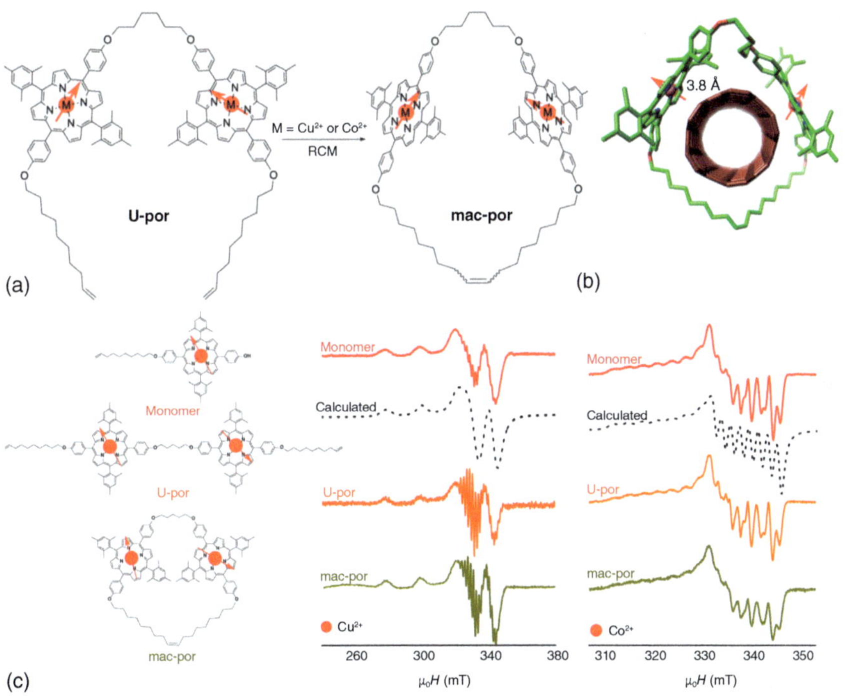

Figure 3.10 (a) Chemical structure of the bisporphyrin U-shape precursor and macrocycle. (b) Relaxed mMINT structure obtained by DFT calculations. The magnetic centers are placed at van der Waals distance from the SWNT surface. (c) X-band electron paramagnetic resonance (EPR) spectra measured at $T = 20$ K on the Cu and Co porphyrin monomers, U-shapes, and Macrocycles. The dotted lines are simulations of the monomer spectra. Source: Reproduced with permission from Moreno-Da Silva et al. [50] / American Chemical Society.

changing the type of molecule, which will allow the exploitation of SWNT properties as reinforcing fillers [74, 75].

More recently, a last example of potential applications of MINTs was described in collaboration with the groups of Moonshiram and Burzurí [50]. Magnetic porphyrin (Cu^{2+} and Co^{2+}) dimeric rings were attached to SWNTs following the clipping strategy and leading to the magnetic MINTs (mMINTs, Figure 3.10a,b). The authors used the SWNTs as a one-dimensional intermediate channel for introducing and wiring magnetic molecules on functional devices. Magnetic porphyrin dimers (Co^{2+} and Cu^{2+}) were mechanically interlocked with SWNTs and fully characterized. The metal centers in both the mac-por and mMINT porphyrins preserved the coordination sphere, the structure, and the spin geometry in the Cu-mMINT (Figure 3.10a). The interaction between spin and conduction electrons can be modulated by selecting the metal in the mMINTs design, which can be used to read the molecular spin and make the mMINTs good candidates for molecular spintronics and molecular spin qubits.

3.6 Conclusions

The mechanical bond brought the possibility for the connection of (sub)molecular fragments in a stable way without establishing covalent bonds between them and for the observation and control of their motion with respect to one another. Exploiting these features has led to some of the most exciting developments in chemistry in recent years, as manifested in the 2016 Nobel Prize for Chemistry, which was awarded to Profs. Sauvage and Stoddart, for the development of MIM-based molecular machinery, with Ben L. Feringa completing the picture with his covalently connected molecular motors.

The interesting characteristics that have made the mechanical bond a transformative force in small-molecule chemistry make it a very relevant tool for carbon nanotube chemistry. In this review, we have focused on the particular case of SWNTs. We have discussed the synthetic strategies toward MINTs and their potential advantages for applications compared to the covalent and non-covalent functionalization of SWNTs.

In particular, we believe that the fields where MINTs are most likely to find success are those where a modification of the surface properties of SWNTs is required, such as the fabrication of SWNT–polymer composites, or those in which there is a need for an intimate yet respectful connection between molecular fragments and SWNTs, for instance, catalysis. The formation of MINTs has already been shown to be useful in several instances, which suggests that the correct combination of molecules with the desired properties with SWNTs will probably be successful in other specific cases too. However, the scalability of their synthesis is still an obstacle. So far, the vast majority of the synthetic methods reported have focused on chirality-enriched SWNTs, which are expensive and work only with one part of the SWNT samples, which requires a purification step before MINT preparation. Plausible approaches to make more and cheaper MINTs should focus on diameter-adaptable macrocycles, for instance, monomers that can form cyclic dimers, trimers, tetramers, etc., around SWNTs under suitable conditions or rigid macrocycles that can functionalize specific chiralities. In principle, the RCM, disulfide or Cu-nanobracket methods already reported for the synthesis of MINTs would be suitable for this.

References

1 Qiu, Y., Song, B., Pezzato, C. et al. (2020). A precise polyrotaxane synthesizer. *Science* 368 (6496): 1247–1253.

2 Amano, S., Fielden, S.D.P., and Leigh, D.A. (2021). A catalysis-driven artificial molecular pump. *Nature* 594 (7864): 529–534.

3 Wasserman, E. (1960). The preparation of interlocking rings: a catenane. *J. Am. Chem. Soc.* 82: 4433–4434.

4 Harrison, I.T. and Harrison, S. (1967). Synthesis of a stable complex of a macrocycle and a threaded chain. *J. Am. Chem. Soc.* 89 (22): 5723–5734.

5 Dietrich-Buchecker, C.O., Sauvage, J.P., and Kintzinger, J.P. (1983). New family of molecules: the metallo-catenanes. *Tetrahedron Lett.* 24 (46): 5095–5098.

6 Dietrich-Buchecker, C.O., Sauvage, J.P., and Kern, J.M. (1984). Templated synthesis of interlocked macrocyclic ligands: the catenands. *J. Am. Chem. Soc.* 106 (10): 3043–3045.

7 Gatti, F.G., Leigh, D.A., Nepogodiev, S.A. et al. (2001). Stiff, and sticky in the right places: the dramatic influence of preorganizing guest binding sites on the hydrogen bond-directed assembly of rotaxanes. *J. Am. Chem. Soc.* 123 (25): 5983–5989.

8 Chichak, K.S., Cantrill, S.J., Pease, A.R. et al. (2004). Molecular borromean rings. *Science* 304 (5675): 1308–1312.

9 Ashbridge, Z., Kreidt, E., Pirvu, L. et al. (2022). Vernier template synthesis of molecular knots. *Science* 375 (6584): 1035–1041.

10 Perez, E.M. and Martin, N. (2015). π–π interactions in carbon nanostructures. *Chem. Soc. Rev.* 44 (18): 6425–6433.

11 Thostenson, E.T., Ren, Z., and Chou, T.-W. (2001). Advances in the science and technology of carbon nanotubes and their composites: a review. *Compos. Sci. Technol.* 61 (13): 1899–1912.

12 Blackburn, J.L., Ferguson, A.J., Cho, C., and Grunlan, J.C. (2018). Carbon-nanotube-based thermoelectric materials and devices. *Adv. Mater.* 30 (11): 1704386.

13 Schroeder, V., Savagatrup, S., He, M. et al. (2019). Carbon nanotube chemical sensors. *Chem. Rev.* 119 (1): 599–663.

14 Yang, Z., Tian, J., Yin, Z. et al. (2019). Carbon nanotube- and graphene-based nanomaterials and applications in high-voltage supercapacitor: a review. *Carbon* 141: 467–480.

15 Yan, Y., Miao, J., Yang, Z. et al. (2015). Carbon nanotube catalysts: recent advances in synthesis, characterization and applications. *Chem. Soc. Rev.* 44 (10): 3295–3346.

16 Dyke, C.A. and Tour, J.M. (2004). Covalent functionalization of single-walled carbon nanotubes for materials applications. *J. Phys. Chem. A* 108 (51): 11151–11159.

17 Herrero, M.A. and Prato, M. (2008). Recent advances in the covalent functionalization of carbon nanotubes. *Mol. Cryst. Liq. Cryst.* 483 (1): 21–32.

18 Singh, P., Campidelli, S., Giordani, S. et al. (2009). Organic functionalisation and characterisation of single-walled carbon nanotubes. *Chem. Soc. Rev.* 38 (8): 2214–2230.

19 Mallakpour, S. and Soltanian, S. (2016). Surface functionalization of carbon nanotubes: fabrication and applications. *RSC Adv.* 6 (111): 109916–109935.

20 Zhao, Y.-L. and Stoddart, J.F. (2009). Noncovalent functionalization of single-walled carbon nanotubes. *Acc. Chem. Res.* 42 (8): 1161–1171.

21 Matiyani, M., Pathak, M., Bohra, B.S., and Sahoo, N.G. (2020). Noncovalent functionalization of carbon nanotubes. In: *Handbook of Carbon Nanotubes* (ed. J. Abraham, S. Thomas, and N. Kalarikkal), 1–28. Cham: Springer International Publishing.

22 Chen, F., Wang, B., Chen, Y., and Li, L.-J. (2007). Toward the extraction of single species of single-walled carbon nanotubes using fluorene-based polymers. *Nano Lett.* 7 (10): 3013–3017.

23 Nish, A., Hwang, J.-Y., Doig, J., and Nicholas, R.J. (2007). Highly selective dispersion of single-walled carbon nanotubes using aromatic polymers. *Nat. Nanotechnol.* 2 (10): 640–646.

24 Ju, S.-Y., Doll, J., Sharma, I., and Papadimitrakopoulos, F. (2008). Selection of carbon nanotubes with specific chiralities using helical assemblies of flavin mononucleotide. *Nat. Nanotechnol.* 3 (6): 356–362.

25 Lei, T., Pochorovski, I., and Bao, Z. (2017). Separation of semiconducting carbon nanotubes for flexible and stretchable electronics using polymer removable method. *Acc. Chem. Res.* 50 (4): 1096–1104.

26 Chortos, A., Pochorovski, I., Lin, P. et al. (2017). Universal selective dispersion of semiconducting carbon nanotubes from commercial sources using a supramolecular polymer. *ACS Nano* 11 (6): 5660–5669.

27 Gao, T.Z., Sun, Z., Yan, X. et al. (2020). Engineering supramolecular polymer conformation for efficient carbon nanotube sorting. *Small* 16 (26): 2000923.

28 Kim, D., Lee, T., Kwon, M. et al. (2020). Polymer wrapping-induced dispersion of single walled carbon nanotubes in ethylene glycol under mild sonication. *RSC Adv.* 10 (44): 26262–26267.

29 Zheng, M., Jagota, A., Semke, E.D. et al. (2003). DNA-assisted dispersion and separation of carbon nanotubes. *Nat. Mater.* 2 (5): 338–342.

30 Zheng, M., Jagota, A., Strano, M.S. et al. (2003). Structure-based carbon nanotube sorting by sequence-dependent DNA assembly. *Science* 302 (5650): 1545–1548.

31 Tu, X., Manohar, S., Jagota, A., and Zheng, M. (2009). DNA sequence motifs for structure-specific recognition and separation of carbon nanotubes. *Nature* 460 (7252): 250–253.

32 Maune, H.T., Han, S.-p., Barish, R.D. et al. (2010). Self-assembly of carbon nanotubes into two-dimensional geometries using DNA origami templates. *Nat. Nanotechnol.* 5 (1): 61–66.

33 Penzo, E., Palma, M., Chenet, D.A. et al. (2016). Directed assembly of single wall carbon nanotube field effect transistors. *ACS Nano* 10 (2): 2975–2981.

34 McMorrow, J., Freeley, M., and Palma, M. (2017). DNA-wrapped single-walled carbon nanotube assemblies. *Ind. Eng. Chem. Res.* 56 (18): 5302–5308.

35 Wu, S.-J., Schuergers, N., Lin, K.-H. et al. (2018). Restriction enzyme analysis of double-stranded dna on pristine single-walled carbon nanotubes. *ACS Appl. Mater. Interfaces* 10 (43): 37386–37395.

36 Clément, P., Trinchera, P., Cervantes-Salguero, K. et al. (2019). A one-step chemical strategy for the formation of carbon nanotube junctions in aqueous solution: reaction of DNA-wrapped carbon nanotubes with diazonium salts. *ChemPlusChem* 84 (9): 1235–1238.

37 Star, A., Liu, Y., Grant, K. et al. (2003). Noncovalent side-wall functionalization of single-walled carbon nanotubes. *Macromolecules* 36 (3): 553–560.

38 Ortiz-Acevedo, A., Xie, H., Zorbas, V. et al. (2005). Diameter-selective solubilization of single-walled carbon nanotubes by reversible cyclic peptides. *J. Am. Chem. Soc.* 127 (26): 9512–9517.

39 Chichak, K.S., Star, A., Altoé, M.V.P., and Stoddart, J.F. (2005). Single-walled carbon nanotubes under the influence of dynamic coordination and supramolecular chemistry. *Small* 1 (4): 452–461.

40 Hijazi, I., Bourgeteau, T., Cornut, R. et al. (2014). Carbon nanotube-templated synthesis of covalent porphyrin network for oxygen reduction reaction. *J. Am. Chem. Soc.* 136 (17): 6348–6354.

41 de Juan, A., Pouillon, Y., Ruiz-González, L. et al. (2014). Mechanically interlocked single-wall carbon nanotubes. *Angew. Chem. Int. Ed.* 53 (21): 5394–5400.

42 Romero-Nieto, C., García, R., Herranz, M.Á. et al. (2012). Tetrathiafulvalene-based nanotweezers—noncovalent binding of carbon nanotubes in aqueous media with charge transfer implications. *J. Am. Chem. Soc.* 134 (22): 9183–9192.

43 Blanco, M., Nieto-Ortega, B., de Juan, A. et al. (2018). Positive and negative regulation of carbon nanotube catalysts through encapsulation within macrocycles. *Nat. Commun.* 9 (1): 2671.

44 Calbo, J., López-Moreno, A., de Juan, A. et al. (2017). Understanding noncovalent interactions of small molecules with carbon nanotubes. *Chem. Eur. J.* 23 (52): 12909–12916.

45 de Juan, A., Lopez-Moreno, A., Calbo, J. et al. (2015). Determination of association constants towards carbon nanotubes. *Chem. Sci.* 6 (12): 7008–7014.

46 de Juan, A., Mar Bernal, M., and Pérez, E.M. (2015). Optimization and insights into the mechanism of formation of mechanically interlocked derivatives of single-walled carbon nanotubes. *ChemPlusChem* 80 (7): 1153–1157.

47 López-Moreno, A. and Pérez, E.M. (2015). Pyrene-based mechanically interlocked SWNTs. *Chem. Commun.* 51 (25): 5421–5424.

48 Leret, S., Pouillon, Y., Casado, S. et al. (2017). Bimodal supramolecular functionalization of carbon nanotubes triggered by covalent bond formation. *Chem. Sci.* 8 (3): 1927–1935.

49 de Juan-Fernández, L., Münich, P.W., Puthiyedath, A. et al. (2018). Interfacing porphyrins and carbon nanotubes through mechanical links. *Chem. Sci.* 9 (33): 6779–6784.

50 Moreno-Da Silva, S., Martínez, J.I., Develioglu, A. et al. (2021). Magnetic, mechanically interlocked porphyrin–carbon nanotubes for quantum computation and spintronics. *J. Am. Chem. Soc.* 143 (50): 21286–21293.

51 Zhang, W., Guillén-Soler, M., Moreno-Da Silva, S. et al. (2022). Mechanical interlocking of SWNTs with N-rich macrocycles for efficient ORR electrocatalysis. *Chem. Sci.* 13 (33): 9706–9712. https://doi.org/10.1039/D2SC02346F.

52 Paolesse, R., Nardis, S., Monti, D. et al. (2017). Porphyrinoids for chemical sensor applications. *Chem. Rev.* 117 (4): 2517–2583.

53 Magna, G., Mandoj, F., Stefanelli, M. et al. (2021). Recent advances in chemical sensors using porphyrin-carbon nanostructure hybrid materials. *Nanomaterials* 11 (4): 997.

54 Negut, C.C., van Staden, R.-I.S., and van Staden, J.F. (2020). Porphyrins-as active materials in the design of sensors. an overview. *ECS J. Solid State Sci. Technol.* 9 (5): 051005.

55 Yang, H., Shi, R., Shang, L., and Zhang, T. (2021). Recent advancements of porphyrin-like single-atom catalysts: synthesis and applications. *Small Struct.* 2 (6): 2100007.

56 Maurin, A. and Robert, M. (2016). Noncovalent immobilization of a molecular iron-based electrocatalyst on carbon electrodes for selective, efficient CO_2-to-CO conversion in water. *J. Am. Chem. Soc.* 138 (8): 2492–2495.

57 Wei, P.-J., Yu, G.-Q., Naruta, Y., and Liu, J.-G. (2014). Covalent grafting of carbon nanotubes with a biomimetic heme model compound to enhance oxygen reduction reactions. *Angew. Chem. Int. Ed.* 53 (26): 6659–6663.

58 Mahmood, A., Hu, J.-Y., Xiao, B. et al. (2018). Recent progress in porphyrin-based materials for organic solar cells. *J. Mater. Chem. A* 6 (35): 16769–16797.

59 Pengxia, L., Du, Z., Wang, D. et al. (2014). Optoelectronic and self-assembly properties of porphyrin derivatives with click chemistry modification. *ChemPhysChem* 15 (16): 3523–3529.

60 Villalva, J., Nieto-Ortega, B., Melle-Franco, M., and Pérez, E.M. (2020). Quasi-barrierless submolecular motion in mechanically interlocked carbon nanotubes. *J. Phys. Chem. C* 124 (28): 15541–15546.

61 Balakrishna, B., Menon, A., Cao, K. et al. (2020). Dynamic covalent formation of concave disulfide macrocycles mechanically interlocked with single-walled carbon nanotubes. *Angew. Chem. Int. Ed.* 59 (42): 18774–18785.

62 Miki, K., Saiki, K., Umeyama, T. et al. (2018). Unique tube–ring interactions: complexation of single-walled carbon nanotubes with cycloparaphenyleneacetylenes. *Small* 14 (26): 1800720.

63 Chamorro, R., de Juan-Fernández, L., Nieto-Ortega, B. et al. (2018). Reversible dispersion and release of carbon nanotubes via cooperative clamping interactions with hydrogen-bonded nanorings. *Chem. Sci.* 9 (17): 4176–4184.

64 Mann, F.A., Horlebein, J., Meyer, N.F. et al. (2018). Carbon nanotubes encapsulated in coiled-coil peptide barrels. *Chem. Eur. J.* 24 (47): 12241–12245.

65 López-Moreno, A., Ibáñez, S., Moreno-Da Silva, S. et al. Single-walled carbon nanotubes encapsulated within metallacycles. *Angew. Chem. Int. Ed.* 61 (35): e202208189.

66 Cheng, G., Hayashi, T., Miyake, Y. et al. (2022). Interlocking of single-walled carbon nanotubes with metal-tethered tetragonal nanobrackets to enrich a few hundredths of a nanometer range in their diameters. *ACS Nano* 16 (8): 12500–12510.

67 Liu, G., Wang, F., Chaunchaiyakul, S. et al. (2013). Simultaneous discrimination of diameter, handedness, and metallicity of single-walled carbon nanotubes with chiral diporphyrin nanocalipers. *J. Am. Chem. Soc.* 135 (12): 4805–4814.

68 Peng, X., Komatsu, N., Bhattacharya, S. et al. (2007). Optically active single-walled carbon nanotubes. *Nat. Nanotechnol.* 2 (6): 361–365.

69 Liu, G., Rahman, A.F.M.M., Chaunchaiyakul, S. et al. (2013). Bis(tert-butylpyrene) nanotweezers and nanocalipers: enhanced extraction and recognition abilities for single-walled carbon nanotubes. *Chem. Eur. J.* 19 (48): 16221–16230.

70 Liu, G., Saito, Y., Nishio-Hamane, D. et al. (2014). Structural discrimination of double-walled carbon nanotubes by chiral diporphyrin nanocalipers. *J. Mater. Chem. A* 2 (44): 19067–19074.

71 Martínez-Periñán, E., de Juan, A., Pouillon, Y. et al. (2016). The mechanical bond on carbon nanotubes: diameter-selective functionalization and effects on physical properties. *Nanoscale* 8 (17): 9254–9264.

72 Wielend, D., Vera-Hidalgo, M., Seelajaroen, H. et al. (2020). Mechanically interlocked carbon nanotubes as a stable electrocatalytic platform for oxygen reduction. *ACS Appl. Mater. Interfaces* 12 (29): 32615–32621.

73 López-Moreno, A., Nieto-Ortega, B., Moffa, M. et al. (2016). Threading through macrocycles enhances the performance of carbon nanotubes as polymer fillers. *ACS Nano* 10 (8): 8012–8018.

74 Mena-Hernando, S., Eaton, M., Fernández-Blázquez, J.P. et al. (2023). Mechanical interlocking to unlock the reinforcing potential of carbon nanotubes. *Chem. Eur. J.* 29 (58): e202301490. https://chemistry-europe.onlinelibrary.wiley.com/doi/full/10.1002/chem.202301490.

75 Villalva, J., Rapakousiou, A., Monclús, M.A. et al. (2023). Interlocking matrix and filler for enhanced individualization and reinforcement in polymer–single-walled carbon nanotube composites. *ACS Nano.* 17 (17): 16565–16572. https://pubs.acs.org/doi/full/10.1021/acsnano.3c02255.

4

Concepts of Molecular Motors in Solution and on Surfaces

*Monika Schied**

Formerly: University of Graz, Institute of Chemistry, Heinrichstr. 28, 8043 Graz, Austria
Present: Elettra-Sincrotrone Trieste, Strada Statale 14 - km 163.5 in AREA Science Park, 34149 Basovizza, Trieste, Italy

The central part of any machine is the motor that drives it. The same holds true for machines on the nanometer scale: in the process of miniaturization, one can go down to the limit of motors made up of only a single molecule.

In nature, molecular motors are ubiquitous: biological organisms heavily rely on naturally occurring molecular motors or machines. While biological motors are intricate machines capable of performing complex tasks efficiently and in a cooperative manner, they also have drawbacks regarding their operational conditions. Most biological systems are restricted to ambient conditions, i.e. specific temperatures or chemical environments. Synthetic molecular motors can be designed in a more versatile, robust way that allows operation in various environmental conditions, thereby enabling new applications such as in nanotechnology or medicine. As research on naturally occurring motors is a highly complex topic that by itself can fill whole books [1–3], this chapter focuses exclusively on artificial motors, i.e. systems that have been designed and synthesized in a bottom-up approach.

However, before one can discuss the specific concepts of molecular motors, it is important to clarify what constitutes them. The American Heritage® Dictionary of the English Language defines the term "motor" as follows [4]:

1. Something, such as a machine or an engine, that produces or imparts motion.
2. A device that converts any form of energy into mechanical energy, especially an internal-combustion engine or an arrangement of coils and magnets that converts electric current into mechanical power.
3. A motor vehicle, especially an automobile

While these definitions are clearly tailored for macroscopic motors, such as car engines, they can be partly applied to motors made up of a single molecule as well. A motor takes energy of any kind – chemical reactions, thermal energy, light, voltage, concentration gradients … – and converts it into useful mechanical work.

* Contact Details: monika.schied@alumni.uni-ulm.de

Mechanically Interlocked Materials: Polymers, Nanomaterials, MOFs, and more, First Edition.
Edited by Emilio M. Pérez.

For continuous operation, the motion should be ***repetitive***, which requires the motor to cycle through a number of states and then return to its original state. But a simple switch is not a motor: despite its ability to change its state and to return to the original state, doing so reverts any previously performed work. Thus, a key property of a molecular motor is its ***unidirectional*** movement.

The major difficulty in the design and operation of a molecular motor does not lie in inducing a rotational or translational motion, but rather in controlling it and giving it a preferred directionality. The world of single molecules is different from what we experience in our macroscopic surroundings. Instead of gravity, the main driving force in the nano world is random Brownian motion [5]. Under these conditions, moving along a defined pathway is like trying to "walk in a hurricane" [6]. In order to perform useful work, the molecular motors have to either overcome Brownian motion or make use of it [7].

One famous theoretical concept is the Brownian motor, which exploits the random forces not by enhancing forward motion, but instead by preventing backward motion. Astumian compared the concept of Brownian motors to a car in hailstorm [6]. The hailstones hit the car randomly from all directions, pushing it forward and backward with – on average – equal rates. These random forces can still be exploited to move the car forward, even up a small slope, by replacing the car brake with a ratchet, i.e. an asymmetric gear and a pawl that blocks the movement in any direction. The energy landscape is schematically shown in Figure 4.1. As long as the break

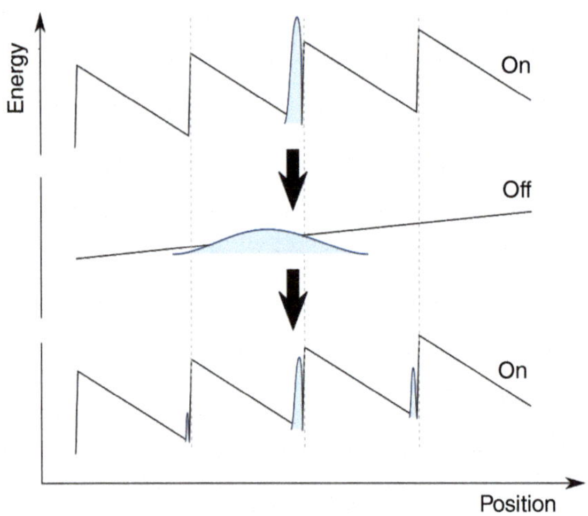

Figure 4.1 Schematic energy landscape of a Brownian motor. A particle is initially confined in one position in an asymmetric ratchet potential (top). Switching the sawtooth modulation off lets the particle move, driven by the slope and by random Brownian motion in both directions. Switching the sawtooth potential back on (bottom), there is a higher chance to find the particle on the right (uphill) than on the left, due to its asymmetry.

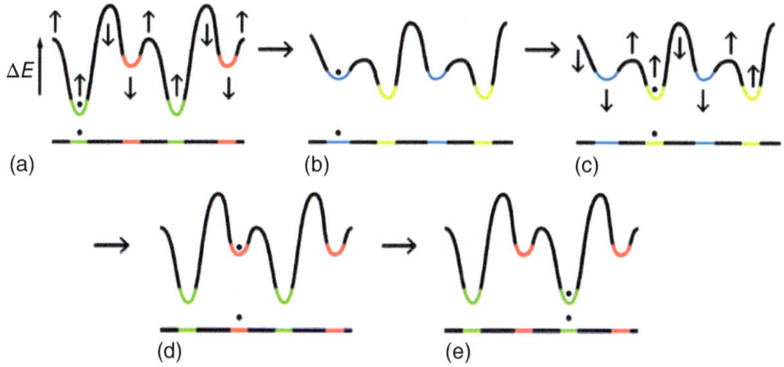

Figure 4.2 Schematic energy landscape of an energy ratchet. The particle (black dot) is moved from the left (a) to the right (e) via periodically switching the potential energy landscape. The modulations of the energy landscape are independent of the particle position (details in the main text). Reproduced with permission from [9]/Royal Society of Chemistry; permission conveyed through Copyright Clearance Center, Inc.

is engaged, the car will stay in its position. When the break is open, the car is free to move and is pushed forward or backward by the hailstones. Due to the asymmetry of the ratchet, however, reengaging the break is more likely to advance the gear by one tooth to the right than going in the opposite direction.

The main ingredients for a functional Brownian motor are

- a potential energy landscape with asymmetry in the dimension of the desired motion
- a randomizing element, such as the break being applied and released
- an external energy input [8].

Generally, two types of ratchets can be distinguished, namely the energy ratchet and the information ratchet [9]. The energy ratchet is schematically shown in Figure 4.2. The particle, shown as small black dot, is initially located in the green potential minimum. Modifying the energy landscape in a way that raises the green minimum higher than the red minimum and simultaneously lowers the energy barrier leads to a movement to the right (a–c). Changing back to the original energies, the particle is located in an unfavorable red minimum and relaxes to the next green global minimum (e). The energy ratchet is characterized by a modulation of the energy landscape irrespective of the position of the particle.

In contrast, in an information ratchet, the energy landscape is modified by the location of the particle. In the scheme in Figure 4.3, the energy barrier to the right of the particle is selectively lowered, allowing transport in this direction only. Alternatively, the barrier in the opposite direction could be raised to prevent backward movement. As the modification of the potential energy is based on the particle position, information transfer is needed.

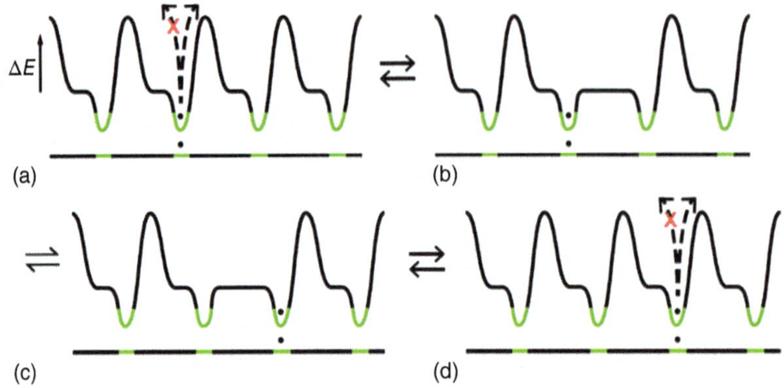

Figure 4.3 Schematic energy landscape of an information ratchet. The position of the particle (black dot) affects the energy landscape, in this example, selectively lowering the energy barrier to its right. Information transfer is indicated by dashed arrows in (a) and (c). Reproduced with permission from [9]/Royal Society of Chemistry.

4.1 Light-driven Overcrowded Alkenes

A ratchet-like system has been presented by Kelly et al. [10]. Based on chemical reactions, a molecular ratchet - a helicene connected to a triptycene - is rotated by 120° in clockwise direction from the perspective view in Figure 4.4. The single bond between the helicene and the triptycene form the "axle." The energy barrier for a rotation around a C—C single bond is typically in the order of 3–5 kcal mol^{-1}.

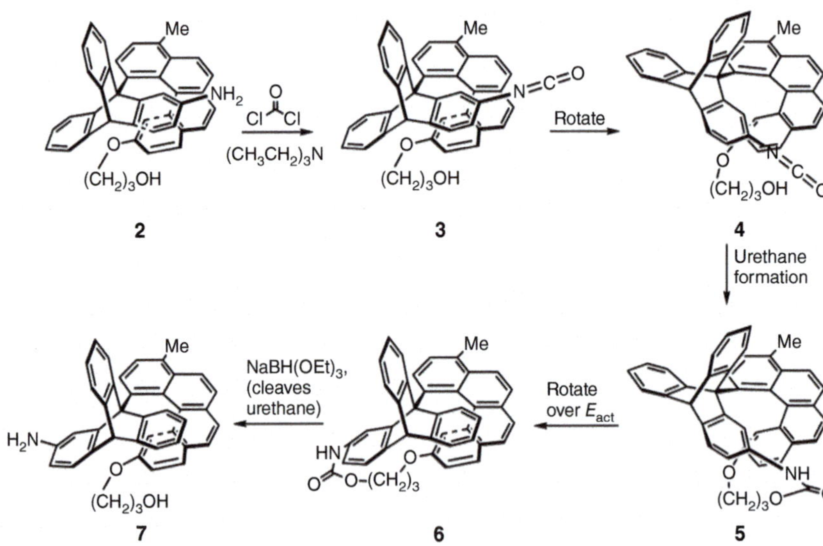

Figure 4.4 Ratchet-like system capable of 120° unidirectional rotation. A chemical reaction defines the sense of rotation. However, repetitive motion is not achieved. Reproduced with permission from [10]/Springer Nature.

However, in this case, the helicene acting as a "friction brake" increases this barrier to about 25 kcal mol^{-1} due to steric hindrance. This brake inhibits anti-clockwise rotation – if not thermally activated, in which case the rotations would proceed randomly in either direction, whereas the clockwise rotation is enabled by energy released from a chemical reaction powered by carbonyl dichloride. While this system has an energy intake (from the chemical reaction) and proceeds directionally, it lacks the repetitive motion necessary to drive it as continuous motor.

The first artificial molecular capable of full, repetitive 360° unidirectional rotation was developed in the group of Bernard Feringa in 1999 [11]. For this design, he was awarded the Nobel Prize in Chemistry in 2016, together with Fraser Stoddart and Jean-Pierre Sauvage "for the design and synthesis of molecular machines" [12].

Feringa's motor is based on sterically overcrowded alkenes. In a four-step process, one half of the molecule (the "rotor") rotates unidirectionally with respect to the other half (the "stator") around the central double bond connecting the two. This sequence of steps is shown schematically for a second generation motor in Figure 4.5a [13]. A "top view" of the molecule is shown above each of the four chemical structures to clarify the relative positions of the different chemical moieties with respect to each other. The colors in the chemical structures correspond to the same colors in the top view. The bottom thioxanthene half (with the carbon rings marked in blue and red) is defined as the stator here.

Starting in ⓐ, the molecule is excited by ultraviolet (UV) light, leading to an isomerization of the double bond and therefore to a 180° clockwise rotation of the rotor

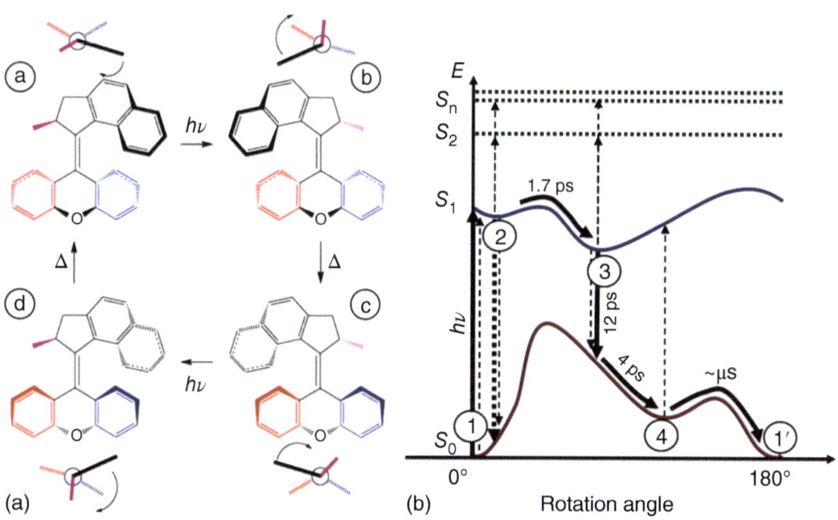

Figure 4.5 Second-generation molecular motor based on overcrowded alkenes. By an alternating sequence of two light-induced double bond isomerizations and two thermal helix inversion steps, unidirectional rotation of the rotor with respect to the stator is achieved. The irreversible helix inversion step (from ④ to ①′ in the energy scheme) dictates the unidirectionality. (a) Chemical structure and unidirectional rotation cycle. (b) Potential energy landscape of the motor cycle. Reproduced from [13] with permission from WILEY-VCH Verlag GmbH & Co. KGaA, Weinheim.

with respect to the stator. In the energy scheme presented in Figure 4.5a, this step is represented by ① (corresponding to the ground state ⓐ) to ④ (corresponding to state ⓑ). Due to steric hindrance within the molecule, this new conformation ⓑ is energetically unfavorable. Thus, in a thermally induced helix inversion step, the molecule overcomes the energy barrier and relaxes to state ⓒ (or ①'). Due to the symmetry of the stator, ⓐ and ⓒ, and therefore ① and ①' are chemically identical. Another set of double bond isomerization (DBI) and helix inversion leads to the full 360° rotation of the rotor back to state ⓐ.

4.1.1 General Concept and Development

The development of the unidirectional molecular motor is based on earlier studies on nondirectional chiroptical molecular switches, which are very similar in structure [14–16]. Their development was inspired by the chemical processes involved in vision, which is based on a light-induced *cis–trans* isomerization of an alkene group [17]. The light-induced *cis–trans* isomerization together with a helicity change leads to conformational changes.

Further refinement of this molecule lead to the development of the unidirectional motor [11]. This molecule consists of two equal halves connected by an overcrowded olefinic C=C bond [18]. Compared to the chiroptical switch, it has two stereogenic centers bearing methyl groups. In the energetically favored (*P*,*P*)-*trans* conformation, these methyl groups adapt an axial orientation, minimizing steric repulsion. Upon *trans–cis* isomerization by light, which is a fast and reversible process, the molecule is converted to the (*M*,*M*)-*cis* form, which is energetically less favored, due to steric interaction between the aromatic regions of the two halves. The methyl groups adopt equatorial conformations. The steric hindrance is released in the thermal helix inversion step, leading to the (*P*,*P*)-*cis* isomer. In a repetition of light-induced *cis–trans* isomerization to (*M*,*M*)-*trans* and another thermal helix inversion step the original (*P*,*P*)-*trans* form is recovered. The authors showed with experiments in solution that the two thermal helix inversion steps are irreversible, and no other pathway between the different isomers has been observed.

The ingenious design of this first repetitive artificial unidirectional motor has some limitations. For once, the two thermal helix inversion steps require different temperatures (20 and 60 °C), which makes cycling through the process slow, because the temperature needs to be changed during the process. Furthermore, the two equal halves render individual functionalization of only the rotor or the stator chemically difficult. This is also important in regard of attaching motor molecules to surfaces.

Soon after the first publication, a second-generation molecular motor (Figure 4.5 shows an example) was presented [19, 20]. Besides addressing the mentioned issues, it was shown that one single stereocenter is sufficient for unidirectional motion. By exchanging the different substituents, as shown in Figure 4.6, the energy barriers of the thermal steps – the same barriers within the complete cycle, thus requiring the same temperature – could be tuned between 22 and 25 kcal mol^{-1}. Furthermore,

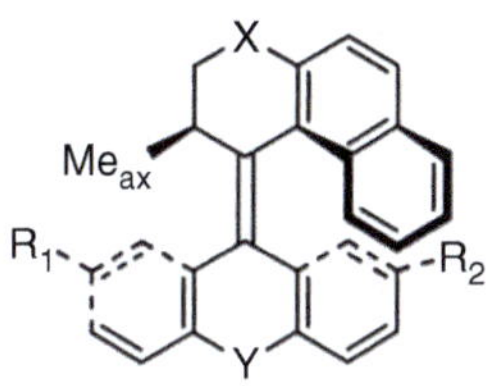

(2′*R*)-(*M*)-*trans*-**1a**: X = S, Y = S, R_1 = OMe, R_2 = H
(2′*R*)-(*M*)-*cis*-**2a**: X = S, Y = S, R_1 = H, R_2 = OMe
(2′*R*)-(*M*)-**3a**: X = S, Y = S, R_1 = H, R_2 = H
(2′*R*)-(*M*)-**4a**: X = S, Y = S, R_1 = OMe, R_2 = OMe
(2′*R*)-(*M*)-**5a**: X = S, Y = O, R_1 = H, R_2 = H
(2′*R*)-(*M*)-**6a**: X = S, Y = $C(CH_3)_2$, R_1 = H, R_2 = H
(2′*S*)-(*M*)-**7a**: X = CH_2, Y = S, R_1 = H, R_2 = H
(2′*S*)-(*M*)-**8a**: X = CH_2, Y = $C(CH_3)_2$, R_1 = H, R_2 = H
(2′*S*)-(*M*)-**9a**: X = CH_2, Y = CHCH, R_1 = H, R_2 = H

Figure 4.6 Second generation of unidirectional molecular motors. In contrast to the symmetric first generation, the rotor and stator have different chemical structures, allowing for the functionalization of the individual parts and thereby facilitating attachment to surfaces. Reprinted with permission from [20]. Copyright 2002 American Chemical Society.

Table 4.1 High-speed second-generation molecular motors.

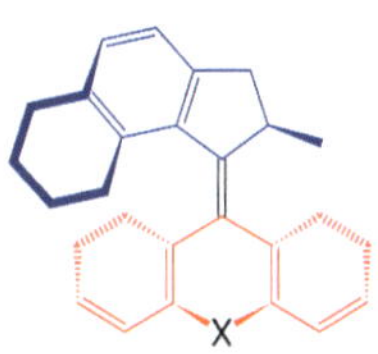

X	$t_{1/2}$ (s)
S	1.1×10^{-7}
O	2.8×10^{-6}
NBoc	5.7×10^{-7}
$C(CH_3)_2$	8.0×10^{-9}
CH_2CH_2	2.4×10^{-3}

Note: The introduction of a cyclopentane-based rotor increased the rotational speed by orders of magnitude. Depending on the substituent X the half-life time was reduced down to 8.0×10^{-9} s [24–26].

motors 7 and 8 reached almost perfect photoequilibria[1] of 1 : 99 in their photochemical conversion [20].

Great effort went into the increase of the rotational speed [21, 22]. A plethora of chemical modifications have been employed in order to vary the flexibility of the rotor and/or the stator, alter the relative energy levels and barriers and modify the electronic states. A detailed summary of the steps taken in this direction can be found in [23]. In 2008 motors with rotation frequencies in the MHz range have been reported [24], which was achieved by introducing a cyclopentane-based rotor. This type of fast motor can be seen in Table 4.1 for a variety of substituents in the stator.

The motors presented so far are excited by ultraviolet (UV) light, which is harmful for biological tissues. Medical applications of motors require excitation by visible or even infrared (IR) light instead [27–29].

One possibility for red-shifting the required irradiation wavelength is the direct modification of the gap between the highest occupied molecular orbital (HOMO) and lowest unoccupied molecular orbital (LUMO). This was first successfully achieved by van Delden et al. by incorporating a nitro-acceptor and a dimethylamine-donor into the stator [30]. The motor corresponds to the structure shown in Figure 4.6 where X = Y = S, $R_11 = Me_2N$, and $R_2 = NO_2$. With this structural modification the excitation wavelength was shifted to 435 nm. A similar

1 Photoequilibrium: the equilibrium state of a photochemical reaction, i.e. equal rates for both reaction directions.

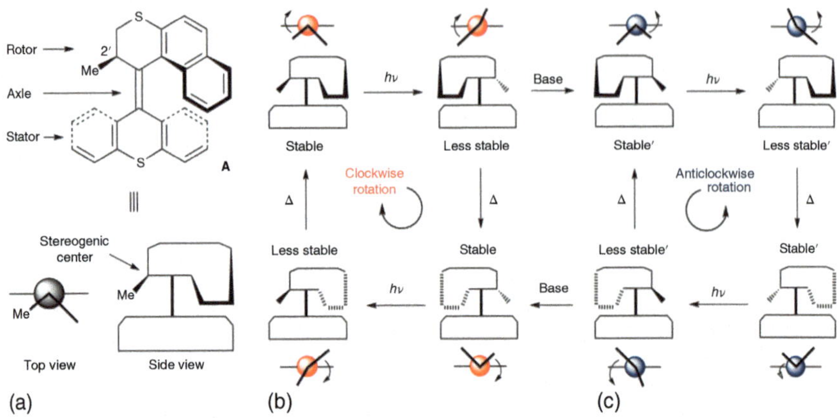

Figure 4.7 Reversible molecular motor. A base-induced epimerization leads to a reversal in directionality of the motor. Reproduced with permission from [34]/Springer Nature.

approach involves the extension of the aromatic system of the stator [31]. An excitation wavelength of up to 490 nm was achieved.

Alternatively, strategies based on metal complexes have been successfully demonstrated [32, 33]. Palladium tetraphenyl-porphyrin was used as a triplet sensitizer, which drives the isomerization of the molecular motor upon irradiation with wavelengths of ≈ 530 nm [32]. On the other hand, incorporating the molecular motor as a ligand in a ruthenium(II)-bipyridine complex increases the necessary wavelength to 450 nm [33].

Unidirectionality, on the one hand, means operation of the system in one way only. However, for application, it might make sense to change the direction of motion as needed. This has been achieved by Ruangsupapichat et al., employing the scheme in Figure 4.7 [34]. Epimerization[2] of the motor from the less stable state with a base leads yields the stable state of the epimer and hence to a reversal of the rotational direction.

Moving from the first generation to the second generation motors reduced the number of stereocenters from two to one. The same holds true for the transition from second to third generation motors: an achiral molecule – with a pseudo-asymmetric center – can still perform unidirectional rotation [36]. The new design of molecular motors consists of two joint motor units, i.e. a stator with a rotor attached on either side (cf. Figure 4.8).

4.1.2 Applications

A variety of applications are possible, and many have been realized already. This section covers a selection of applications in several disciplines.

2 Epimerization: Change of configuration at one asymmetric carbon atom of a molecule with at least two asymmetric carbon atoms. The configuration of the other asymmetric carbon atoms remains the same. Epimers are diastereomers with the same configuration at all but (exactly) one stereogenic centers [35].

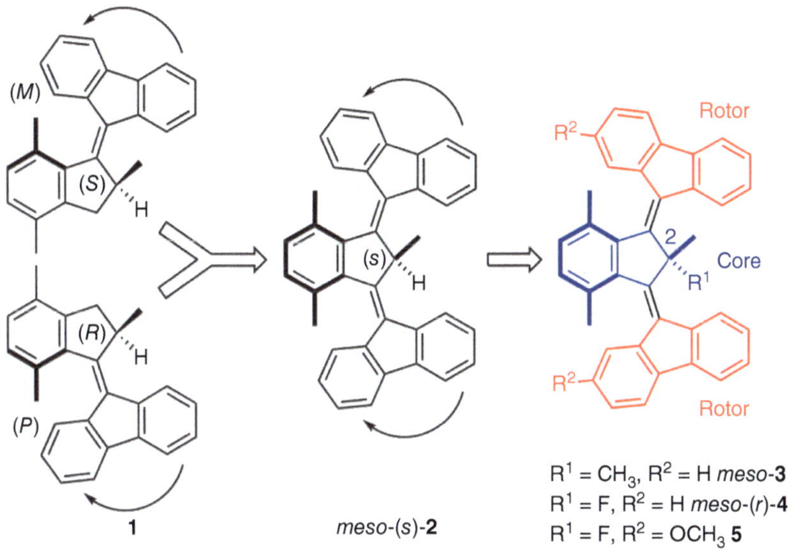

Figure 4.8 Third generation of unidirectional molecular motors. Merging two second generation motors yields an achiral molecule with a pseudo-asymmetric center, which is sufficient to provide unidirectional motion of both rotor units. Reprinted by permission from Springer Nature Customer Service Centre GmbH: Springer Nature [36].

4.1.2.1 Molecular Motors in Liquid Crystals

Implementing molecular motors into liquid crystal media has been shown to successfully modify their properties. For example, doping of nematic liquid crystal with first-generation molecular motors leads to a re-organization of the liquid crystal molecules [37]. The change of the cholesteric phase leads to a visible color change, i.e. a visually observable change of the material. By adjusting the duration, intensity or wavelength of the illumination, the observed wavelength could be tuned throughout the whole visible range, as demonstrated in Figure 4.9.

Also the rotation of a macroscopic object, a glass rod of $5 \times 28\,\mu m^2$ size on a liquid crystal doped with 1% (by weight) enantiopure molecular motors has been observed [38]. The rotated object is by a factor of 10 000 larger than the molecules. Furthermore, a change in the surface relief is connected to the change in the helicity of the motor molecules. Therefore, during the isomerization process, the rotation proceeds

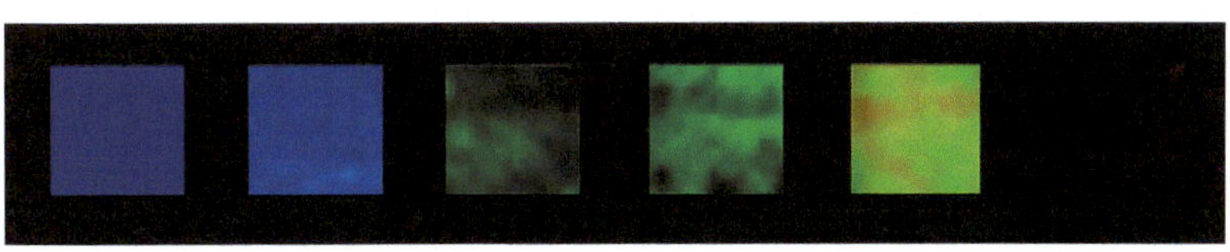

Figure 4.9 Photographs of a molecular motor doped liquid crystal. Depending on the illumination duration (from left to right: 0, 10, 20, 30, 40, and 80 s) the color visibly changes. Reproduced from [37]. © (2002) National Academy of Sciences.

in one way, and is reversed in the helix inversion step. The other enantiomeric form of the motor shows opposite directionalities.

In a more recent study, a continuous, regular and unidirectional rotation of large supramolecular chiral structures, tens of μm in size, has been observed [39]. By illuminating a localized area, molecules are free to diffuse into and out of the photoactive zone and it is therefore more difficult to reach the photostationary state.

4.1.2.2 Self-assembly of Molecular Motors

Amphiphilic molecular motors, functionalized with hydrophobic and hydrophilic groups, are found to self-assemble into nanotubes and vesicles [40]. A hydrophilic group is attached to the rotor and two hydrophobic groups are connected to the stator of a fast and a slow motor. In water with additional phospholipids, the fast amphiphilic motor assembles into toroided nanotubes. The half-time of the helix inversion is too short to reach different aggregates upon illumination. The slow motor, on the other hand, initially forms straight nanotubes. The light-induced isomerization leads to a reorganization into vesicles. The original tubular form is restored upon heating.

A similar system based on a switch is used to form a photo-responsive nano-container [41]. Here, the switching between the two forms changes the

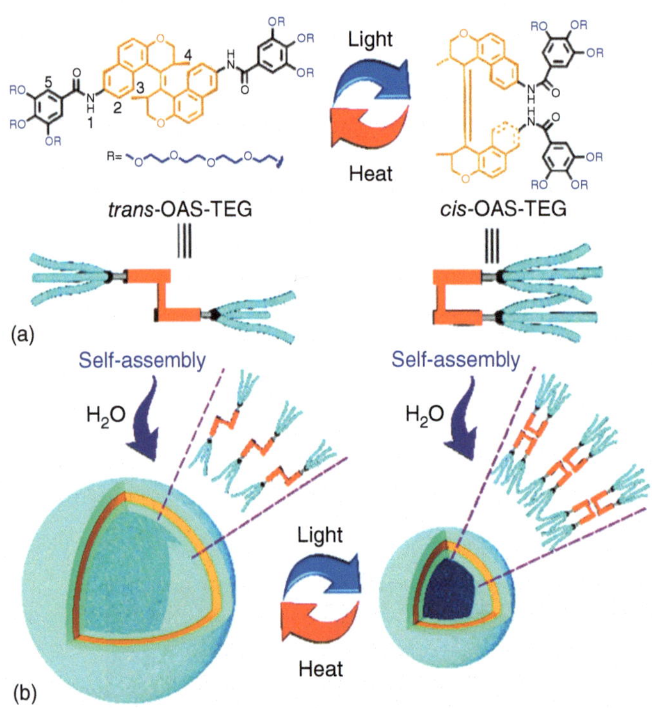

Figure 4.10 Photo-responsive nano-container. Depending on the state of the amphiphilic switch, the volume of the self-assembled vesicle changes. Reproduced with permission from [9]/Royal Society of Chemistry.

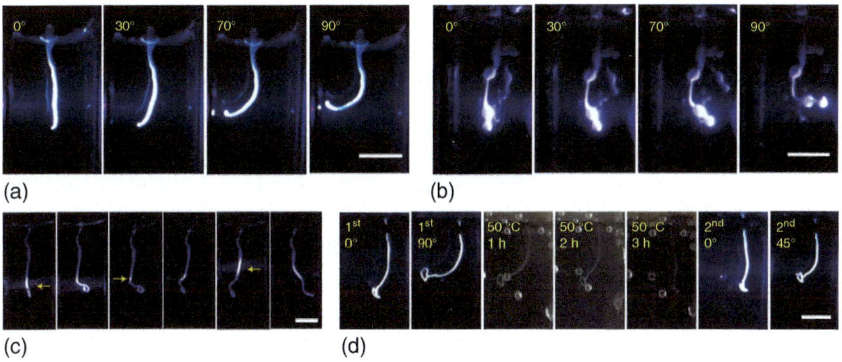

Figure 4.11 Self-assembled artificial muscles made from molecular motors. (a) Upon illumination, the artificial muscle bends up to an angle of 90° within 60 s. (b) The artificial muscle with a ball-shaped end is able to bend to an angle of 90° within 90 s. (c) Selective illumination of parts of the muscle leads to multiple kinks. (d) Upon illumination, the muscle bends, which is fully reversed by heating. The bending of the muscle can subsequently be induced again (all scale bars: 0.5 cm). Reproduced by permission from [42], Springer Nature.

volume of the self-assembled vesicles. The molecule and the proposed structures of the vesicles are shown in Figure 4.10b. A release of the enclosed model cargo calcein, due to the light-induced contraction, has been demonstrated.

In a similar vein, artificial muscles made from self-assembled amphiphilic molecular motors with additional dodecyl chains first assemble into nanofibers, which further congregate into centimeter long macroscopic strings [42]. This is similar to natural muscle tissue consisting of a hierarchical supramolecular assembly of protein molecules, for example myosin and actin. Their coordinated movement leads to an amplification from the molecular to the macroscopic level. The artificial muscles made from molecular motors are capable to lift small cargo (0.4 mg of paper). Figure 4.11 demonstrates the light-induced bending and the reversibility of the motion by heat. In particular the noninvasive activation by light and the operability both in water and air make this a promising candidate for future bio-related applications.

4.1.2.3 Macroscopic Contraction of Gels

The twisting of the molecular motor can be used to entangle polymers connected to both rotor and stator [43]. A simple system consisting of two polymers connected to a motor is shown in Figure 4.12. Upon illumination, the polymers coil up and therefore reduce the size of the system. A more elaborate motor-polymer network is shown in Figure 4.13a, which consists of many motors and polymer strands. This gel significantly shrinks upon UV illumination [43–45].

Going one step further, motor units that spin in the opposite direction triggered by visible light can be used to disentangle the network and thereby restore the original size of the gel [46]. Alternating between UV and visible light reversibly shrinks and enlarges the gel, respectively, as shown in Figure 4.13b.

4.1.2.4 Cancer Treatment

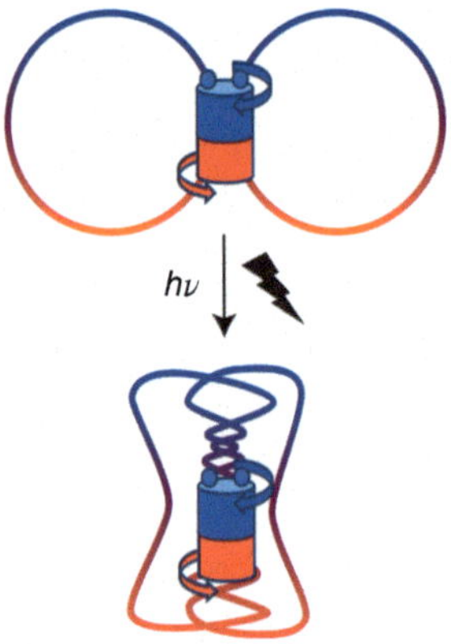

Figure 4.12 Eight-shaped polymer system twisted by a molecular motor. Two strands of polymer are attached to both the rotor and the stator of a molecular motor. The rotation of the motor leads to twisting of the polymer and a significant reduction of its volume. Reproduced with permission from [43]/Springer Nature.

A possible application of molecular motors in the medical field is the prospect of cancer treatment. Molecular motors can physically adsorb on lipid bilayers and, upon activation by UV illumination, drill holes into cell membranes [47]. Control experiments with molecules without rotor, suppressed helix inversion, or nondirectional switches have shown that the unidirectionality is important for this process. By appropriate functionalization with side groups, the motors can be targeted to specific cell-surface recognition sites.

As briefly addressed earlier, the usage of UV light is harmful for biological systems. Therefore, in order to use the molecular motors in a medical treatment, excitation with longer wavelengths, preferably in the red or infrared (IR) regime, is desirable. This goal has been achieved by using two-photon activation with IR light [48]. Besides removing the need of UV light, this allows a very localized application and therefore reduces secondary damage of surrounding tissue by the motor activation. In a step toward actual medical applications, an *in vivo* study of three eukaryotic species has been conducted [49]. The effectiveness of inducing cell necrosis in *Caenorhabditis elegans*, Daphnia, and mice has been demonstrated.

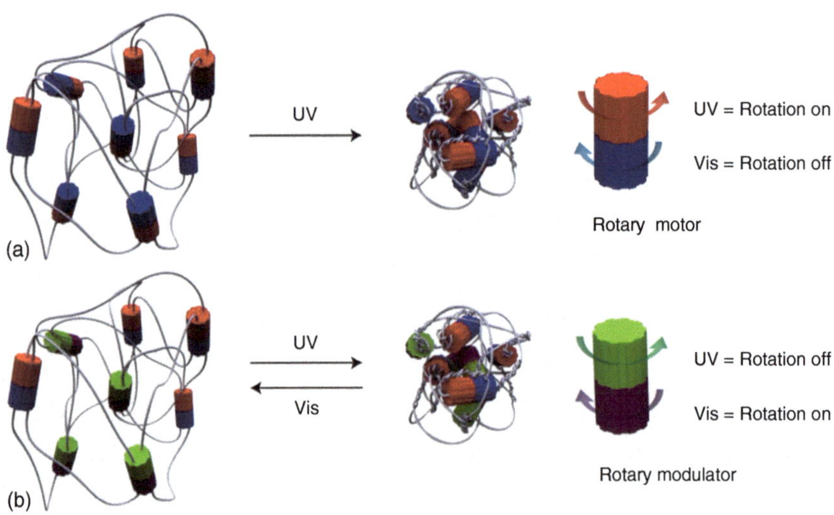

Figure 4.13 Contraction of a polymer gel by molecular motors. (a) UV-light actuated motors irreversibly shrink the network of motors and polymer strands. (b) Additionally, visible light actuated motors rotating in the opposite direction are incorporated in the system and can restore the original volume of the gel. Reproduced with permission from [46]/Springer Nature.

4.2 Molecular Motors Based on Catenanes

Since Feringa's seminal work [11], many other concepts for molecular motors have been developed. They convert energy from light, heat, chemical reactions, other external stimuli, or combinations thereof into directed motion. Catenanes present

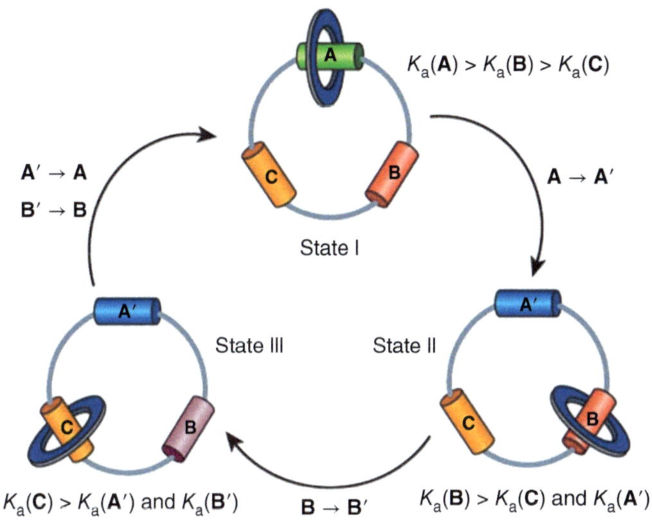

Figure 4.14 Nondirectional [2]-catenane system. By varying the thermodynamical environment, the smaller macrocycle can be positioned on one of three chemical recognition sites on the larger ring. However, the direction of motion cannot be controlled in this system. Reproduced with permission from [50]/Springer Nature.

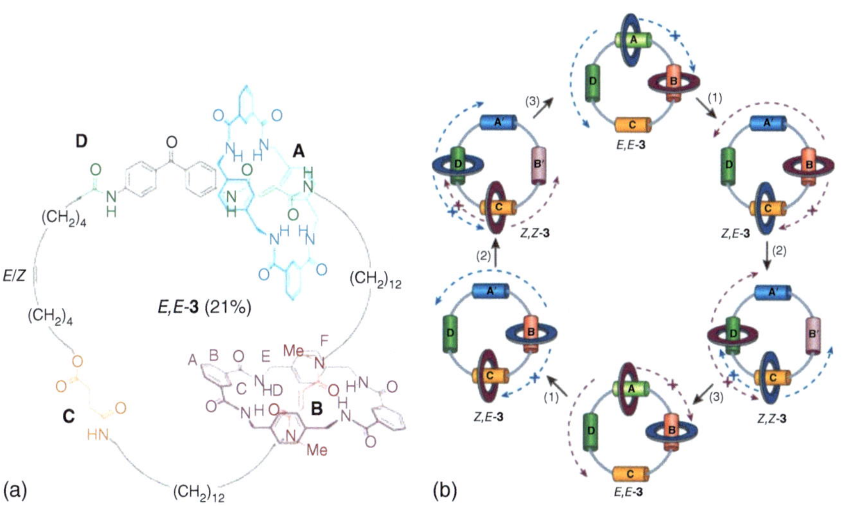

Figure 4.15 Unidirectional rotation of two macrocycles in a [3]-catenane. The two small macrocycles mutually block the backwards direction of rotation. In a sequence of three steps, the rings have effectively swapped places. Repeating the same steps leads to one full 360° rotation along the large ring. (a) Chemical structure. (b) Motion sequence. Reproduced with permission from [50]/Springer Nature.

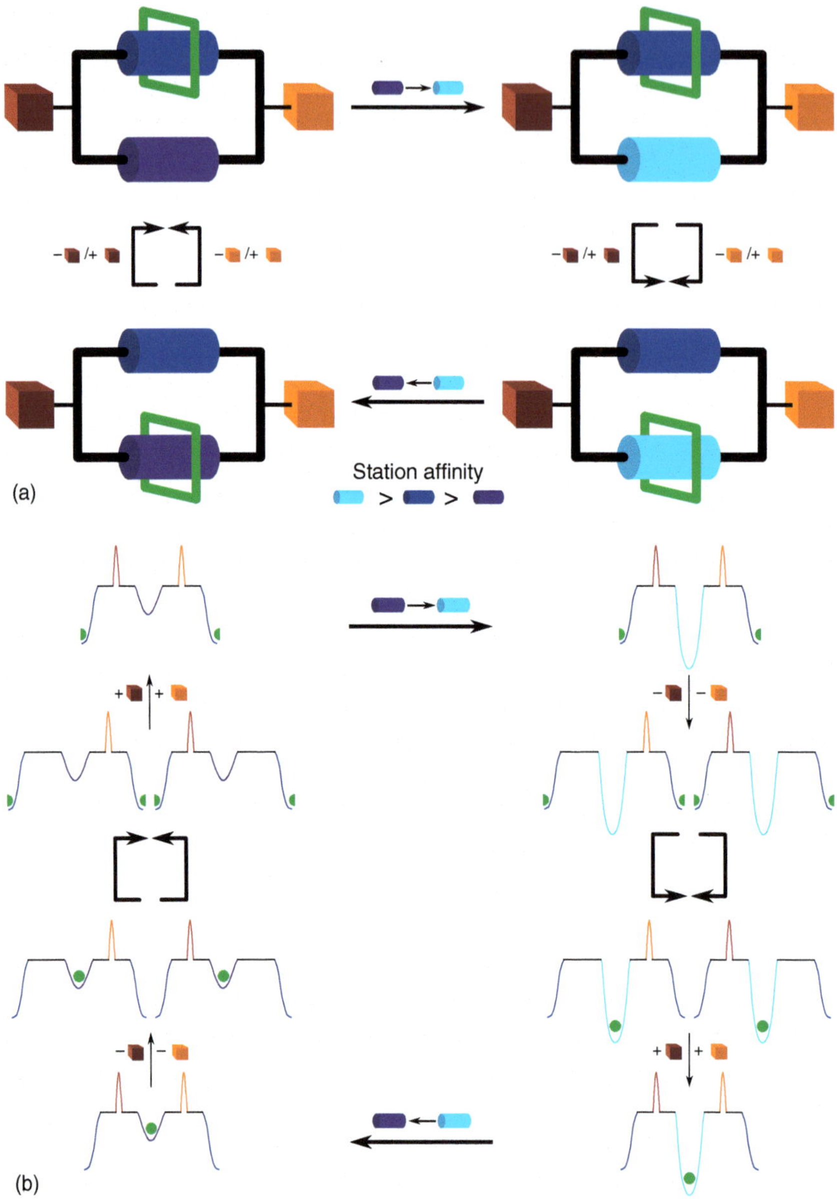

Figure 4.16 Reversible molecular motor based on a [2]-catenane with two detachable chemical stoppers. The lower chemical recognition site can undergo an *E-Z* isomerization and is energetically more or less favorable (symbolized by teal or purple, respectively) than the other site (blue). Removing the stopper (red and orange cubes) on one side dictates the direction of motion of the small ring. (a) Rotation scheme of the small ring around the larger ring. Both directions are possible. (b) Switching of the lower recognition sites alters the energy landscape. The stoppers allow energetical relaxation in one direction only. Adapted from [8].

a promising basis for molecular motors, in which one macrocycle can go around another macrocycle in a unidirectional fashion. Some of these concepts are highlighted here.

In 2003, Leigh et al. presented a motor based on [2]- and [3]-catenanes [50]. Driven by chemical stimuli, light and heat, one or more smaller macrocycles can be moved between different chemical recognition sites on the larger ring. This is achieved by altering the thermodynamics of the binding sites and therefore favoring one recognition site over the others.

In Figure 4.14 the simplest concept with a [2]-catenane, i.e. two interconnected rings, is shown. The smaller ring can be precisely positioned at three distinct positions on the larger ring. However, the direction of movement between these stations cannot be controlled. For example, the ring can move from station A to B in the direct (clockwise) direction, or in the opposite direction via station C.

For directional motion of the small ring around the larger ring, one of the pathways has to be blocked. This is achieved by the introduction of a second macrocycle (i.e. a [3]-catenane) and four recognition sites A–D, as shown in Figure 4.15a. The two rings mutually block the backwards direction. In a cycle, schematically shown in Figure 4.15b, both small macrocycles move around the larger ring in a total of six steps. After a sequence of three steps, the two rings have effectively swapped places with each other.

A similar concept using [2]-catenanes, which is schematically shown in Figure 4.16a, has been presented by [8]. The corresponding chemical structure is shown in Figure 4.17. The different chemical moieties are represented with different colors, that match the colors of Figure 4.16. A small macrocycle can stay at two chemical recognition sites on the larger ring. The lower site can be photochemically *E*-*Z* isomerized, which influences its thermodynamic stability of hosting the small ring. The relative binding abilities toward the small ring on the different positions of the larger ring are symbolized with colors, teal being the most and purple the least stable. The blue recognition site at the top of the ring is not changed throughout the

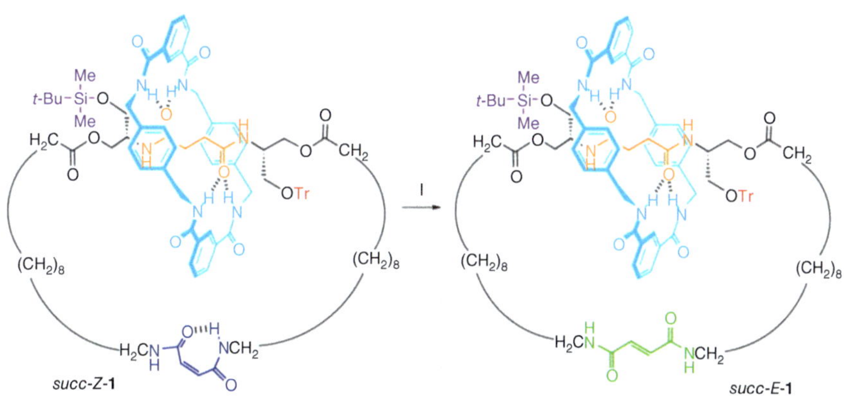

Figure 4.17 Chemical structure of the motor shown in Figure 4.16. Adapted from [8].

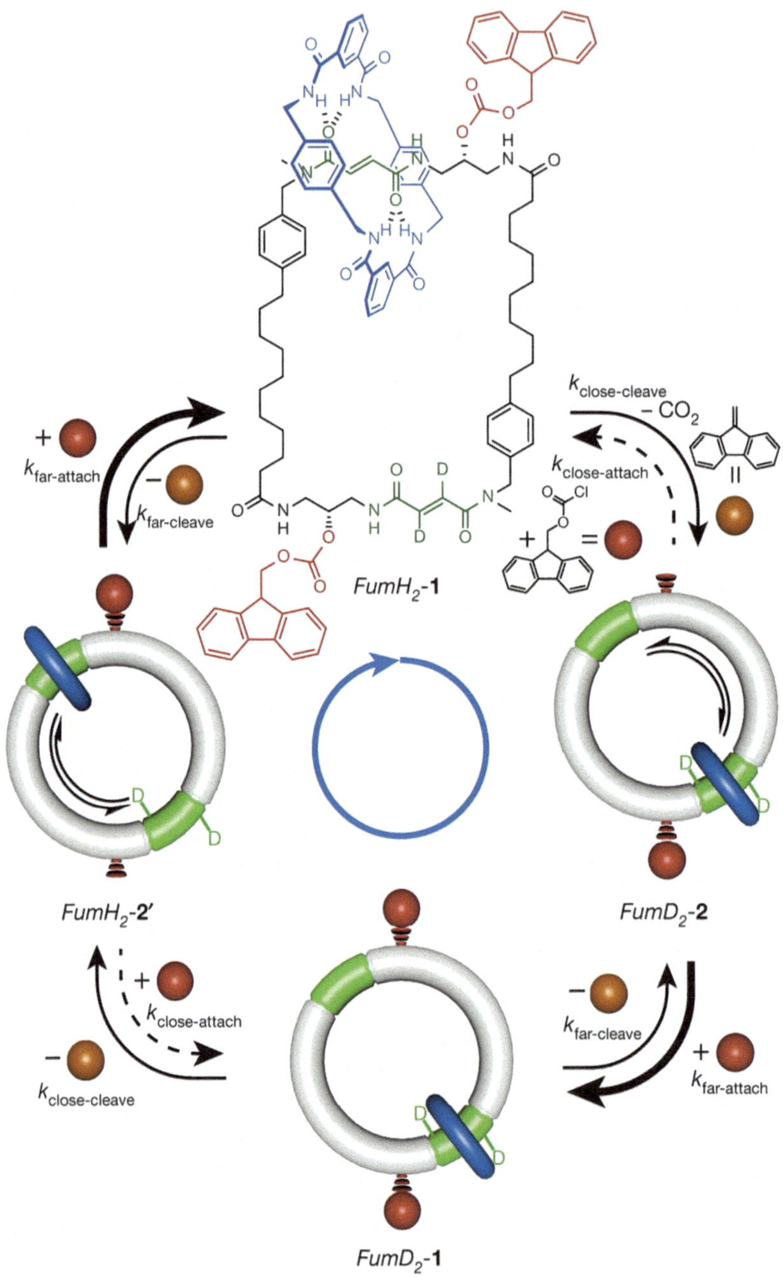

Figure 4.18 Symmetric [2]-catenane based molecular motor. The rate of removal of the stoppers (marked red) is independent from the position of the blue macrocycle. The rate of attachment is higher when the ring is on the opposite site, leading to a preferred sense of rotation. Reproduced with permission from [51]/Springer Nature.

process and lies energetically between teal and purple. Therefore, depending on the state of the lower recognition site, the macrocycle is more likely to be found in one or the other position.

The directionality of the motion is given by two chemically different stopper groups (marked in red and orange) which can be removed or attached individually. For a full clockwise cycle starting from the top left position in Figure 4.16a, the orange stopper is removed and the lower recognition group is switched to be energetically favorable for the small macrocycle. Subsequently, the orange stopper is reattached and the red stopper is removed. By switching the lower binding site back, the macrocycle returns to its original position at the top site. After the re-attachment of the red stopper, the molecule is in its original state. All of these steps require specific chemical reactions and environments. The direction of the cycle can be reversed by inverting the order of removing and attaching the stoppers. A sketch of the energy landscape is shown in Figure 4.16b. Switching of the lower recognition site alters the relative energies, but due to the stoppers, only one pathway for relaxation is open.

Yet another concept involving [2]-catenanes with stopper groups has been introduced by [51]. Instead of different recognition sites and different stoppers, a symmetrical concept as shown in Figure 4.18 is used here. The macrocycle (blue) can freely move between the two recognition sites (green), unless the pathway is blocked by a stopper (red). The two stoppers are located right next to each binding site and can be chemically attached and cleaved under identical chemical conditions. While the rate of cleavage of the stopper groups is independent of the position of the small ring, the rate of attachment is higher when the ring is located on the far site. With this mechanism of an information ratchet (Figure 4.3), a preferred sense of movement of the small macrocycle along the larger ring is achieved. A more detailed description of the reaction kinetics and thermodynamics is out of the scope of this chapter and can be found in Ref. [52].

4.3 Other Concepts of Molecular Motors

Besides the aforementioned concepts of molecular motors based on overcrowded alkenes and catenanes, a plethora of other implementations of unidirectional rotation within a single molecule have been designed. One of them is the motor by Fletcher et al. [53]. Their motor, shown in Figure 4.19, consists of a rotor and a stator connected by a single C—C bond, that acts as axis of rotation. Driven by chemical reactions, the molecule can assume four distinct states, which are stabilized against random Brownian motion by chemical bonds or by nonbonding interactions. By employing specific chemical reactions, involving well-established protection or deprotection steps, a directionality of 90–100 % over the full 360° cycle is achieved.

This motor, as well as the ones based on catenanes, operate with chemical fuel, as is also the case for most natural molecular motors. However, a continuous supply of fresh reactants is required. Furthermore, in general, waste products are formed that must be removed for the operation of the device. Light-driven motors do not

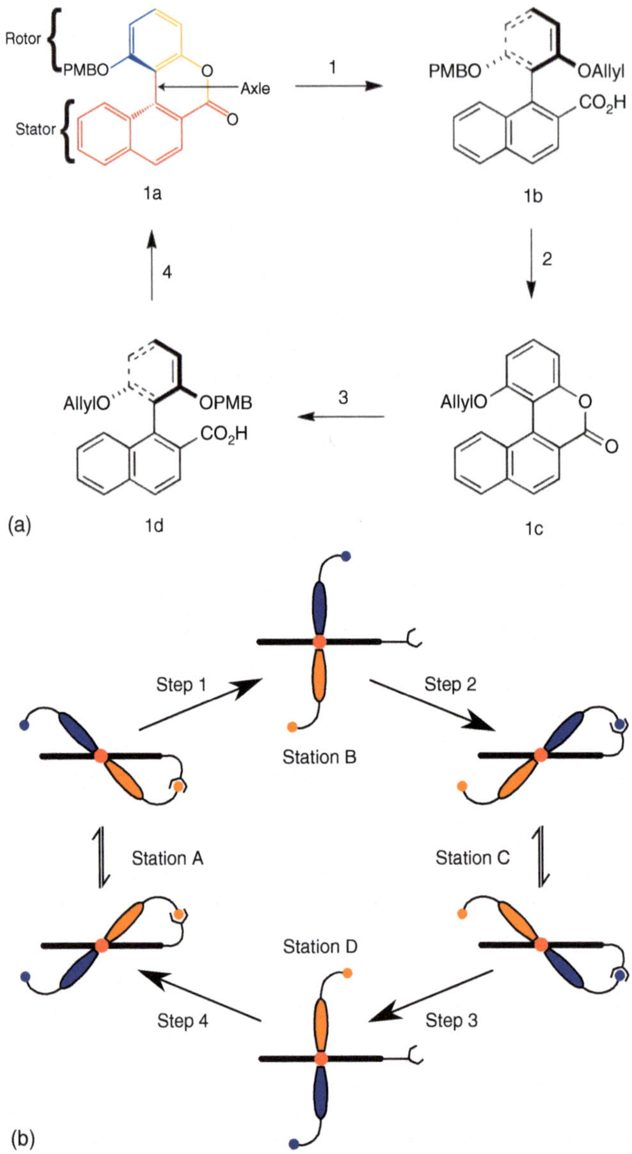

Figure 4.19 Reversible, unidirectional motor driven by chemical fuel. The directionality is given by the specific chemical reactions by which either the yellow or the blue part of the rotor are bound to the stator (red). (a) Chemical structure of the motor in each station. (b) Schematical depiction of the cycle. Adapted from [7, 53].

have these requirements: photons can trigger energonic and reversible reactions in a "clean" fashion without creating any waste product. The light source can be placed far from the place of reactions, and therefore triggers many molecules at the same time. Furthermore, as the wavelength and intensity of the light can be easily adjusted for the specific needs, the energy input is generally very well controllable. Also other properties, such as the polarization, can be adapted. Finally, the illuminated area or volume as well as the illumination duration can be tuned from nanometer resolution (with near-field optics) and extremely short laser pulses to simultaneous excitation of large scale areas or volumes with continuous photon flux [54]. For these reasons, a multitude of light-driven molecular motors have been developed.

In 2006, Lehn conjectured that imines can be used as basis for unidirectional photo-driven molecular motors [55]. Subsequently, these motors were successfully synthesized and their functionality proven [56, 57]. Instead of the C=C double bond of the Feringa type motors [11] (cf. 4.1), imine-based motors rotate around the C=N imine bond. The rotation is triggered by light-induced isomerization, followed by one of two thermal processes (pathways A and B), as sketched in Figure 4.20. It has been shown that both thermal processes follow a different route than the photo-isomerization, hence not simply undoing the work of the first step. The irreversible character of the thermal processes leads to the desired unidirectionality.

Pathway A is a N-inversion, which returns the motor back into its original state, i.e. 360° rotation is achieved in a two-step process. The alternative route B proceeds via a ring inversion. A repetition of the isomerization and the ring inversion leads to a full 360° rotation. Which route is taken is determined by the choice of the substituents that change the conformational flexibility of the molecule. The speed of the motor can be tuned in the same way.

A hemithioindigo chromophore-based molecular motor was introduced by Guentner et al. [58, 59]. It consists of a thioindigo and a stilbene fragment, connected by a double bond, and a sulfur-based stereocenter. The steric crowding

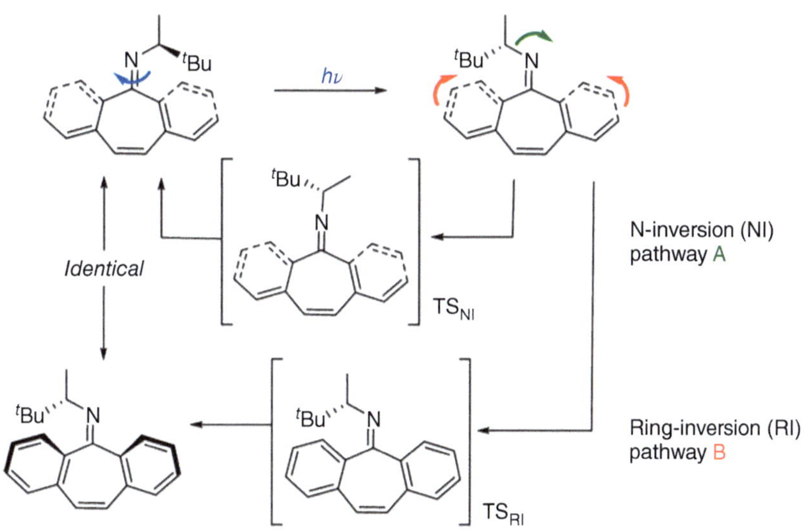

Figure 4.20 Imine-based light-driven molecular motor. A light-induced isomerization of the C=N imine bond is followed by one of two irreversible thermal relaxation processes. The two thermal processes are different from each other, as well as from the light-induced isomerization. Reprinted with permission from [56]. Copyright 2014 American Chemical Society.

at the stilbene fragment leads to helical twisting around the double bond. This leads to four diastereomeric structures with either *R*- or *S*-configuration and *P*- or *M*-helicity. Similar to the Feringa-type motor based on overcrowded alkenes [11], the rotation is based on an alternating sequence of light-induced double-bond isomerizations and thermal helix inversion steps. The four different diastereomeric structures and the reaction route are shown in Figure 4.21. However, instead of UV light, the hemithioindigo motor is operated with visible light up to 500 nm, which is therefore more suitable for biological systems. At a temperature of 20 °C, a rotation frequency of 1 kHz is measured.

In the previously described motors, the thermally induced relaxation steps usually limit the rotational speed of the motor. If the temperature is too low, the helix inversion cannot take place, and the motor turns into a simple photo-switch. On the other hand, at too high temperatures, the directionality of the rotation is lost.

The hemithioindigo chromophore-based molecular motor [58] has then been further improved to be driven by light alone, eliminating the thermal helix inversion step [60]. This structurally modified molecule is driven exclusively by light-induced processes and an enhanced directionality and efficiency at lower temperatures is observed.

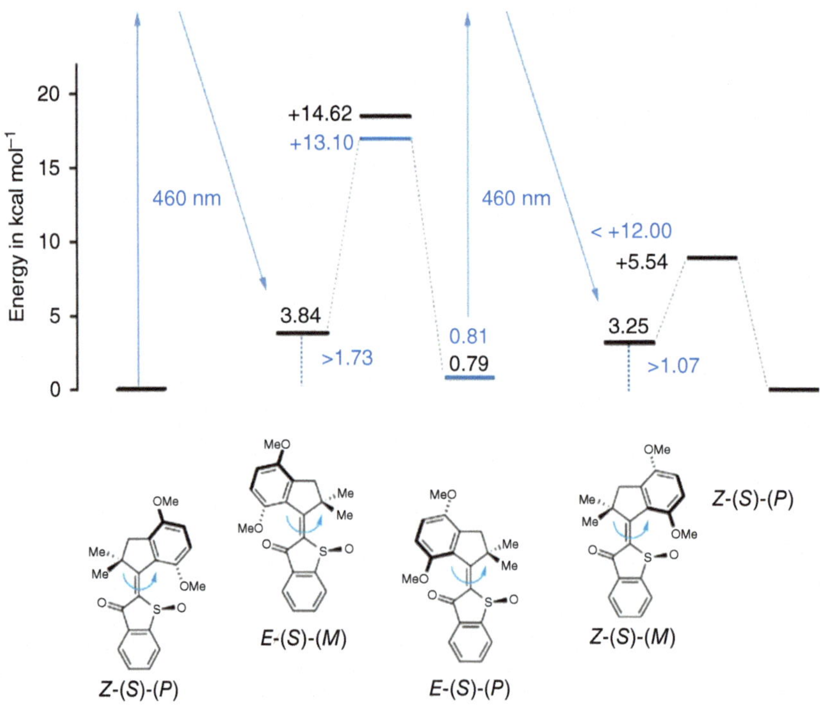

Figure 4.21 Visible light driven hemithioindigo chromophore-based molecular motor. The four diastereomeric structures of the *S*-enantiomer and the respective energies (black: theoretical calculations, blue: experimental data) are shown. A light-induced double-bond isomerization is followed by a thermal helix inversion. Another isomerization and helix inversion complete the 360° cycle. [58]/Springer Nature/CC BY.

The motor has four diastereomeric forms (A–D). They are interconverted via different photoreactions: single-bond rotation (SBR), DBI, and hula twist (HT). For each interconversion (e.g. A → B, B → A, B → C, etc.) the quantum yield has been determined at 20 and −50 °C. From the yields of the individual pathways, the probability of completing a certain cycle (e.g. ABC, ABD, etc.) is computed for 20 and −50 °C, respectively. At room temperature (RT), mainly two competing cycles, ABC with a probability of (81 ± 7 %) and ABD (16 ± 6 %), are found. Both cycles proceed with a very high degree of unidirectionality close to 100%. At low temperatures, the cycle ABD is strongly suppressed and the motor cycles through ABC with a probability of (98 ± 2 %) and 99 % unidirectionality. Furthermore, the average number of absorbed photons necessary to complete one full cycle of one molecule is reduced from 3400 at 20 °C to 1900 at −50 °C. This is in stark contrast to other motors, in which the efficiency is decreasing at lower temperatures.

Another modification of the hemithioindigo chromophore-based motor does not rotate in a simple loop, but instead follows an eight-shaped path [61]. The molecule, shown in Figure 4.22a, has a permanent sulfoxide stereocenter. The nonsymmetric julolidine unit additionally introduces a chiral axis. At slightly elevated temperatures (27–60 °C) predominantly conversions from D to A and from C to B are observed. Under illumination with green light (520 nm), transitions of A to C as well as B to D are found to be at least one order of magnitude more efficient than any other photoreaction. Combining thermal and photochemical processes, the cycle ACBD is achieved. The right part of Figure 4.22b shows the relative position of the methyl group with respect to the thioindigo fragment, highlighting the eight-shaped cycle.

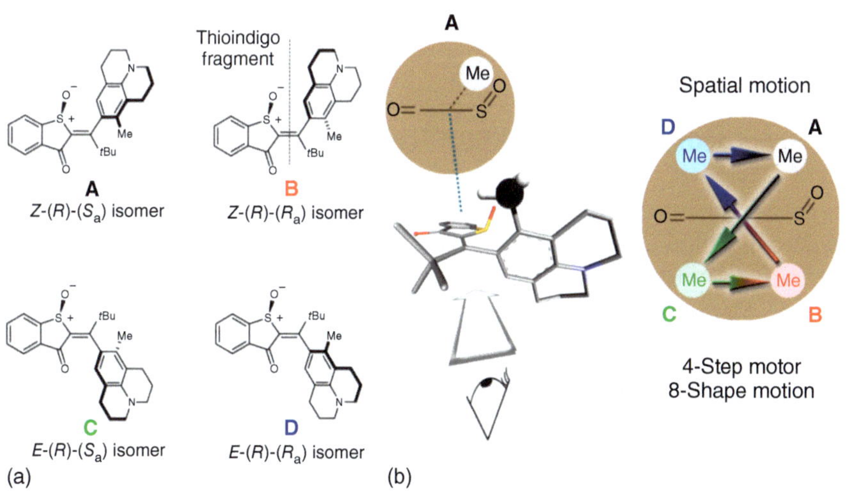

Figure 4.22 Hemithioindigo chromophore-based motor undergoing an eight-shaped motion. If the solvent is changed between each step, 82 % directionality is achieved. (a) Chemical structure of the four diastereomeric states of the motor. (b) Left: Perspective view of the molecule. Right: Relative position of the methyl group, following an eightshape path. [61]/Springer Nature/CC BY.

Each step of the cycle has different optimal operating conditions. If the solvent is changed between each motor step, a directionality of 82 % is achieved. In case of a single solvent – 1,2-dichlorobenzene – for all steps, which is desirable for continuous operation of the motor, the directionality is reduced to 47 %, as the thermal steps become less selective.

4.4 Computationally Designed Light-driven Molecular Motors

In addition to the experimentally synthesized and characterized motor molecules described above, a number of motor designs have been proposed based on computational calculations. For instance, Wang et al. proposed a molecular motor incorporating a cyclohexenylidene moiety into a protonated Schiff-base framework [62]. In contrast to most other designs, this motor does not require point chirality in order to operate in a unidirectional manner, as confirmed through minimum energy path (MEP) calculations and non-adiabatic molecular dynamics (NAMD) simulations based on multiconfigurational quantum chemistry.

Other examples of computationally designed molecular motors are:

- Two-step photoactive molecular motor (without thermal steps) based on hydrogen bond induced chiral environment [63].
- Synthetic indanylidene-pyrrolinium molecular frameworks [64].
- Imine-based motor with bulky chiral stator [65].
- Fulgides based rotary motor [66].

To the best knowledge of the authors, the unidirectional nature of these molecules has not yet been experimentally investigated.

4.5 Molecular Motors on Surfaces

The motors presented so far in this chapter have typically been synthesized and characterized in solution. However, in the liquid phase, the unidirectional motion of the rotor with respect to the stator competes with the Brownian motion of the whole molecule. Moreover, a fixed point of reference for translation is missing. In order to generate useful work and for many possible applications, adsorption on a solid surface, which provides a point of reference for the motion, is preferred.

4.5.1 Tethering of Molecular Motors

An effort toward this goal was the grafting of molecular motors onto gold nanoparticles [67, 68]. In order to prevent uncontrolled thermal rotation of the whole molecule (not to be confused with the rotation of the rotor against the stator within the molecule), the authors chose two points of attachment on the surface, as indicated in Figure 4.23. Furthermore, the tethers should be sufficiently long

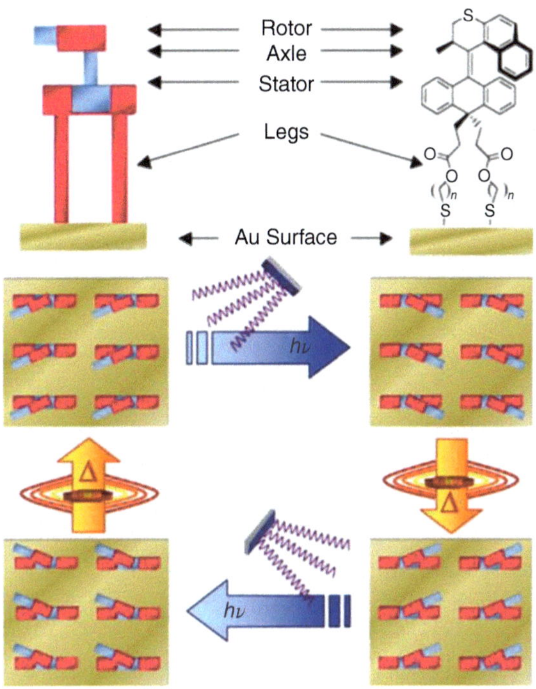

Figure 4.23 Molecular motors grafted onto a thin gold film on a quartz substrate. The motors are lifted from the surface by atom chains of length 8 or 16. Upon illumination, the motors tethered with the longer chains undergo unidirectional rotation in a plane approximately parallel to the surface. Reproduced with permission from [69]/Royal Society of Chemistry.

to provide enough conformational flexibility of the motor unit. An important side effect is the relatively large distance between the motor unit and the metal, thus minimized quenching of electronic states by the gold nanoparticle surface, allowing the photoisomerization to take place efficiently. For these reasons, tethers consisting of eight carbon atoms are chosen. A sketch of a similar system on a thin gold film (see below) is shown in Figure 4.23. The axis of rotation of the rotor is perpendicular of the surface of the nanoparticle, leading to an in-plane rotation of the rotor.

While unidirectional motion of the rotor with respect to the stator – and hence the whole nanoparticle – is confirmed, the rotation speed of this system is decreased by a factor of approximately two as compared to solution. This is assigned to the reduced degrees of freedom within the motors due to intermolecular interactions [68], but the motor functionality could also be affected by an energy transfer to the gold surface [23].

While the nanoparticles provide a frame of reference for the rotation of the motor unit, the particles themselves are subject to Brownian motion. Therefore, a similar approach has been followed for a gold thin film on a quartz substrate [69]. Molecular motors adsorb on the surface in a self-assembled monolayer [70]. In this case, with tethers of eight atoms length, the motor activity is suppressed. This is in line with previous reports of different amounts of quenching on nanoparticles as compared to films [71, 72]. Accordingly, with longer tethers of 16 atoms, the functionality of the motor molecules is regained.

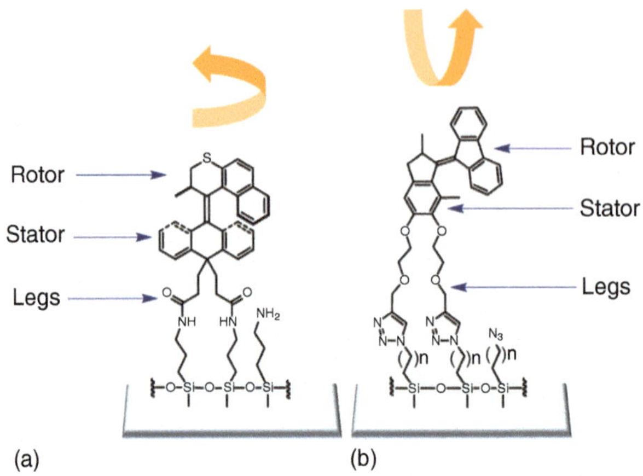

Figure 4.24 (a) Azimuthal [73] and (b) altitudinal [75] molecular motors on a quartz surface. Reproduced with permission from [23]/American Chemical Society.

An insulating substrate, such as quartz, has the strong advantage of not quenching the excited states necessary for the motor rotation. Similar to the tethering on gold nanoparticles of thin films, motor molecules have been attached to a quartz film by Pollard et al. [73]. Also here, the rotation axis of the motor is perpendicular to the surface, leading to in-plane rotation, as shown in Figure 4.24a. It has been found that the molecular motors in this setting still maintain their unidirectionality.

By changing the point of attachment of the motor, an altitudinal motor can be built on a surface [74]. This is shown in Figure 4.24b: the new rotor unit rotates around an axis parallel to the surface. The rotational speed depends on the coverage of the motors. High coverage hampers the rotation, while high speed is achieved at low coverages. Therefore, most likely the intermolecular interaction affects the rotary speed.

4.5.1.1 Tunable Wettability of Surfaces

As one possible application, the altitudinal motors have been used to change the macroscopic surface wettability [76, 77]. One side of the (altitudinal) rotor is functionalized with a functional group (FG). This group – which for example could be hydrophobic – points either away from the surface or toward the surface and is therefore exposed to or screened from the water droplet on the surface. Hence, depending on the state of the motor, the properties of the surface are altered.

In a first attempt the wettability of a quartz surface could be modified by functionalization with different isomers of the motor. However, switching of motors on the surface was not possible [76]. By modifying the attachment of the stator to the surface – in this case gold on mica – to a tripodal structure the motor functionality on the surface is restored and *in situ* switching of the surface wettability is possible [77]. The tripodal structure increases the spacing between adjacent motor molecules, leading to less intermolecular interactions and an increased conformational flexibility. Furthermore, sufficient decoupling from the gold substrate is provided. Figure 4.25b

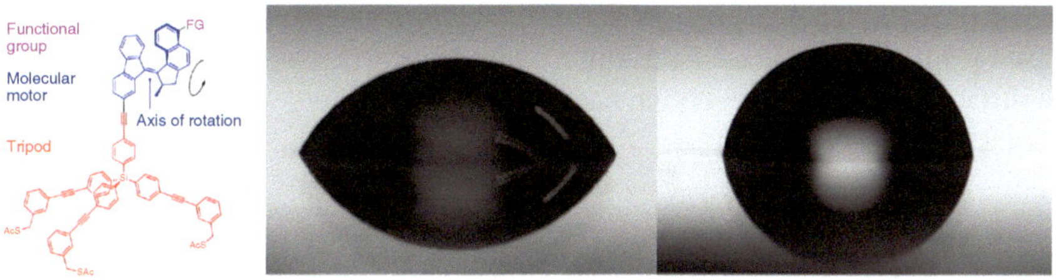

(a) (b)

Figure 4.25 Tunable wettability of a gold on mica surface. (a) Attaching the stator to the surface with a tripodal structure allows for a rotation of the rotor with the hydrophobic FG. (b) Depending on the state of the motor, the hydrophobic group is either screened by the motor unit (left) or fully exposed to the surface (right), which changes the water contact angle up to 16°. Reproduced with permission from [77]. © 2014, American Chemical Society.

shows a water droplet on this surface before and after switching the motor unit with light. A change of the water contact angle of up to 16° is observed.

4.5.2 Molecular Motors on Surfaces Without Tethers

Besides tethering of motors to the surface to keep them in a defined orientation with respect to the surface, it is also possible to deposit motor molecules without any side groups onto surfaces. The motor shown in Figure 4.26a is found to adsorb in three conformations on the Cu(111) surface [78, 79]. After depositions onto the cold surface (5 K), most molecules are in conformations A and B. Conformation C is only reached after heating the sample to roughly RT. This indicates that C is the most stable conformation on the surface, but it requires thermal energy to reach this state. In comparison to the motor states in solution, A and B are similar to the – in solution more stable – (*M*)-1 isomer, while C corresponds to the less stable (*P*)-1. Thus, the relative stability of the motor states is reversed on the surface.

Upon excitation with voltage pulses, the molecular motors in conformation C exhibit rotational motion [78, 80]. In contrast to the motorized nanocar by

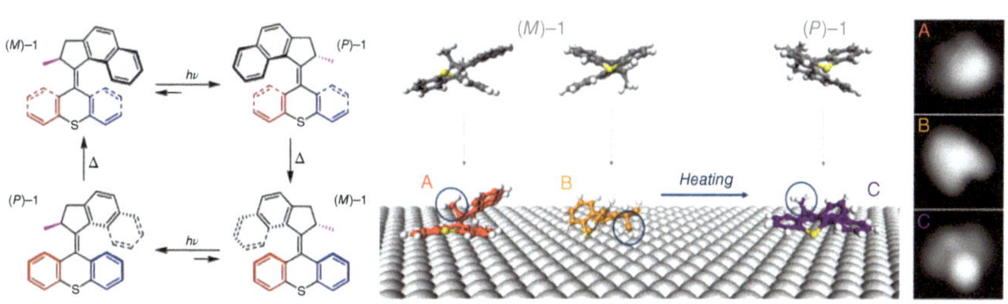

(a) (b) (c)

Figure 4.26 Molecular motors on the Cu(111) surface. (a) Full 360° rotary cycle of the molecular motor in solution. (b) Scheme of the molecular adsorption from density functional theory (DFT) calculations. The most stable conformation C is reached after heating the sample. (c) STM images of all three conformations ($1.8 \times 1.8\ nm^2$, A: 9 pA, 200 mV, B: 10 pA, 100 mV, C: 450 pA, −50 mV). Reproduced from [79], American Chemical Society/CC BY 4.0.

Kudernac et al. [81], which moves across the surface, the motor is rotating around a fixed pivot point at or near the sulfur atom, which is strongly interacting with the surface. This rotational motion has a high directionality $\xi \approx 80 - 90\%$, which is defined as

$$\xi = \frac{n_{\leftarrow} - n_{\rightarrow}}{n_{\leftarrow} + n_{\rightarrow}}, \tag{4.1}$$

where $n_{\leftarrow}$ and $n_{\rightarrow}$ are the movements (translation, rotation) in forward and backward direction, respectively. A directionality ξ of 100 or -100 % indicate fully directional rotation in either direction. A completely nondirectional system, i.e. having equal probability of either direction, corresponds to a directionality of 0 %.

4.5.3 Unidirectional Motion on Surfaces

Besides motors consisting of just one molecule, systems of adsorbates on surfaces open up many opportunities for building and tuning unidirectional systems. The controlled manipulation – i.e. induced motion – of adsorbates with the scanning tunneling microscope (STM) tip via the electric field, tunneling electrons, or mechanical pushing or pulling with the tip is well established [82–84]. Furthermore, tip-induced rotations of adsorbates have been studied for decades, but often no directionality has been observed [85].

Simpson et al. achieved directional control of the rotation of a single molecule on a surface by the appropriate choice of the tip position and the parameters of the applied voltage pulse [86]. This excellent control about the rotational orientation of the molecule enabled the successful participation in the first nanocar race [87, 88], in which a molecule – the nanocar – had to be navigated through a predefined course without mechanically dragging it with the tip [89].

The second edition of the nanocar in March 2022 saw two winners with very different approaches of gaining control of the motion on the singe-molecule level [90, 91]. The NANOHISPA team from IMDEA Nanociencia in Madrid, Spain designed their molecule similar to an actual car, reducing the interaction of the "chassis" by the addition of "wheels." In contrast, the NIMS-MANA team from Tsukuba, Japan chose a very flat molecule with a stronger interaction with the surface. But similar to graphene nano-ribbons, superlubricity was employed to allow for fast motion along the surface.

One of the most important ingredients of unidirectionality is an asymmetric potential energy landscape. However, this alone is not sufficient. The symmetric dithioanthracene (DTA) molecule moves along one axis on the Cu(111) surface, with equal probability for forward and backward movement [92]. The addition of two methyl groups on one side of the molecule, forming dimethyl-DTA, decreases the surface mobility by a factor of 100, but the symmetry of the movement along remains unchanged [93]. The authors attribute this finding to the principle of microscopic reversibility stating that in thermodynamic equilibrium reactions rates in the forward and reverse direction are equal [94].

One of the first molecule-surface systems with a slight preferential directionality was butyl methyl sulfide (BuSMe) on Cu(111) [96]. This molecule is protochiral and becomes chiral only upon adsorption on the surface, as seen in Figure 4.27b [97].

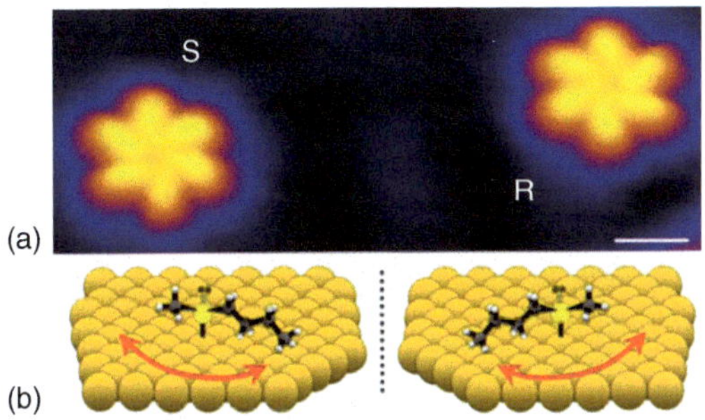

Figure 4.27 BuSMe on Cu(111) as electric motor. (a) Due to rotation of the molecule during imaging, all six orientations on the surface are seen as superposition in the scanning tunnelling microscope (STM) image. (b) Depending on the tip, a small directionality of up to 5%, where the sense of rotation depends on the enantiomer, is observed. Reproduced with permission from [95]. © 2011/American Chemical Society.

The molecule can adopt six orientations on the surface. Imaging with the STM with sufficiently large bias induces a rapid rotation of the molecule around the central sulfur-metal bond and all six orientations are seen simultaneously (Figure 4.27). Using a low bias, the molecule can be imaged without changing its orientation. Upon voltage pulses, single rotational steps can be induced. From these observations, a small directionality of ≤ 5 % is determined, whereby the sense of rotation depends on the chirality of the molecule on the surface. Furthermore, it has been found that the state of the tip influences the rotational speed and the directionality. This is attributed to the chirality of the tip apex [95].

A different molecule, a porphyrin derivative with a platinum atom coordinated in the center of the macrocycle, does not rotate unidirectionally by itself, but only in a dimer assembled on a Au(111) surface [98]. The chemical structures and the dimer arrangement on the surface are shown in Figure 4.28a,b, respectively. The dimer, roughly resembling an S shape in the STM images, as shown in Figure 4.29a, can adapt two chiralities, which are seen in the first and last panels of Figure 4.29d.

Typically, one of the molecules is pinned on an adsorbate at an elbow site of the herringbone reconstruction of the Au(111) surface, while the other one is free to move. Two series of STM images in Figure 4.29b,c show the preferred directionality of each enantiomer. Furthermore, the dimers can be temporarily deformed (e.g. in the second to last image of Figure 4.29c) or their chirality can be reversed (panel d).

For better illustration, the relative position of the mobile monomer with respect to the pinned center in a series of voltage pulses is graphically represented in Figure 4.28c, where the numbers indicate the sequence of the observed position. Between the red and the blue series with a pulse voltage of +2 V, the chirality was switched using a negative bias of −1.9 V (green dots).

The relative probabilities of "forward" or "backward" movement (defined as the preferred sense of rotation for each enantiomer), deformation or no change are listed in Figure 4.28d for positive and negative pulse voltage. While deformation is only observed for negative voltage, the preferred direction of rotation is independent of the polarity. The authors report probabilities of forward movement of 89% for the

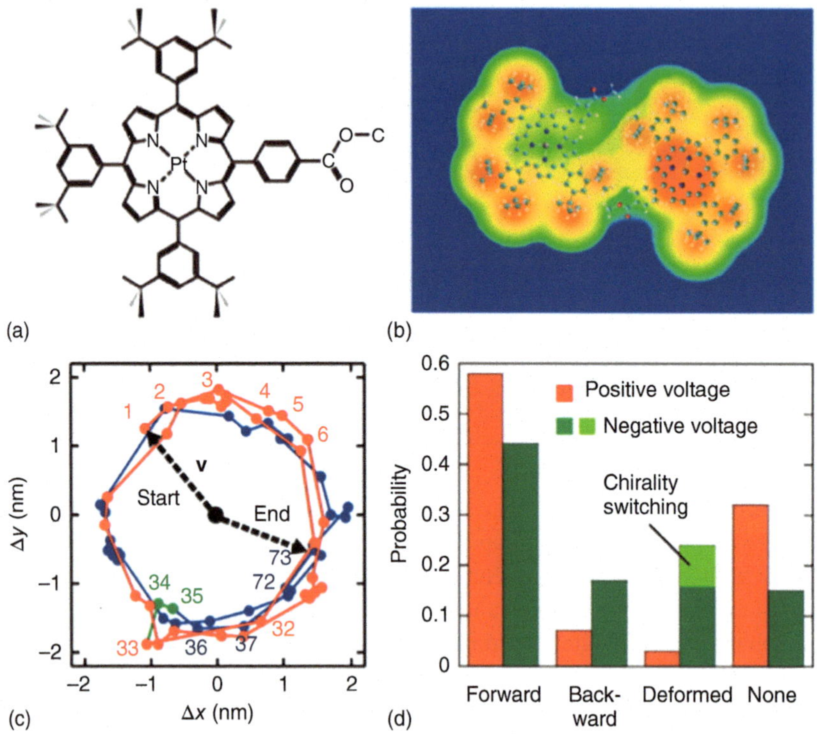

Figure 4.28 Chemical structure and rotation of a platinum-porphyrin-based supramolecular assembly on Au(111). (a) Chemical structure of the single molecule with a platinum atom coordinated in the center of the macrocycle. (b) Suggested dimer arrangement on Au(111) superimposed on a simulated STM image. (c) Relative position of the two dimers in a sequence of voltage pulses (red, blue: +2 V, green: −1.9 V). The chirality of the dimer was reversed in the pulses marked in green. (d) Probability of forward, backward or no movement or deformation of the dimer separated by pulse polarity. Reproduced with permission from [98]. © 2015/American Chemical Society.

positive voltage and 72% for the negative bias (if deformation events and pulses without movement are excluded). This corresponds to directionalities ξ of 78 % and 44 % for positive and negative voltage, respectively.

Another molecule pinned to the surface is 2-(2-methoxyphenyl)-1,3-dimethyl-1*H*-benzoimidazol-3-ium (o-MeO-DMBI), which is chemisorbed on the Au(111) surface through a lone pair of the oxygen [99]. Similar to BuSMe, the molecule gains its chirality only upon adsorption on the surface. Voltage pulses induce 60° rotations, where the directionality depends on the chirality of the molecule. No values for the directionality of the motion have been given.

Zhang et al. reported a multicomponent molecular propeller adsorbed on an Au(111) surface [100]. The stator, resembling a five-teethed gear, is coordinated with a ruthenium atom to the rotor, which consists of three indazole groups [101, 102]. Figure 4.30a,b show the chemical structure and a top view of the molecule. On the surface, the rotor is found in two chiral forms, which are highlighted by circles

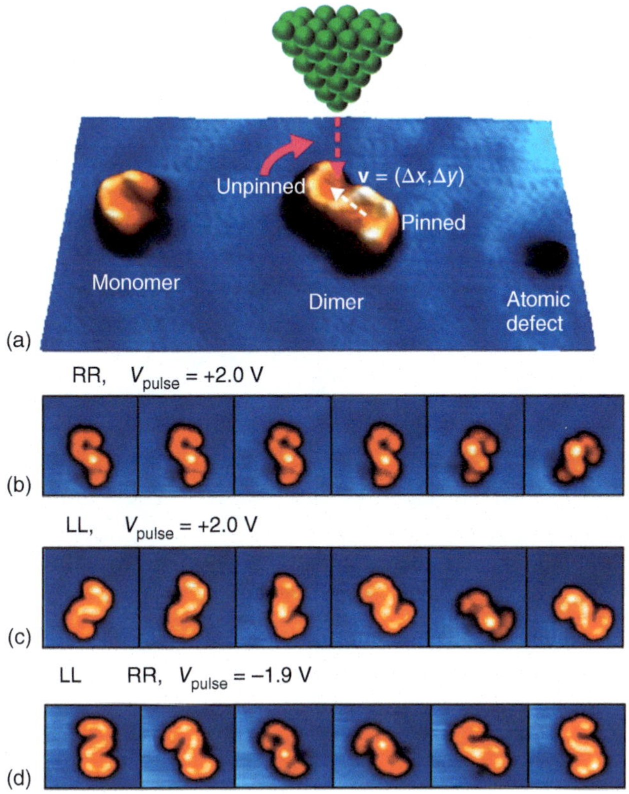

Figure 4.29 Voltage pulse-induced rotation or deformation of supramolecular motors on Au(111). (a) Pseudo-3D representation of the surface. (b, c) Unidirectional rotation sequences for the two enantiomeric forms. (d) Change of chirality. Reproduced with permission from [98]. © 2015, American Chemical Society.

in Figure 4.30c. This is in contrast to the motor in solution at RT, where the two atropisomers are interconverted with a fast exchange rate.

The rotation of the rotor unit can be induced by STM imaging with a bias of more than ± 1 V. Given the three blades of the rotor and five teeth of the stator, full 360° rotation proceeds in 15 steps of $\approx$ 24° each. Also, double-steps of 48° are observed. In some cases, not all blades of the rotor move to a new position, but remain bent. A directionality of about 57% is reported, counting only complete rotation steps of all three blades.

Furthermore, STM manipulation using the electric field as well as excitation via inelastic electron tunneling [103, 104] lead to a preferred sense of rotation of the motor. Also, mechanical manipulation with the STM tip is possible. While the rotor can be rotated in both directions, the reverse direction often leads only to partial movement of the rotor, i.e. with some of the blades in a bent position. This further corroborates that the system has an intrinsically preferred sense of rotation, which depends only on the chirality of the rotor.

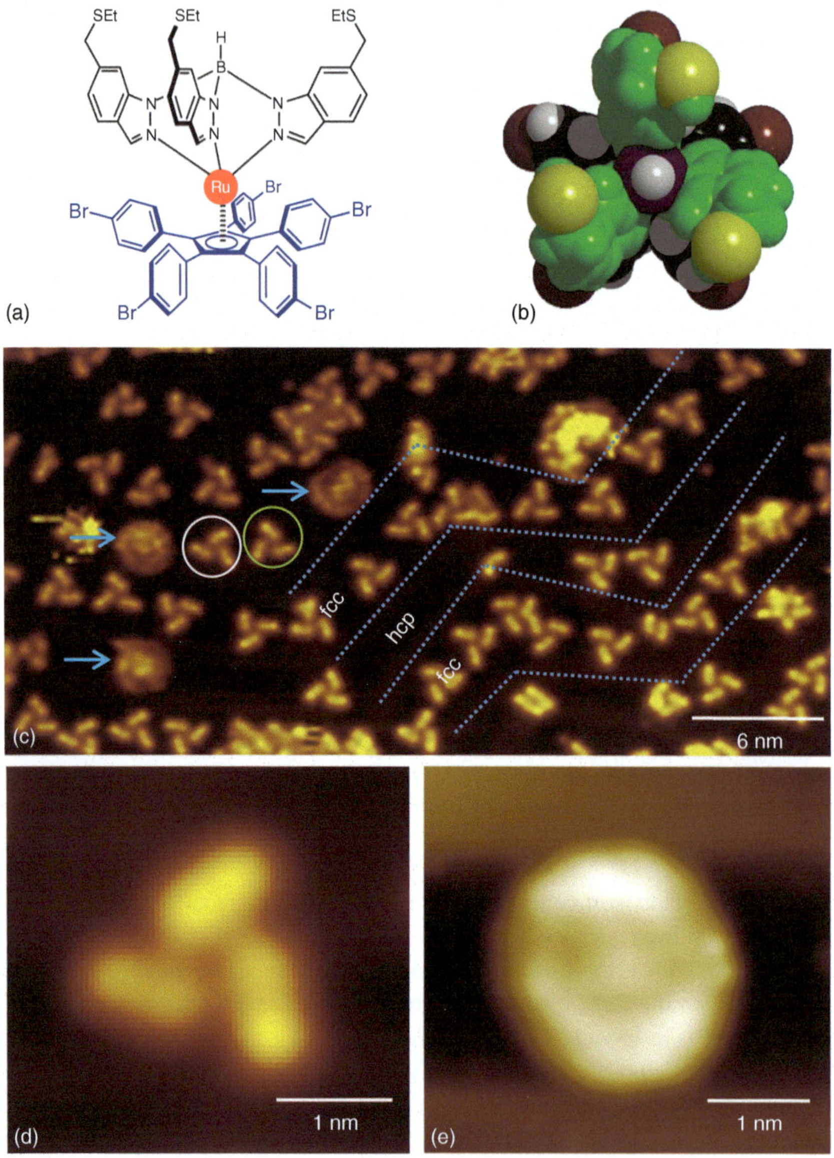

Figure 4.30 Multicomponent molecular propeller on Au(111). (a, b) A three-bladed rotor is connected to a five-teethed stator via a ruthenium atom. (c) The molecule appears as a tripod structure in one of two chiralities. (d) Some molecules are found to rotate under the STM tip. Reproduced from [100], Springer Nature/CC BY 4.0.

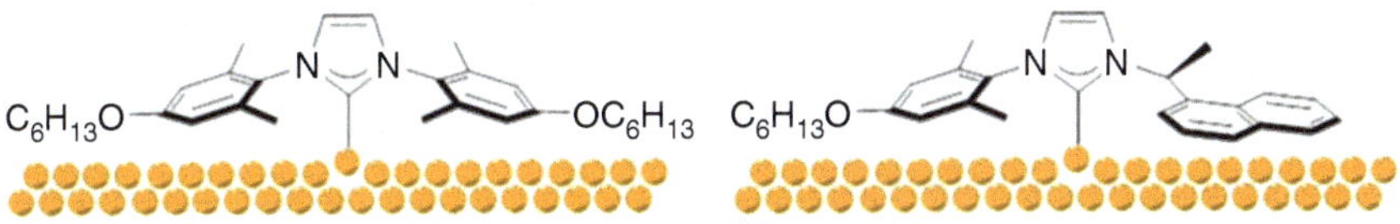

Figure 4.31 Chemical structures of IoMes (a) and IRotate (b). Both the symmetric IoMes as well as the asymmetric IRotate are bound to a single atom of the Ag(111) surface. This bond acts as the axis of rotation. Reproduced with permission from [105]. © 2020/American Chemical Society.

Surface-anchored N-heterocyclic carbene (NHC)-based rotors can be tuned by varying the nitrogen substituents and therefore optimized for the operation as motors on surfaces [105]. Ren et al. optimized the molecular design by simulations, before carrying out experiments with the most promising candidates. Two molecules are presented here: the symmetric IoMes, shown on the left in Figure 4.31, and the asymmetric IRotate on the right in the same figure.

Both molecules are characterized by STM on a Ag(111) surface. As sketched in Figure 4.31, they are bound to a single surface atom via a vertical C-metal bond. The molecules can be laterally manipulated on the surface and are moved together with the surface atom they are bound to. This bond furthermore acts as axis of rotation. The free energy along the rotation angle coordinate φ is shown to be approximately symmetric for IoMes (Figure 4.32d) and sawtooth-like for IRotate (Figure 4.32e).

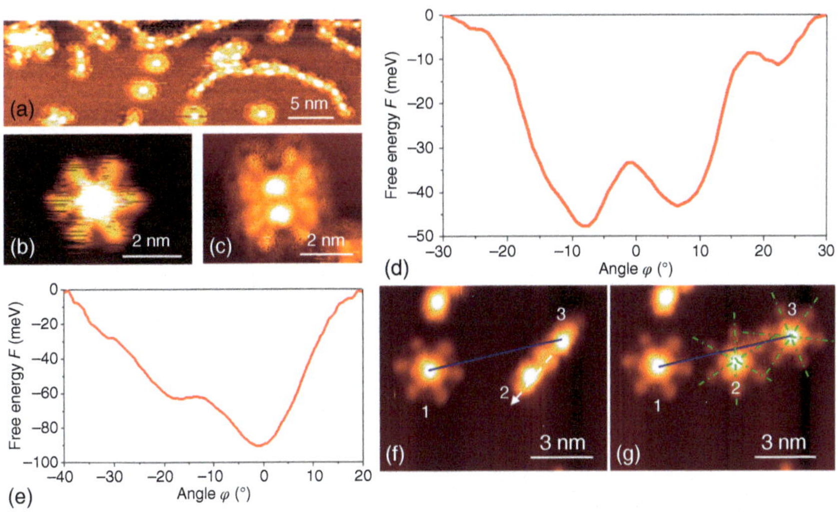

Figure 4.32 STM images (at 78 K) and energy landscape of IoMes (a–d) and IRotate (e–g). Single molecules are free to rotate, but the rotation can be blocked by neighboring molecules in close vicinity (IoMes: c, IRotate: 2, 3 in f). The symmetric IoMes still undergoes a swinging type motion, indicating nondirectional rotation. The blocked IRotate is not moving, evidencing unidirectional rotation. The free energy landscape is roughly symmetric for IoMes (d) and sawtooth-shaped for IRotate (e). Reproduced with permission from [105]. © 2020, American Chemical Society.

At 78 K, the two molecules rotate quickly, leading to a superposition of all possible adsorption positions in the STM images, as shown in Figure 4.32.

Bringing two molecules close to each other blocks the rotational movement. In the case of the symmetric IoMes, the molecules are not lying still, but can still perform a swinging type of motion, which is shown in Figure 4.32c. In contrast, two asymmetric IRotate molecules next to each other are stabilized in one specific orientation, as shown in panel f, labeled as 2 and 3. Moving them apart restores the rotation of both molecules (panel g). This indicates that the rotation was blocked by the neighboring molecules. While the symmetric IoMes rotates in both directions, leading to the swinging motion, the unidirectional rotation of the asymmetric counterpart IRotate is effectively blocked.

At 5 K, the rotation of molecules is frozen and they are imaged statically, as shown in Figure 4.33. Single rotation steps are induced by inelastic electron tunneling and the sense of rotation is recorded. In the examples shown in Figure 4.32, both the IRotate, as well as the IoMes molecule rotate anti-clockwise by 60° and 30°, respectively. The directionality is not quantified, but a clear preference for anti-clockwise rotation for IRotate is observed: after more than 60 consecutive pulses, the molecule rotated by a cumulative angle of ca. +1800°, as shown in panel b. In contrast, the rotation of IoMes does not lead to a significant net rotation (panel d), i.e. IoMes does not show a preferential sense of rotation.

All previously described systems are based on *(proto)chiral* molecules on *achiral* surfaces. In the work of Stolz et al. this is reversed [106]: an achiral molecule,

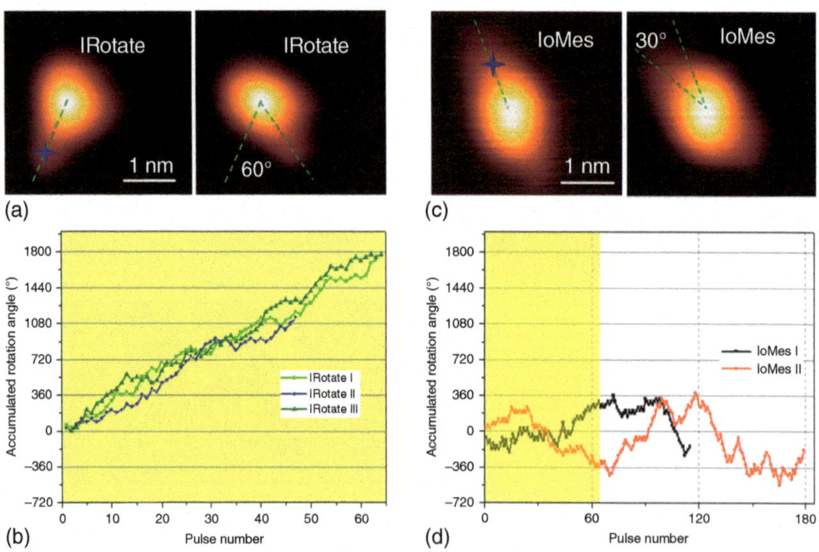

Figure 4.33 Directionality of the rotation of IRotate and IoMes on Au(111). (a, c) At 5 K, the rotation is frozen and the molecules are imaged stably. Examples of anti-clockwise rotation of 60° (IRotate) and 30° (IoMes) after −6.5 V pulses at the position marked with a blue cross are shown. (b, d) Cumulative rotation angle after multiple voltage pulses. The yellow ranges in (b) and (d) are the same. Reproduced with permission from [105]. © 2020, American Chemical Society.

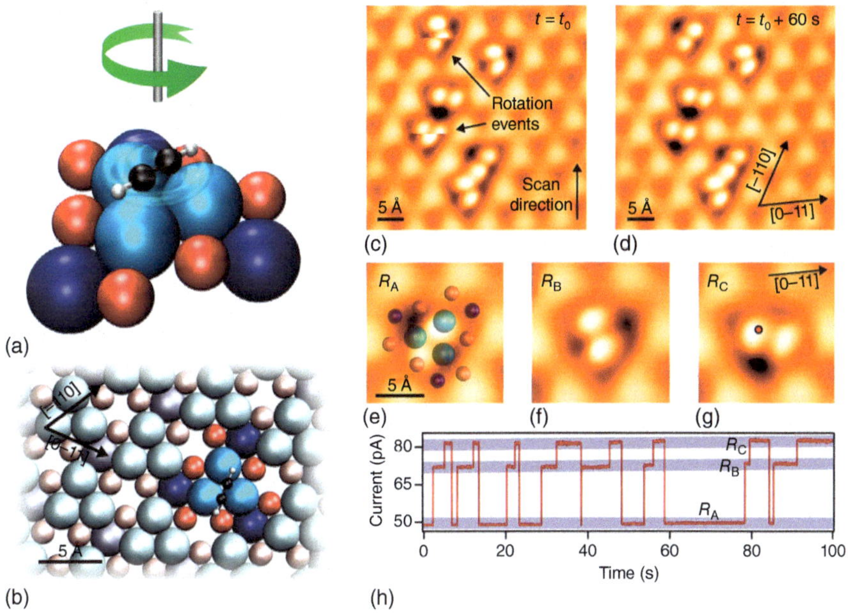

Figure 4.34 Achiral acetylene on the chiral PdGa($\overline{111}$) surface. The surface and the adsorbed molecule are schematically shown in (a) and (b). (c, d) Overview image with multiple molecules, two of which are moving below the tip in (c). (e, f) Three adsorption positions. (h) Tunneling current measured during a voltage pulse. The molecule cycles through three adsorption positions (each leading to a different current), showing the unidirectionality of the process. Reproduced with permission from [106]/National Academy of Science.

acetylene (H—C≡C—H) is adsorbed on a chiral PdGa surface, specifically the PdGa($\overline{111}$) A enantiomorph [107]. Each trigonal surface unit cell contains one Pd trimer (Pd_3). The neighboring six Ga atoms and three Pd atoms lift the local inversion symmetry of the trimer as sketched in Figure 4.34B. Panel A of the same figure shows an acetylene molecule adsorbed on the Pd_3 trimer. It is firmly anchored and typically dissociates rather than being moved away with the STM tip.

STM images reveal three energetically equivalent adsorption positions of the acetylene molecule. The molecule is rotated in 60° steps by voltage pulses, exemplarily shown in Figure 4.34e–g. The current trace in panel (h) shows the cyclic motion through the three adsorption positions, indicating the unidirectionality of the process. The authors identify two regimes; the tunneling regime with no pronounced parameter dependence of directionality or rotation rate and the classical regime with a strong dependence on temperature, pulse bias, and tunneling current. Overall, a very high directionality ξ of up to 98% is observed.

The tunneling nature of the process is confirmed by replacing one or both hydrogen atoms with the heavier deuterium isotope (C_2DH, C_2D_2). Despite the small change of the moment of inertia (1:1.08:1.2), a strong relative decrease of the rotation frequency is observed (1:0.56(11):0.24(5)). In the case of the asymmetric

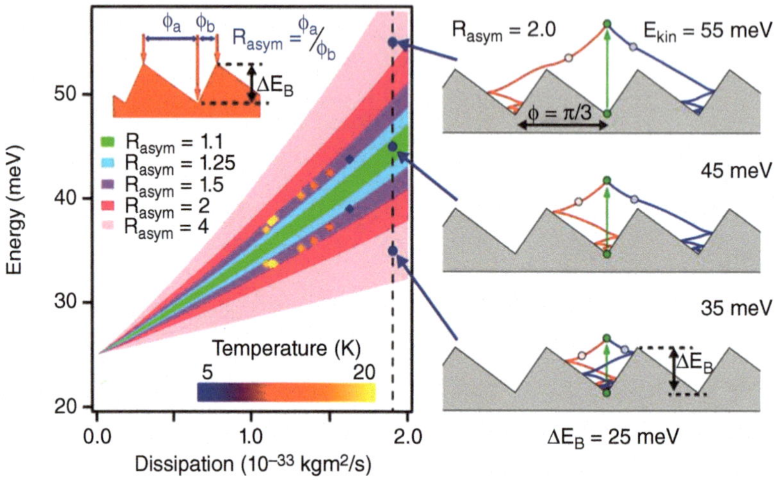

Figure 4.35 Suggested model for the classical regime: only with an adequate energy unidirectional motion is observed. Too low energy cannot overcome the barrier, and too high energy leads to rotations in both directions. Reproduced with permission from [106]/National Academy of Science.

C_2DH six distinct adsorption positions are observed, confirming that indeed six 60° steps are necessary for a full 360° rotation.

To explain the unidirectional motion in the classical regime, the authors propose the model shown in Figure 4.35. Excitation from a sawtooth potential (right image) leads to no motion if the energy is too low, and to nondirectional motion if the energy is too high and the potential barrier can be overcome. Only a narrow range of the excitation energy leads to motion with a preferred sense of rotation.

4.6 Conclusions and Outlook

Since the first successful demonstration of a unidirectional molecular motor by Bernard Feringa, defying random Brownian motion, its design has been refined and improved in many ways. Various aspects of the motor's properties have been addressed, such as the rotational speed, the excitation wavelength or even the presence of stereocenters in the molecule. Besides variations of the initial Feringa motor, much effort went into developing other motor concepts. Many are similar to the Feringa motor by achieving unidirectional rotation around a specific chemical bond. But also entirely different solutions have been presented, e.g. by propelling one macrocycle of a catenane around the other in a unidirectional fashion.

Many applications already exist, ranging from medical cures to functional surface coatings. Some systems exist purely on the nanoscale, such as nanocars, which can move along predefined tracks across the surface. Others bridge the gap from the nano- to the macro-world by exploiting the collective movement of self-organized hierarchical structures, e.g. in artificial muscles. But there are many possibilities for other uses of molecular motors. In the field of medicine, systems based on molecular motors could deliver and release pharmaceuticals, targeted down to the cellular level, thereby reducing side effects and increasing treatment efficacy. Nanorobots could repair biological tissue or clean blocked arteries without requiring surgeries. In nano-technology, more elaborate "nano-machines" comprised not only of motors, but also switches, wheels, axles, gears, propellers, tweezers, etc., could enable the assembly of complex nanostructures with high precision and thus the creation of new advanced materials and devices. For example, one could think of advanced sensors, actuators, or analytical tools. Even molecular computing seems achievable by utilizing the molecular motors' work for data processing and storage on the nano-scale.

Compared to the intricate and highly specialized biological molecular motors, artificial motors still require much refinement and further research efforts. A fundamental understanding of the processes involved in the motor rotation is required in order to make use of their intriguing properties. Among the largest challenges to be tackled is the ability to precisely steer the integration of molecular motors into specific devices. Furthermore, it is necessary to externally operate and control the molecular motors, possibly in complex environments like the human body. For any applications, high stability and durability are essential. Tackling these challenges will open up countless exciting possibilities and can reshape our technological landscape.

Acknowledgments

I would like to thank Leonhard Grill for proof-reading the manuscript. Special thanks go to Petra Rudolf for recommending me as an author. Funding through the MEMO project is gratefully acknowledged.

References

1 Credi, A. (ed.) (2014). *Molecular Machines and Motors*, 342 pp. Springer-Verlag GmbH. ISBN: 9783319086781.

2 Kolomeisky, A.B. (2015). *Motor Proteins and Molecular Motors*, 222 pp. Taylor & Francis Ltd. ISBN: 1482224763.

3 Lavelle, C. (2018). *Molecular Motors: Methods and Protocols*. Totowa, NJ: Humana Press. ISBN: 9781493985562.

4 American Heritage® Dictionary of the English Language (2011). Motor. https://www.thefreedictionary.com/motor (visited on 15 April 2021).

5 Einstein, A. (1905). Über die von der molekularkinetischen Theorie der Wärme geforderte Bewegung von in ruhenden Flüssigkeiten suspendierten Teilchen. *Ann. Phys.* 322 (8): 549–560.

6 Astumian, R.D. (2001). Making molecules into motors. *Sci. Am.* 285 (1): 56–64.

7 Browne, W.R. and Feringa, B.L. (2006). Making molecular machines work. *Nat. Nanotechnol.* 1 (1): 25–35.

8 Hernández, J.V., Kay, E.R., and Leigh, D.A. (2004). A reversible synthetic rotary molecular motor. *Science* 306 (5701): 1532–1537.

9 Kassem, S., van Leeuwen, T., Lubbe, A.S. et al. (2017). Artificial molecular motors. *Chem. Soc. Rev.* 46 (9): 2592–2621.

10 Kelly, T.R., De Silva, H., and Silva, R.A. (1999). Unidirectional rotary motion in a molecular system. *Nature* 401 (6749): 150–152.

11 Koumura, N., Zijlstra, R.W.J., van Delden, R.A. et al. (1999). Light-driven monodirectional molecular rotor. *Nature* 401 (6749): 152–155.

12 NobelPrize.org. Nobel Media AB (2021). Press Release: The Nobel Prize in Chemistry 2016. https://www.nobelprize.org/prizes/chemistry/2016/press-release/ (visited on 09 April 2021).

13 Augulis, R., Klok, M., Feringa, B.L., and van Loosdrecht, P.H.M. (2009). Light-driven rotary molecular motors: an ultrafast optical study. *Phys. Status Solidi C* 6 (1): 181–184.

14 Feringa, B.L., Jager, W.F., De Lange, B., and Meijer, E.W. (1991). Chiroptical molecular switch. *J. Am. Chem. Soc.* 113 (14): 5468–5470.

15 Jager, W.F., de Jong, J.C., de Lange, B. et al. (1995). A highly stereoselective optical switching process based on donor-acceptor substituted dissymmetric alkenes. *Angew. Chem. Int. Ed. Engl.* 34 (3): 348–350.

16 Huck, N.P.M., Jager, W.F., De Lange, B., and Feringa, B.L. (1996). Dynamic control and amplification of molecular chirality by circular polarized light. *Science* 273 (5282): 1686–1688.

17 Feringa, B.L., van Delden, R.A., and ter Wiel, M.K.J. (2003). In control of switching, motion, and organization. *Pure Appl. Chem.* 75 (5): 563–575.

18 Harada, N., Koumura, N., and Feringa, B.L. (1997). Chemistry of unique chiral olefins. 3. Synthesis and absolute stereochemistry of trans- and cis-1,1′,2,2′,3,3′,4,4′- octahydro-3,3′-dimethyl-4,4′-biphenanthrylidenes. *J. Am. Chem. Soc.* 119 (31): 7256–7264.

19 Koumura, N., Geertsema, E.M., Meetsma, A., and Feringa, B.L. (2000). Light-driven molecular rotor: unidirectional rotation controlled by a single stereogenic center. *J. Am. Chem. Soc.* 122 (48): 12005–12006.

20 Koumura, N., Geertsema, E.M., van Gelder, M.B. et al. (2002). Second generation light-driven molecular motors. Unidirectional rotation controlled by a single stereogenic center with near-perfect photoequilibria and acceleration of the speed of rotation by structural modification. *J. Am. Chem. Soc.* 124 (18): 5037–5051.

21 Vicario, J., Walko, M., Meetsma, A., and Feringa, B.L. (2006). Fine tuning of the rotary motion by structural modification in light-driven unidirectional molecular motors. *J. Am. Chem. Soc.* 128 (15): 5127–5135.

22 Pollard, M., Klok, M., Pijper, D., and Feringa, B.L. (2007). Rate acceleration of light-driven rotary molecular motors. *Adv. Funct. Mater.* 17 (5): 718–729.

23 García-López, V., Liu, D., and Tour, J.M. (2019). Light-activated organic molecular motors and their applications. *Chem. Rev.* 120 (1): 79–124.

24 Klok, M., Boyle, N., Pryce, M.T. et al. (2008). MHz unidirectional rotation of molecular rotary motors. *J. Am. Chem. Soc.* 130 (32): 10484–10485.

25 Kulago, A.A., Mes, E.M., Klok, M. et al. (2009). Ultrafast light-driven nanomotors based on an acridane stator. *J. Org. Chem.* 75 (3): 666–679.

26 Bauer, J., Hou, L., Kistemaker, J.C.M., and Feringa, B.L. (2014). Tuning the rotation rate of light-driven molecular motors. *J. Org. Chem.* 79 (10): 4446–4455.

27 Bléger, D. and Hecht, S. (2015). Visible-light-activated molecular switches. *Angew. Chem. Int. Ed.* 54 (39): 11338–11349.

28 Lerch, M.M., Hansen, M.J., van Dam, G.M. et al. (2016). Emerging targets in photopharmacology. *Angew. Chem. Int. Ed.* 55 (37): 10978–10999.

29 Russew, M.-M. and Hecht, S. (2010). Photoswitches: from molecules to materials. *Adv. Mater.* 22 (31): 3348–3360.

30 van Delden, R.A., Koumura, N., Schoevaars, A. et al. (2003). A donor-acceptor substituted molecular motor: unidirectional rotation driven by visible light. *Org. Biomol. Chem.* 1 (1): 33–35.

31 van Leeuwen, T., Pol, J., Roke, D. et al. (2017). Visible-light excitation of a molecular motor with an extended aromatic core. *Org. Lett.* 19 (6): 1402–1405.

32 Cnossen, A., Hou, L., Pollard, M.M. et al. (2012). Driving unidirectional molecular rotary motors with visible light by intra-and intermolecular energy transfer from palladium porphyrin. *J. Am. Chem. Soc.* 134 (42): 17613–17619.

33 Wezenberg, S.J., Chen, K.Y., and Feringa, B.L. (2015). Visible-light-driven photoisomerization and increased rotation speed of a molecular motor acting as a ligand in a ruthenium (II) complex. *Angew. Chem. Int. Ed.* 54 (39): 11457–11461.

34 Ruangsupapichat, N., Pollard, M.M., Harutyunyan, S.R., and Feringa, B.L. (2011). Reversing the direction in a light-driven rotary molecular motor. *Nat. Chem.* 3 (1): 53–60.

35 Gold, V. (ed.) (2019). *The IUPAC Compendium of Chemical Terminology*. International Union of Pure and Applied Chemistry (IUPAC).

36 Kistemaker, J.C.M., Štacko, P., Visser, J., and Feringa, B.L. (2015). Unidirectional rotary motion in achiral molecular motors. *Nat. Chem.* 7 (11): 890–896.

37 van Delden, R.A., Koumura, N., Harada, N., and Feringa, B.L. (2002). Unidirectional rotary motion in a liquid crystalline environment: Color tuning by a molecular motor. *Proc. Natl. Acad. Sci. U.S.A.* 99 (8): 4945–4949.

38 Eelkema, R., Pollard, M.M., Vicario, J. et al. (2006). Nanomotor rotates microscale objects. *Nature* 440 (7081): 163.

39 Orlova, T., Lancia, F., Loussert, C. et al. (2018). Revolving supramolecular chiral structures powered by light in nanomotor-doped liquid crystals. *Nat. Nanotechnol.* 13 (4): 304–308.

40 van Dijken, D.J., Chen, J., Stuart, M.C.A. et al. (2016). Amphiphilic molecular motors for responsive aggregation in water. *J. Am. Chem. Soc.* 138 (2): 660–669.

41 Yu, J.-J., Cao, Z.Q., Zhang, Q. et al. (2016). Photo-powered stretchable nano-containers based on well-defined vesicles formed by an overcrowded alkene switch. *Chem. Commun.* 52 (81): 12056–12059.

42 Chen, J., Leung, F.K.C., Stuart, M.C.A. et al. (2017). Artificial muscle-like function from hierarchical supramolecular assembly of photoresponsive molecular motors. *Nat. Chem.* 10 (2): 132–138.

43 Li, Q., Fuks, G., Moulin, E. et al. (2015). Macroscopic contraction of a gel induced by the integrated motion of light-driven molecular motors. *Nat. Nanotechnol.* 10 (2): 161–165.

44 Colard-Itté, J.R., Li, Q., Collin, D. et al. (2019). Mechanical behaviour of contractile gels based on light-driven molecular motors. *Nanoscale* 11 (12): 5197–5202.

45 Mariani, G., Colard-Itté, J.R., Moulin, E. et al. (2020). Structural properties of contractile gels based on light-driven molecular motors: a small-angle neutron and X-ray study. *Soft Matter* 16 (16): 4008–4023.

46 Foy, J.T., Li, Q., Goujon, A. et al. (2017). Dual-light control of nanomachines that integrate motor and modulator subunits. *Nat. Nanotechnol.* 12 (6): 540–545.

47 García-López, V., Chen, F., Nilewski, L.G. et al. (2017). Molecular machines open cell membranes. *Nature* 548 (7669): 567–572.

48 Liu, D., García-López, V., Gunasekera, R.S. et al. (2019). Near-infrared light activates molecular nanomachines to drill into and kill cells. *ACS Nano* 13 (6): 6813–6823.

49 Gunasekera, R.S., Galbadage, T., Ayala-Orozco, C. et al. (2020). Molecular nanomachines can destroy tissue or kill multicellular eukaryotes. *ACS Appl. Mater. Interfaces* 12 (12): 13657–13670.

50 Leigh, D.A., Wong, J.K.Y., Dehez, F., and Zerbetto, F. (2003). Unidirectional rotation in a mechanically interlocked molecular rotor. *Nature* 424 (6945): 174–179.

51 Wilson, M.R., Solà, J., Carlone, A. et al. (2016). An autonomous chemically fuelled small-molecule motor. *Nature* 534 (7606): 235–240.

52 Astumian, R.D. (2017). How molecular motors work – insights from the molecular machinist's toolbox: the Nobel prize in Chemistry 2016. *Chem. Sci.* 8 (2): 840–845.

53 Fletcher, S.P., Dumur, F., Pollard, M.M., and Feringa, B.L. (2005). A reversible, unidirectional molecular rotary motor driven by chemical energy. *Science* 310 (5745): 80–82.

54 Balzani, V., Credi, A., and Venturi, M. (2009). Light powered molecular machines. *Chem. Soc. Rev.* 38 (6): 1542.

55 Lehn, J.-M. (2006). Conjecture: imines as unidirectional photodriven molecular motors – motional and constitutional dynamic devices. *Chem. Eur. J.* 12 (23): 5910–5915.

56 Greb, L. and Lehn, J.M. (2014). Light-driven molecular motors: imines as four-step or two-step unidirectional rotors. *J. Am. Chem. Soc.* 136 (38): 13114–13117.

57 Greb, L., Eichhöfer, A., and Lehn, J.M. (2015). Synthetic molecular motors: thermal N inversion and directional photoinduced C–N bond rotation of camphorquinone imines. *Angew. Chem.* 127 (48): 14553–14556.

58 Guentner, M., Schildhauer, M., Thumser, S. et al. (2015). Sunlight-powered kHz rotation of a hemithioindigo-based molecular motor. *Nat. Commun.* 6 (1): 8406.

59 Wilcken, R., Schildhauer, M., Rott, F. et al. (2018). Complete mechanism of hemithioindigo motor rotation. *J. Am. Chem. Soc.* 140 (15): 5311–5318.

60 Gerwien, A., Mayer, P., and Dube, H. (2018). Photon-only molecular motor with reverse temperature-dependent efficiency. *J. Am. Chem. Soc.* 140 (48): 16442–16445.

61 Gerwien, A., Mayer, P., and Dube, H. (2019). Green light powered molecular state motor enabling eight-shaped unidirectional rotation. *Nat. Commun.* 10 (1): 4449.

62 Wang, J., Oruganti, B., and Durbeej, B. (2017). Light-driven rotary molecular motors without point chirality: a minimal design. *Phys. Chem. Chem. Phys.* 19 (10): 6952–6956.

63 Garcia-Iriepa, C., Marazzi, M., Zapata, F. et al. (2013). Chiral hydrogen bond environment providing unidirectional rotation in photoactive molecular motors. *J. Phys. Chem. Lett.* 4 (9): 1389–1396.

64 Marchand, G., Eng, J., Schapiro, I. et al. (2015). Directionality of double-bond photoisomerization dynamics induced by a single stereogenic center. *J. Phys. Chem. Lett.* 6 (4): 599–604.

65 Amatatsu, Y. (2018). Computational design of a light-driven imine-based motor with bulky chiral stator. *Chem. Phys.* 508: 51–60.

66 Filatov, M., Paolino, M., Min, S.K., and Kim, K.S. (2018). Fulgides as light-driven molecular rotary motors: computational design of a prototype compound. *J. Phys. Chem. Lett.* 9 (17): 4995–5001.

67 van Delden, R.A., ter Wiel, M.K.J., Pollard, M.M. et al. (2005). Unidirectional molecular motor on a gold surface. *Nature* 437 (7063): 1337–1340.

68 Pollard, M.M., ter Wiel, M.K.J., van Delden, R.A. et al. (2008). Light-driven rotary molecular motors on gold nanoparticles. *Chem. Eur. J.* 14 (36): 11610–11622.

69 Carroll, G.T., Pollard, M.M., van Delden, R., and Feringa, B.L. (2010). Controlled rotary motion of light-driven molecular motors assembled on a gold film. *Chem. Sci.* 1 (1): 97.

70 Ulman, A. (1996). Formation and structure of self-assembled monolayers. *Chem. Rev.* 96 (4): 1533–1554.

71 Evans, S.D., Johnson, S.R., Ringsdorf, H. et al. (1998). Photoswitching of azobenzene derivatives formed on planar and colloidal gold surfaces. *Langmuir* 14 (22): 6436–6440.

72 Dulkeith, E., Morteani, A.C., Niedereichholz, T. et al. (2002). Fluorescence quenching of dye molecules near gold nanoparticles: radiative and nonradiative effects. *Phys. Rev. Lett.* 89 (20): 203002.

73 Pollard, M.M., Lubomska, M., Rudolf, P., and Feringa, B.L. (2007). Controlled rotary motion in a monolayer of molecular motors. *Angew. Chem.* 119 (8): 1300–1302.

74 Carroll, G.T., London, G., Landaluce, T.F. et al. (2011). Adhesion of photon-driven molecular motors to surfaces via 1,3-dipolar cycloadditions: effect of interfacial interactions on molecular motion. *ACS Nano* 5 (1): 622–630.

75 London, G., Carroll, G.T., Landaluce, T.F. et al. (2009). Light-driven altitudinal molecular motors on surfaces. *Chem. Commun.* (13): 1712.

76 London, G., Chen, K.-Y., Carroll, G.T., and Feringa, B.L. (2013). Towards dynamic control of wettability by using functionalized altitudinal molecular motors on solid surfaces. *Chem. Eur. J.* 19 (32): 10690–10697.

77 Chen, K.-Y., Ivashenko, O., Carroll, G.T. et al. (2014). Control of surface wettability using tripodal light-activated molecular motors. *J. Am. Chem. Soc.* 136 (8): 3219–3224.

78 Schied, M. (2021). Molecular motors on surfaces studied by scanning tunnelling microscopy. Ph.D. dissertation. University of Graz. https://unipub.uni-graz.at/obvugrhs/download/pdf/6286427.

79 Schied, M., Prezzi, D., Liu, D. et al. (2022). Inverted conformation stability of a molecular motor on a metal surface. *J. Phys. Chem. C* 126 (21): 9034–9040.

80 Schied, M., Prezzi, D., Liu, D. et al. (2023). Chirality-specific unidirectional rotation of molecular motors on Cu(111). *ACS Nano* 17 (4): 3958–3965.

81 Kudernac, T., Ruangsupapichat, N., Parschau, M. et al. (2011). Electrically driven directional motion of a four-wheeled molecule on a metal surface. *Nature* 479 (7372): 208–211.

82 Eigler, D.M. and Schweizer, E.K. (1990). Positioning single atoms with a scanning tunnelling microscope. *Nature* 344 (6266): 524–526.

83 Bartels, L., Meyer, G., and Rieder, K.H. (1997). Basic steps of lateral manipulation of single atoms and diatomic clusters with a scanning tunneling microscope tip. *Phys. Rev. Lett.* 79 (4): 697–700.

84 Grill, L., Rieder, K.H., Moresco, F. et al. (2007). Rolling a single molecular wheel at the atomic scale. *Nat. Nanotechnol.* 2 (2): 95–98.

85 Stipe, B.C., Rezaei, M.A., and Ho, W. (1998). Coupling of vibrational excitation to the rotational motion of a single adsorbed molecule. *Phys. Rev. Lett.* 81 (6): 1263–1266.

86 Simpson, G.J., García-López, V., Daniel Boese, A. et al. (2019). How to control single-molecule rotation. *Nat. Commun.* 10 (1): 4631.

87 Joachim, C. and Rapenne, G. (2013). Molecule concept nanocars: chassis, wheels, and motors?. *ACS Nano* 7 (1): 11–14.

88 Rapenne, G. and Joachim, C. (2017). The first nanocar race. *Nat. Rev. Mater.* 2 (6): 17040.

89 Simpson, G.J., García-López, V., Petermeier, P. et al. (2017). How to build and race a fast nanocar. *Nat. Nanotechnol.* 12 (7): 604–606.

90 Editorial (2022). 24 hours of toulouse. *Nat. Nanotechnol.* 17 (5): 433–433.

91 Memo Project (2022). Nanocar Race II reveals two winning teams. https://www.memo-project.eu/flatCMS/index.php/News#17 (visited on 21 July 2023).

92 Kwon, K.-Y., Wong, K.L., Pawin, G. et al. (2005). Unidirectional adsorbate motion on a high-symmetry surface: "walking" molecules can stay the course. *Phys. Rev. Lett.* 95 (16): 166101.

93 Pawin, G., Wong, K.L., Kwon, K.Y. et al. (2008). Surface diffusive motion in a periodic and asymmetric potential. *J. Am. Chem. Soc.* 130 (46): 15244–15245.

94 Tolman, R.C. (1925). The principle of microscopic reversibility. *Proc. Natl. Acad. Sci. U.S.A.* 11 (7): 436–439.

95 Tierney, H.L., Murphy, C.J., and Sykes, E.C.H. (2011). Regular scanning tunneling microscope tips can be intrinsically chiral. *Phys. Rev. Lett.* 106 (1): 010801.

96 Tierney, H.L., Murphy, C.J., Jewell, A.D. et al. (2011). Experimental demonstration of a single-molecule electric motor. *Nat. Nanotechnol.* 6 (10): 625–629.

97 Tierney, H.L., Han, J.W., Jewell, A.D. et al. (2010). Chirality and rotation of asymmetric surface-bound thioethers. *J. Phys. Chem. C* 115 (4): 897–901.

98 Mishra, P., Hill, J.P., Vijayaraghavan, S. et al. (2015). Current-driven supramolecular motor with in situ surface chiral directionality switching. *Nano Lett.* 15 (7): 4793–4798.

99 Eisenhut, F., Meyer, J., Krüger, J. et al. (2018). Inducing the controlled rotation of single o-MeO-DMBI molecules anchored on Au(111). *Surf. Sci.* 678: 177–182.

100 Zhang, Y., Calupitan, J.P., Rojas, T. et al. (2019). A chiral molecular propeller designed for unidirectional rotations on a surface. *Nat. Commun.* 10 (1): 3742.

101 Vives, G. and Rapenne, G. (2008). Directed synthesis of symmetric and dissymmetric molecular motors built around a ruthenium cyclopentadienyl tris(indazolyl)borate complex. *Tetrahedron* 64 (50): 11462–11468.

102 Erbland, G., Gisbert, Y., Rapenne, G., and Kammerer, C. (2018). Expedient synthesis of thioether-functionalized hydrotris(indazolyl)borate as an anchoring platform for rotary molecular machines. *Eur. J. Org. Chem.* 2018 (34): 4731–4739.

103 Stipe, B.C., Rezaei, M.A., Ho, W. et al. (1997). Single-molecule dissociation by tunneling electrons. *Phys. Rev. Lett.* 78 (23): 4410–4413.

104 Hla, S.-W. (2008). Scanning tunneling microscope atom and molecule manipulations: realizing molecular switches and devices. *Jpn. J. Appl. Phys.* 47 (7): 6063–6069.

105 Ren, J., Freitag, M., Schwermann, C. et al. (2020). A unidirectional surface-anchored N-heterocyclic carbene rotor. *Nano Lett.* 20 (8): 5922–5928.

106 Stolz, S., Gröning, O., Prinz, J. et al. (2020). Molecular motor crossing the frontier of classical to quantum tunneling motion. *Proc. Natl. Acad. Sci. U.S.A.* 117 (26): 14838–14842.

107 Rosenthal, D., Widmer, R., Wagner, R. et al. (2012). Surface investigation of intermetallic PdGa(111). *Langmuir* 28 (17): 6848–6856.

5

MIMs in MOFs: Designing Mechanically Interlocked Molecules to Function Inside Metal–Organic Frameworks

Benjamin H. Wilson and Stephen J. Loeb

University of Windsor, Department of Chemistry and Biochemistry, 401 Sunset Avenue, Windsor, ON, N9B 3P4, Canada

5.1 Introduction

The two major classes of mechanically interlocked molecule (MIM) are the rotaxane, comprised of dumbbell-shaped rods and encircling rings and the catenane, which involves only interlocked rings [1–4]. Cartoon depictions of these structures, along with the possible rotational and translational motions associated with each, are shown in Figure 5.1. In this chapter, we describe research focused on integrating rotaxane and catenanes into metal–organic framework (MOF) materials as a methodology aimed at organizing and transferring the dynamics of MIMs into the solid state.

Advances in the design, synthesis and application of MOF materials have accelerated exponentially in the past two decades [5–9]. This is, in part, due to the modular nature of their building blocks, which allows a wide variety of components to be integrated into the materials and allows their use for a diverse range of applications such as gas storage and separations [10–16], energy applications [17–19], and drug delivery [20–23]. MOFs are composed of nodes – metal cations or aggregates of metal cations known as secondary building units (SBUs), and rods – bridging organic ligands known as linkers [24, 25]. It is through the organic linkers that MIMs can be incorporated. One of the key properties of MIMs is their ability to exhibit dynamics between two mechanically interlocked components, such as rotation of a ring about an axle in a [2]rotaxane [26–45]. In solution, where the predominant interactions with the MIM are from solvent molecules, the energy barrier for these dynamic processes is often very low, and they readily occur [27, 31, 34, 42, 44]. However, in a solid, close packing of the molecules is thermodynamically favorable, and this introduces significant steric barriers to motion [36, 39, 40, 45, 46]. Therefore, to achieve the large motions associated with MIMs, a solid-state material must contain significant void space, often termed free volume [38, 47, 48]. Since MOFs are three-dimensional (3D) crystalline frameworks that often have low packing density and significant porosity, they are excellent candidates for synthesizing solid-state materials incorporating MIMs.

Mechanically Interlocked Materials: Polymers, Nanomaterials, MOFs, and more, First Edition.
Edited by Emilio M. Pérez.

Figure 5.1 Cartoon depictions of mechanically interlocked molecules (MIMs). (a) A single-station [2]rotaxane; (b) a single-station [2]catenane; (c) a two-station [2]rotaxane known as a molecular shuttle. Basic motions possible – rotation and translation – are indicated by arrows and can occur in the direction shown or in the reverse.

5.2 Coordination Polymers Using [2]Pseudorotaxanes as Linkers

It is important to note that coordination polymers (CPs) are a broad class of polymeric compounds of any periodicity (one-, two-, or three) built using metal–ligand coordination bonds [49–51], while MOFs are a subset of CPs which are usually 3D and possess permanent and verifiable porosity [52, 53]. This distinction is important as it is the porosity of MOFs that facilitates the dynamic behavior of MIMs in the solid state. While MIM-based organic linkers can be readily utilized to synthesize CPs, the resultant material will most likely be densely packed, which precludes any dynamics of the MIM; these differences are illustrated in Figure 5.2.

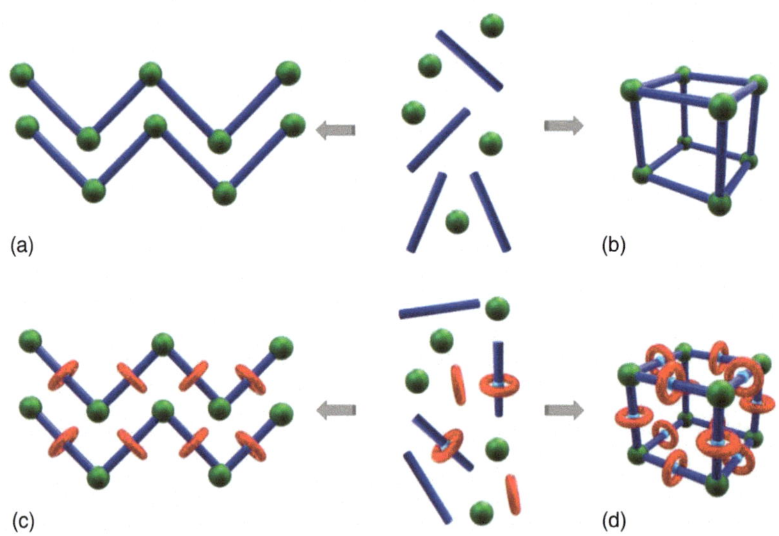

Figure 5.2 The judicious choice of linkers and nodes dictates whether the resulting material is a MOF with permanent porosity or a simple CP without pores. (a) Linkers and nodes produce a well-packed, 1D zigzag CP with no porosity; (b) the linker and nodes give rise to a 3D MOF with porosity; (c) MIM linkers and nodes form a 1D CP with dense packing that precludes any large amplitude motion of the MIM rings; (d) MIM linkers and nodes yield a 3D MOF with porosity which allows for dynamics such as molecular rotation to occur. Color code: metal nodes = green, organic linkers and axles = blue, rotaxane wheels = red, axle recognition sites = light blue.

The predominant approach to creating MIM-based linkers utilizes the axle of a [2]pseudorotaxane or [2]rotaxane to coordinate with the metal SBU [32, 36, 39–41, 47, 54–65]. This mimics, for example, the common use of divergent pyridine or carboxylate derivatives to form CPs [66]. There are, however, examples of MIM linkers in which a rotaxane [67–69] or catenane [64, 70] wheel contains the donor groups as well as daisy chains that function as linkers [71].

Originally, work in this field focused on the proof-of-principle incorporation of MIM components into CPs utilizing [2]pseudorotaxane-based linkers. Initially, Kim employed cucurbituril (CB[*n*]) macrocycles threaded on ammonium axles to synthesize a variety of CPs [72–77] and reported the first 3D polyrotaxane network [72]. An alternative approach by Loeb used crown ether molecules threaded on bis-pyridinium axles [54–59, 78–80], while Sessler employed an inverse approach (in terms of charge) with a cationic "Texas box" macrocycle and anionic dicarboxylate axles [81–84] and Yaghi and Stoddart used metal templated pseudorotaxane linkers [63]. In these early studies, the MIM linkers were [2]pseudorotaxanes since they do not possess bulky stopper groups. Upon coordination of metal ions to both ends of the [2]pseudorotaxane axle and formation of a CP, the linking metal cations become stopper groups, preventing the dethreading of the macrocycles. Examples of these linkers and CPs are depicted in Figures 5.3 and 5.4, respectively. Most significantly, none of these MIM-containing CPs exhibit large amplitude motions (rotation, pirouette, and translation) of their MIM components in the solid state. This work has been summarized in several reviews [56, 57, 59, 75] and a book chapter [85] and the reader is directed to those reports for more detailed descriptions of the syntheses, structures, and properties of these types of CP materials.

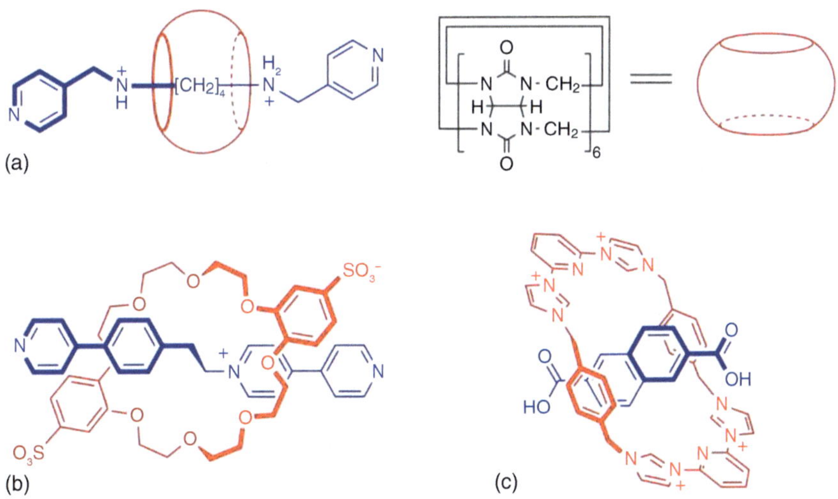

Figure 5.3 Some linker systems used for the preparation of early examples of CPs containing MIM linkers. (a) An example of a [2]pseudorotaxane combining a dipyridine axle and CB[6] wheel from Kim; (b) a [2]pseudorotaxane with a combination of pyridinium axle and 24-crown ether wheel reported by Loeb; (c) Sessler's "Texas box" macrocycle and dicarboxylate axle.

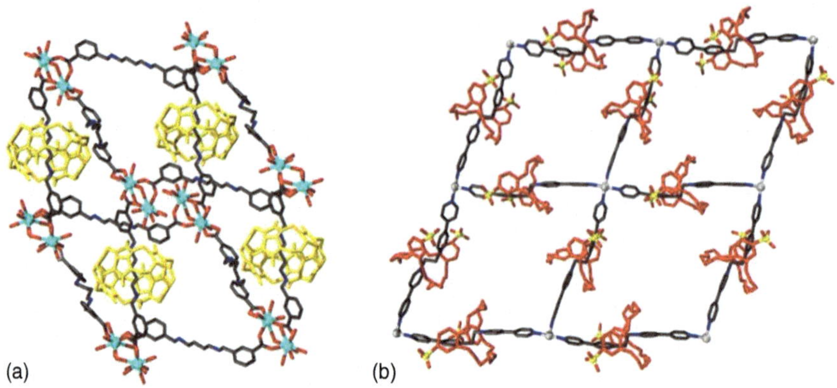

Figure 5.4 Two examples of three-periodic coordination polymers using [2]pseudorotaxanes as linkers. (a) From Kim, a bipyridyl axle and a CB[6] wheel form a [2]pseudorotaxane linker which links Tb(III) ions in a cubic-like arrangement; (b) from Loeb, a pyridinium axle and a crown ether wheel form a [2]pseudorotaxane which is linked by Zn(II) ions in a square net (shown) and pillared by anions (not shown). Source: Reproduced from Vukotic and Loeb [59] with permission from The Royal Society of Chemistry.

5.3 Robust Dynamics

The concept of utilizing the modular assembly of MOF structures for the integration of MIMs led Stoddart and Yaghi to coin the term "robust dynamics." This refers to the idea that for MIMs in MOFs "repeated dynamics of one entity does not affect the integrity of any others linked to it" [38]. MOF materials are ideal candidates for robust dynamics due to their rigid framework structure, which allows permanent porosity with sufficient space for large amplitude dynamic motion, such as the rotation and/or translation of a macrocycle. Ideally, a MOF capable of achieving robust dynamics would have a mobile component, free of encumbrance, attached to a framework that retains its open porous structure during repeated cycles of MIM motion. This pronouncement led to a focus on MIM linkers that mimicked those used to construct known MOFs and, most importantly, the use of permanently interlocked [2]rotaxane or [2]catenane linkers, rather than [2]pseudorotaxane linkers which relied on the axle and wheel components remaining threaded during the sometimes extreme conditions of MOF synthesis.

5.4 Techniques for Elucidating Dynamic Behavior in the Solid State

Determining the design blueprints and synthetic protocols required for creating robust MOFs with MIMs in their pores is paramount for progress in this area, but demonstrating that robust dynamics are actually occurring requires the unambiguous verification and characterization of molecular motion in the solid state. Dynamics in solution can be elucidated by a variety of methods, with the most common being variable temperature (VT) ^{1}H or ^{13}C NMR spectroscopy or exchange spectroscopy (EXSY). However, moving from the solution to the solid state requires a different set of techniques to, not only verify the stability of the material, but also to directly probe dynamic behavior.

Thermal gravimetric analysis (TGA) and gas sorption studies are commonly used to determine the stability and permanent porosity of MOF materials, while experiments such as single-crystal X-ray diffraction (SCXRD) and powder X-ray diffraction (PXRD) are the most common methods for determining the structures of crystalline MOFs. However, these techniques do not really provide insight into dynamic processes. Although VT SCXRD can be used to study MOFs, this approach only provides a snapshot of the molecular conformations adopted at a particular temperature [86–88]. Unfortunately, this does not provide direct information about molecular motion and is often plagued by simultaneous populations of overlapping structures (disorders) that are difficult to deconvolute. PXRD is most often utilized to probe changes in crystallinity, stability, and the occurrence of different phases with changes in temperature, but this does not directly probe molecular motion.

Solid-state nuclear magnetic resonance (SSNMR) has emerged as a highly useful technique for the study of MIM dynamics in crystalline MOF materials. ^{1}H NMR spectroscopy, commonly used to characterize organic molecules in solution, generally produces broad, poorly resolved peaks in the solid state, which are difficult to interpret. On the other hand, ^{2}H SSNMR and cross-polarization magic angle spinning (CP-MAS) ^{13}C SSNMR spectra can provide valuable insight into rotational and translation processes, respectively. The line shape and relaxation patterns of the ^{2}H nuclei make it an ideal probe to study molecular motions in the solid state and, from a practical standpoint, ^{2}H SSNMR is particularly useful as, in many cases, it is relatively straightforward to synthesize the MIM of interest with partial deuteration. ^{2}H SSNMR patterns exhibit characteristic line shapes at the slow-motion regime (SMR), intermediate-motion regime (IMR), and fast-motion regime (FMR) that vary with temperature. Simulating and fitting the dynamic line shapes to suitable models using either multiple independent simple motional modes [32, 36, 39, 60] or a microscopic-order–macroscopic-disorder [89, 90] approach allows for the characterization of different types of motions, including rates of rotation to be elucidated. In comparison, the spectra obtained from CP-MAS ^{13}C experiments are much more like those of their solution counterpart, although with significantly broader peaks. Much like in solution, VT experiments can be used to study the dynamics of a molecular system by collecting spectra when the motion is slower than the NMR timescale at low temperature and faster than the NMR timescale at high temperature. At low temperatures, two or more chemically distinct forms are often observed, while at high temperatures, the chemically averaged form is usually observed. The rates of exchange at different temperatures and the temperature at which the peaks corresponding to the chemically distinct forms coalesce to the chemically averaged form provide important information about the rates of motion and energy barriers associated with the observed dynamics. 2D NMR experiments such as EXSY can also be used to probe exchange rates in the solid state. Readers who are interested in the details of these SSNMR experiments are encouraged to access the supplementary materials associated with the primary literature describing these MIM in MOF materials. Examples of applications of these techniques are illustrated in the following two MIM in MOF case studies.

5.5 Rotational Motion of a MIM Wheel: UWDM-1, a Case Study

The first example of a MIM containing MOF with experimentally verifiable rotational dynamics of a large, interlocked macrocycle was **UWDM-1** (*U*niversity of *W*indsor *D*ynamic *M*aterial) [39]. This highly crystalline material contained a [2]rotaxane-based linker comprised of a 24-crown-6 (**24C6**) ether moiety bound to an aniline station – the remnants of an anilinium recognition site – on a tetracarboxylate axle; see Figure 5.5a. Coordination of the MIM linker to Cu^{II} paddlewheel nodes resulted in a crystalline MOF displaying the **fog** net of NbO topology; see Figure 5.5c. The structure can be described as the linking of Cu_{12} nanoballs to give narrow channels that contain water molecules.

PXRD analysis showed that subsequent activation removes the water molecules from the framework but does not affect crystallinity or change the phase of the material; see Figure 5.6. Importantly, after activation, the material contains sufficient free volume for the motion of the macrocycle to occur.

The use of a partially deuterated **24C6** macrocycle in the MIM component allowed the full range of motions exhibited in the solid state to be studied via VT ^{2}H SSNMR; see Figure 5.7. Three motional regimes are defined based on the relative magnitudes of the ^{2}H quadrupolar frequency (ν_Q) and the exchange rate (ν_{ex}): the SML, where rates are $<10^3$ Hz ($\nu_{ex} \ll \nu_Q$), the IMR, where rates are between 10^4 and 10^7 Hz ($\nu_{ex} \sim \nu_Q$), and the FML where rates are $>10^7$ Hz ($\nu_{ex} \gg \nu_Q$). The powder patterns corresponding to rates in the SML and FML can be modeled using motionally averaged

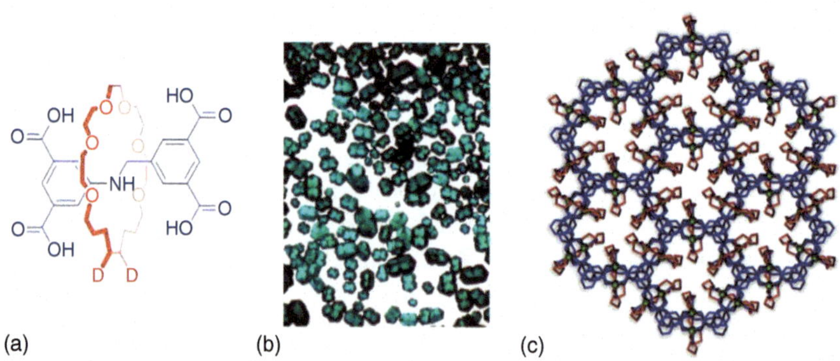

Figure 5.5 The case of **UWDM-1**. (a) The tetracarboxylic acid MIM linker combined with Cu^{II} ions used to make **UWDM-1**, (b) crystals of the as-synthesized material, (c) the structure of **UWDM-1** as determined by SCXRD showing the hexagonal channels of the **fog** net with **24C6**-d_2 wheels in the pores; Color code: MOF framework of nodes and MIM linker axles = blue, **24C6** macrocycles = red. Source: Reproduced from Vukotic et al. [39]/Springer Nature.

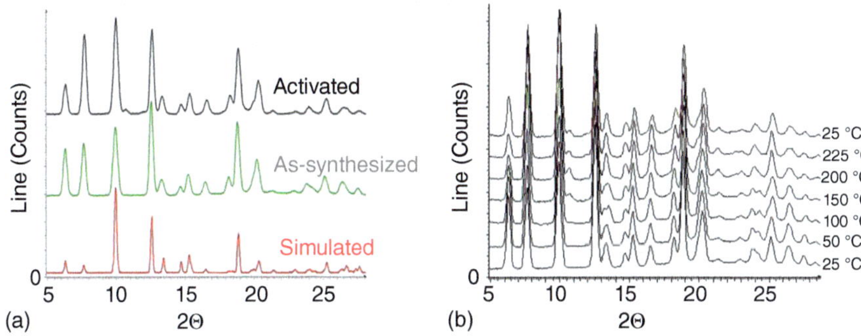

Figure 5.6 The case of **UWDM-1**. (a) PXRD (25 °C) patterns for **UWDM-1** demonstrate the robust nature of the crystalline material, (b) VT PXRD showing that the material is stable at the temperatures used to characterize its motion via SSNMR. Source: Vukotic et al. [39]/Springer Nature.

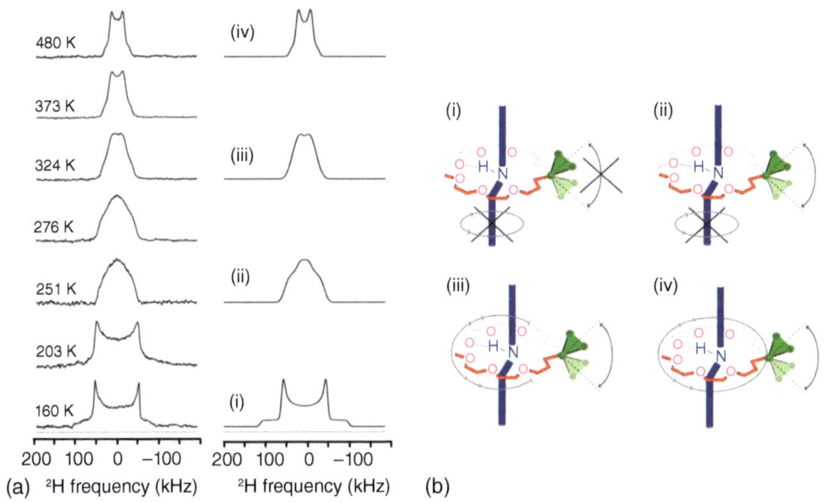

Figure 5.7 The case of **UWDM-1**. (a) VT ^{2}H SSNMR spectra, (b) the various motions of the macrocycle determined from simulations of the experimental spectra; (i) no motion, (ii) flexing of the aliphatic chain, (iii) rocking or partial rotation of the macrocycle around the axle, and (iv) full rotation of the macrocycle around the axle. Source: Vukotic et al. [39]/Springer Nature.

^{2}H electric field gradient (EFG) tensors and are invariant to changes in both echo spacing and temperature. Motions with rates in the IMR produce complex powder patterns that change with echo spacing and temperature. The low-temperature spectrum of **UWDM-1** acquired at 160 K was simulated as a single site with quadrupolar parameters: $C_Q = 160(5)$ kHz and $\eta_Q = 0.0(1)$; these parameters are typical for a deuteron bonded to a carbon atom in an alkyl group. The spectrum was simulated with rates in the SML, and therefore, the motion does not influence the appearance of the powder pattern. Increasing the temperature to 225 K produced no appreciable changes in the powder patterns. The spectrum acquired at 251 K was quite different from the lower temperature spectra and was simulated based on a two-site jump of the deuteron through an angle of 77°. The two-site jump mode of motion consists of the deuterons of the CD_2 groups jumping between two positions that are related by a rotation through an angle β about an axis that is perpendicular to the plane of the CD_2 moiety and produces powder patterns that are dependent on β. The spectrum acquired at 324 K could not be simulated using only the two-site jump model and, instead, was simulated using the partial rotation model, combined with two-site jumps with $\beta = 60°$. The partial rotation model consists of the macrocyclic ring rotating about an axis that is perpendicular to the ring in steps separated by the angle γ. The jumps are made only to adjacent sites that allow for the interaction of the hydrogen bond donor on the axle with the oxygen atoms on the ring. The high-temperature spectrum collected at 480 K is axially symmetric (i.e. $\eta_Q = 0$) and cannot be simulated with the partial rotation model, but instead was simulated as full rotation, combined with two-site jumps with $\beta = 60°$. The macrocyclic ring is no longer constrained to making jumps between adjacent oxygen atoms on the ring, but now has enough energy to jump through the alkyl portion in either direction. The full rotation was simulated with rates within the FML. While it was difficult to comment on the relative rates of the full rotation and two-site jump motional models, it was postulated that the latter is occurring at a higher rate due to its significantly lower onset temperature.

5.6 Translational Motion of a MIM Wheel: UWDM-4, a Case Study

Large amplitude translational motion is associated with [2]rotaxanes that contain two or more recognition sites (stations) at which a single macrocyclic wheel can reside. The ring can then move along the axle in a to-and-fro motion at a rate that depends on the relative interaction energies at each station. These types of MIMs are known as molecular shuttles [4]. Conceptually, translational dynamics are more challenging to integrate into MOFs than rotational dynamics due to the larger free volume required. In the case of rotation, the free volume required for motion is simply that occupied by the macrocycle – rotation can occur as long as there are no close contacts within the structure that block this change in position. However, in the case of a molecular shuttle, the free volume required must include the entire track that the macrocycle will traverse along the axle. The first example of a molecular shuttle imbedded in a MOF, **UWDM-4**, was reported by Loeb and Schurko in 2015 [40]. The MIM linker was comprised of an H-shaped tetracarboxylate axle encircled by a single 24-crown-8 (**24C8**) macrocycle. The resultant structure displayed an **hea** net, which contains sufficient free volume for the crown ether macrocycle to move back and forth between the two benzimidazole recognition sites, which are separated by ca. 7.5 Å. The linker, MOF crystals and MOF structure are shown in Figure 5.8.

Since the recognition sites are equivalent, there is no preference for ring binding to a particular station; the molecular shuttle is degenerate. The thermally induced

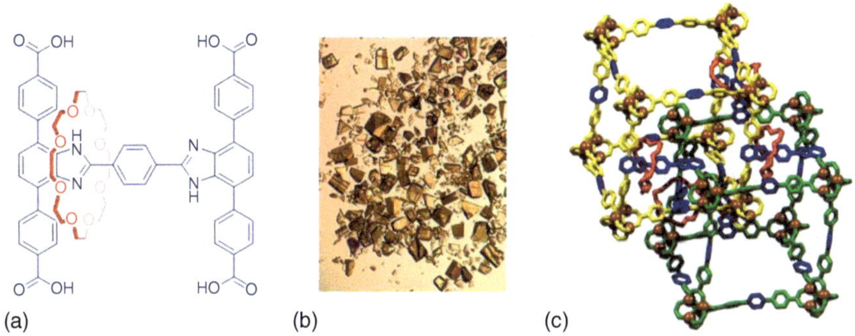

Figure 5.8 The case of **UWDM-4**. (a) The MIM linker was used to create **UWDM-4**, (b) crystals of the as-synthesized material, (c) the structure of **UWDM-4** with the **hea** net as determined by SCXRD. Color code: Zn^{II} ions = brown, axle = blue, macrocycle = red, two different interpenetrated frameworks = yellow and green. Source: Zhu et al. [40]/Springer Nature.

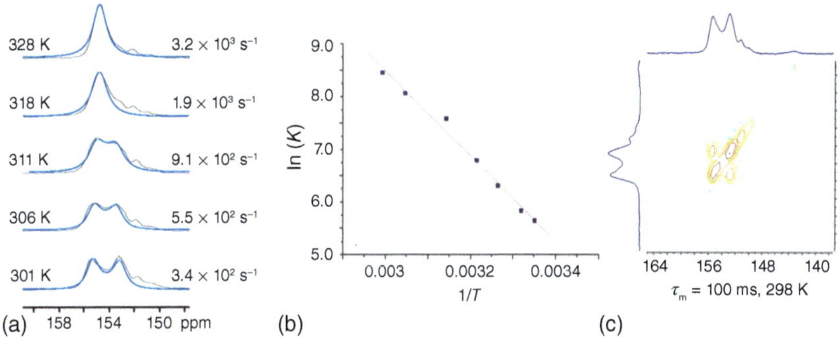

Figure 5.9 The case of **UWDM-4**. (a) Partial VT CP MAS ^{13}C SSNMR showing the ^{13}C enriched 2-position of the benzimidazole rings (black = experimental, blue = simulated), (b) Eyring plot used to calculate the rate of ($283\ s^{-1}$ at 298 K) and energy barrier ($14.1\ kcal\ mol^{-1}$) of shuttling, (c) a ^{13}C EXSY spectrum showing the ^{13}C enriched atoms undergoing slow exchange. Source: Adapted from Zhu et al. [40].

translation was studied using VT CP MAS ^{13}C SSNMR on a sample in which the carbon atoms at the 2-positions of the benzimidazole rings were 50% ^{13}C enriched. This provided sufficient signal strength to clearly identify the benzimidazole C-atom resonance and monitor line shapes as a function of temperature. This data provided the thermodynamic parameters associated with the translation motion of the ring, the salient one being the energy barrier for the translation, $\Delta G^{\ddagger}$; see Figure 5.9.

Since chemical exchange between the occupied and unoccupied recognition sites is slow on the NMR timescale at room temperature, an EXSY spectrum could also be obtained and used to calculate the rate of shuttling; see Figure 5.9. This method gave comparable rates and energy barrier data to those obtained from fitting the VT spectra.

Particularly interesting was the effect of confining the shuttling MIM motion in the pores of the MOF lattice. The energy barrier for translation increased from

7.7 kcal mol^{-1} in solution (for the ester derivative of the linker in toluene) to 14.1 kcal mol^{-1} inside the MOF (*m*-xylene), i.e. the rate of translation of the macrocycle in the MOF was much slower in the solid state than in solution [40]. Despite the porosity of the MOF framework ensuring sufficient free volume for the translation motion to occur, there is a multitude of other considerations when transferring MIM dynamics from the solution to the solid state. In the case of **UWDM-4**, the pores contained *m*-xylene solvent molecules. Since it has been established that, in general, for porous systems such as MOFs with small amounts of confined solvents, the effective viscosity of the solvent is significantly increased [41, 91], it appears that this may influence the molecular dynamics of a MIM, or at least molecular shuttles. Since the number of studies on the motion of MIMs in MOFs was very limited in the early days of this work, it was impossible to unambiguously answer these questions.

However, subsequent studies did shed some light on these observations, and we further address the effect of solvent on rotational and translational motion in Section 5.8 of this chapter.

5.7 MIM Linker Design Strategies

Although **UWDM-1** and **UWDM-4** were built using previously successful MOF building blocks such as a 4-connected linker with isophthalic acid groups, a 4-connected Cu^{II} paddlewheel SBU and a 6-connected basic zinc oxide SBU, serendipity still played a significant role in yielding materials that permitted dynamics of the respective MIM component. Moving forward, it was abundantly clear that a well-thought-out design strategy would be important for the successful synthesis of a MOF with a functioning MIM. Indeed, there are many ways the axle and wheel components of a MIM and the struts and coordinating groups of standard MOF linkers can be arranged to fashion new MIM linkers suitable for synthesizing robust MOFs with MIMs that can potentially display robust dynamics; see Figure 5.10.

One reliable strategy is to select an existing porous MOF with a linker of suitable length which can be modified to contain the MIM component but retain its coordinating groups – a version of reticular synthesis – the process of assembling judiciously designed rigid molecular building blocks into predetermined ordered structures [8, 92]. A simple example of this is linear divergent dipyridyl ligands, which are often used to pillar the paddlewheel SBU's 2D metal-carboxylate layers [93–100]. By using a dipyridyl ligand as an axle and clipping a **24C6** macrocycle around it via Grubb's metathesis, Loeb and Schurko constructed several MOFs with a **pcu** net and varying degrees of interpenetration depending on the length of the dicarboxylate linker, e.g. **UWDM-2** (biphenyldicarboxylate, threefold) [36], **UWDM-3** (diazodicarboxylate, twofold) [36], **UWDM-7** (porphyrintetracarboxylate, twofold); [33] see Figure 5.11a, b for examples of the linker and a pillared MOF respectively. A combination of VT PXRD and VT ^{2}H SSNMR was used to probe the dynamics of the **24C6** macrocycle, and it was shown that all these pillared systems underwent a phase change upon activation [33, 36].

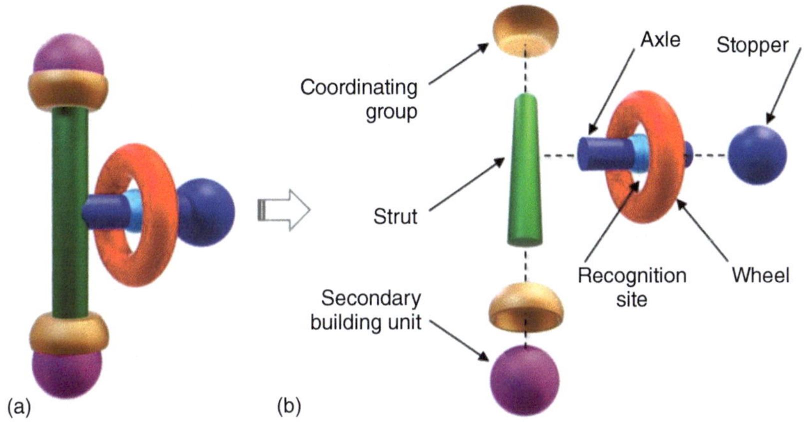

Figure 5.10 The integral parts of [2]rotaxane linker. (a) A cartoon of a T-shaped [2]rotaxane linker coordinated to two SBUs, (b) the exploded version of a T-shaped [2]rotaxane linker with each part labeled. Although sometimes the axle and stopper may be combined covalently to give a semi-rotaxane, or the strut and axle might be the same unit or perhaps a stopper might be replaced by another strut, these labels can be generalized for all of the MIM linkers discussed in this chapter. Color code: pink = SBU, brown = coordinating groups, green = strut, blue = axle, light blue = recognition site, red = macrocyclic wheel.

In a similar vein, Stoddart, Sauvage, and Yaghi utilized a linear Cu^{I} templated [2]rotaxane to mimic the dicarboxylates used as linkers to create the iconic isoreticular metal-organic framework (IRMOF) series of which **MOF-5** is a member [63], see Figure 5.11d. In particular, a 6-connected Zn_4O cluster and terphenyl dicarboxylate linker are known to give **IRMOF-16** with a **pcu** network [101]. Unfortunately, for **MOF-1040** twofold interpenetration of the frameworks resulted in a structure more similar to **IRMOF-15**, and the bulk of the macrocycle combined with the interpretation precluded observation of MIM dynamics.

A different approach to MIM linker design is to append the MIM component onto the side of the MOF linker strut so that, unlike previous designs, the strut of the linker and the axle of the rotaxane are different entities. Once again, this should be a reliable process, as known MOF linkers can be adapted via reticular chemistry. Of course, this approach still requires the original MOF scaffold to possess sufficient porosity to accommodate the appended MIM component.

A systematic study by Loeb and Schurko investigated the use of tetracarboxylate linkers akin to those used to construct **NOTT-101** [102] (University of *Nott*ingham) and **MOF-1** (Mukherjee) [103], but with appended rotaxane moieties capable of rotational dynamics. [45] For the smaller linker shown in Figure 5.12c, while frameworks **UWCM-11** and **UWCM-12** (*U*niversity of *W*indsor *C*rystalline *M*aterial) containing an appended axle could be synthesized and characterized by X-ray diffraction, the addition of a crown ether macrocycle to form the rotaxane resulted in the linkers being too bulky to form the targeted MOF. However, upon changing to the extended version of this linker design shown in Figure 5.12d, the resultant MOF had sufficient free space to accommodate both the axle and macrocycle of the

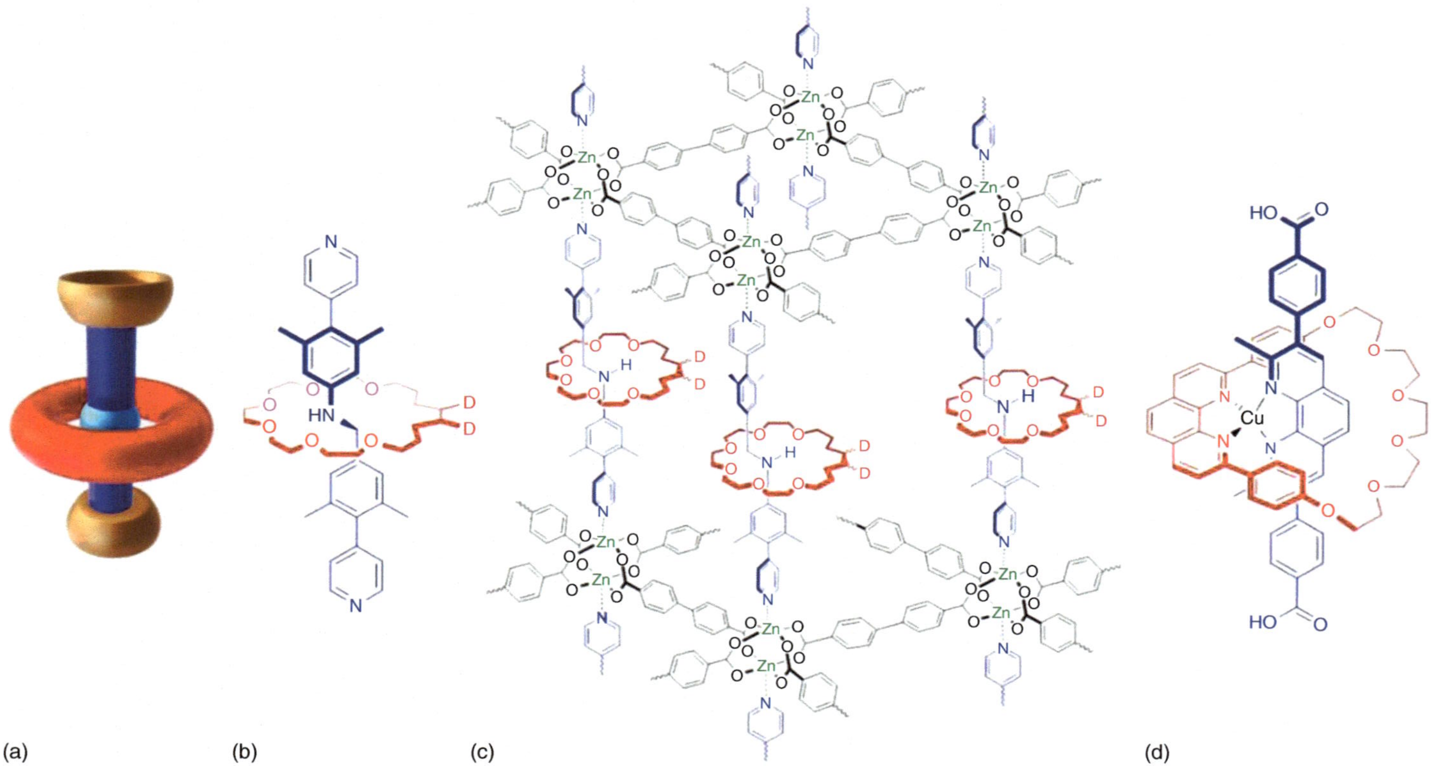

(a) (b) (c) (d)

Figure 5.11 Pillared MIMs in MOFs. (a) A cartoon depiction of a linear [2]rotaxane linker. The axle of the MIM (blue) about which the macrocyclic ring (red) can rotate is built on the strut of the linker (also blue) with terminal ligating groups (brown), (b) the dipyridyl linker used as a pillar in **UWDM-2**, **-3**, and **-7**, (c) the structure of **UWDM-3** showing the pillared layers; note, this material is threefold interpenetrated, (d) the Cu^{I} [2]rotaxane used to make **MOF-1040** with a **pcu** net.

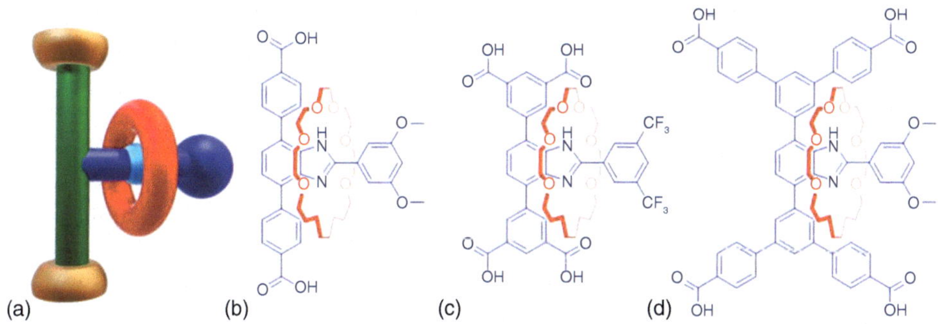

Figure 5.12 Reticular chemistry is used to design different T-shaped MIM linkers capable of rotation. (a) A cartoon depiction of a T-shaped [2]rotaxane. The strut (green) between the ligating groups (brown) is separate and perpendicular to the axle (blue) of the rotaxane about which the ring (red) can rotate, (b) a linear dicarboxylate strut based on **IRMOF-16**, (c) a strut with isophthalate groups based on **NOTT-101**, (d) a strut with extended isophthalate groups based on **MOF-1*** (*Mukherjee, Ref. [104]) used to make **UWDM-13**. Source: Adapted from Krause et al. [104].

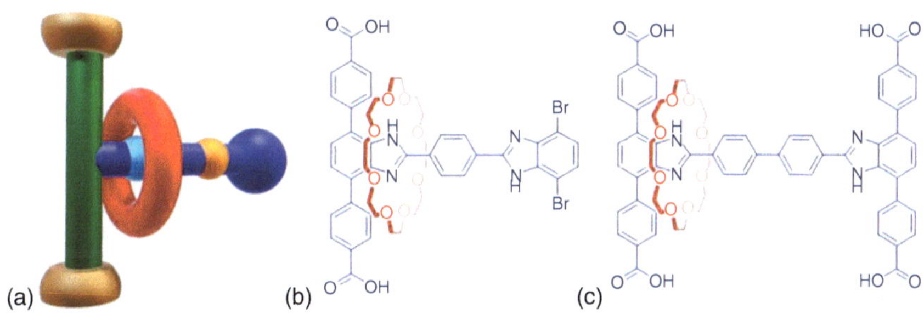

Figure 5.13 Reticular chemistry is used to design different T- or H-shaped MIM linkers capable of translational shuttling motion. (a) A cartoon depiction of a T-shaped [2]rotaxane with two recognition sites accessible to the macrocyclic ring. The strut (green) between the ligating groups (brown) is separate and perpendicular to the axle (blue) of the rotaxane along which the ring (red) can move back and forth between the recognition sites, (b) a T-shaped dicarboxylate strut with a perpendicular molecular shuttle, (c) an H-shaped tetracarboxylate linker with a molecular shuttle as the crossbar between two struts. This is similar to the original H-shaped linker used to make **UWDM-4**.

appended rotaxane and rotation of the macrocycle pores was observed by VT ^{2}H SSNMR for **UWDM-13** [45].

The side group appendage strategy has also successfully been used to incorporate molecular shuttles into the pores of a MOF; see Figure 5.13a. As a [2]rotaxane molecular shuttle requires two spatially separated recognition sites for the macrocycle, the steric bulk and thus free volume required is typically much larger than for a [2]rotaxane displaying molecular rotation. This approach has been applied to an analog of **PCN-57** (*P*orous *C*oordination *N*etwork), a zirconium-based MOF of the **UiO-66** (*Uni*versity of *O*slo) family with a **fcu** net that contains both tetrahedral and octahedral cavities. The large octahedral cavities provide sufficient room for a T-shaped linker based on a linear dicarboxylate with an appended [2]rotaxane molecular shuttle; see Figure 5.13b [41]. In this example, the MOF is multivariate – containing two types of linkers of the same length. This is to ensure that only one molecular shuttle resides in each octahedral cavity while the remaining edges of the cavity are composed of a much simpler terphenyl dicarboxylate linker.

MOF cavities, or entire lattices, can also be cross-linked by using a linker that resembles two individual struts joined by a crossbar; see Figure 5.13c. An example is the joining of two linear terphenyl dicarboxylates to form an H-shaped tetracarboxylate. This strategy was employed for **UWDM-4** to embed a molecular shuttle in a MOF using an H-shaped linker that cross-linked two interpenetrated lattices [40]. **IRMOF-15**, first reported by Yaghi in 2002, is composed of linear terphenyl dicarboxylate linkers bridging Zn_4O nodes to give a framework made up of two interpenetrated **pcu** lattices [101]. The H-shaped linker utilized in **UWDM-4** effectively cross-linked these two interpenetrated lattices such that the translational motion of the molecular shuttle occurs along the phenyl ring of the crosslinking moiety. Changing the length of the crossbar in this H-shaped linker can allow it to be used to crosslink cavities in different MOFs. For example, when a biphenyl crossbar was used, as shown in Figure 5.13c, the axle was of suitable length to cross-link the tetrahedral cavities of the zirconium-based MOF **UiO-68** to give **UWDM-8**, **-9,** and **-10**, with different primary terphenyl dicarboxylate linkers [46, 62].

Each of the three strategies detailed above, strut, T-shaped with appended MIM and H-shaped cross-linking, has its advantages and disadvantages. The strut approach includes most of the bulk of the MIM component onto the MOF linker strut, so there is less steric bulk directed into the pore of the MOF. This seems to be counterintuitive when trying to utilize the void space in MOF cavities, but it still has limited application for axles with a single recognition site and rotational dynamics. For a molecular shuttle with multiple recognition sites, the length of the axle required for the translational motion makes the strut approach less feasible. It is likely to result in lattice interpenetration, which would limit the freedom of motion of any interlocked macrocycle. The strategy of appending the MIM so that it is perpendicular to the strut and potentially oriented into the cavity of the MOF is a good one as long as the cavities are of sufficient size to accommodate the length of the axle. This has worked well for both simple rotaxanes and molecular shuttles and has the advantage that the length of the rotaxane axle does not affect the length of the MOF strut. This approach also allows for varying the structure of the MIM without changing the topology of the MOF, something that should allow for future studies in which the features of the MIM can be fine-tuned within the same framework. While the crosslinking strategy has proven to be a very effective method for imbedding molecular shuttles in MOFs, it is not as easily generalized as the length of the crossbar must exactly span the width of the framework into which it is being inserted. The advantages to this design, when it can be utilized, are that the MIM is reliably centered in the cavity, and the crossbar structure adds to the stability of the MOF framework.

There are also examples of MOFs built using linkers with coordinating groups attached to the macrocycle such that the axle does not participate in coordination with the metal SBU, but the macrocycle does [64, 67, 68, 70]. While the macrocycles

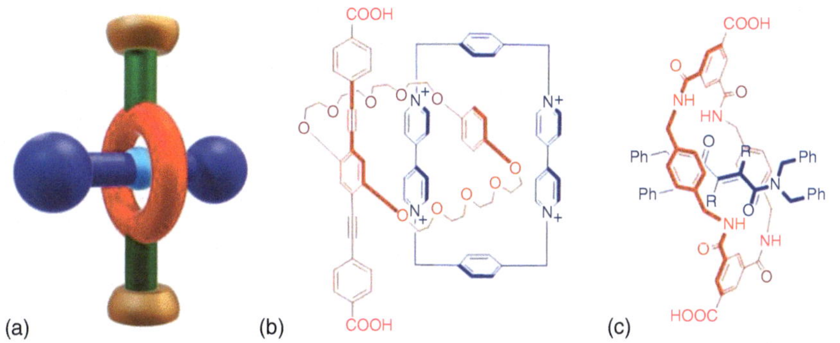

Figure 5.14 Coordinating groups are attached to the macrocyclic wheel rather than the axle or strut. (a) A cartoon depiction of a [2]rotaxane linker with the struts (green) and coordinating groups (brown) attached to the macrocycle (red) with the interlocked dumbbell (blue) free to undergo a pirouetting motion, (b) a [2]catenane linker with coordinating groups extended in a rigid fashion from one of the macrocycles, (c) a [2]rotaxane with coordinating carboxylate groups on the crown ether macrocycle.

used to make MIMs are typically very flexible and thus not ideal for constructing MOF frameworks, the threading of two macrocycles to form a catenane or the threading of an axle through the wheel usually enhances the overall rigidity of the system. This approach of including coordinating moieties on the macrocycle has been applied to both [2]catenanes and [2]rotaxanes. Yaghi and Stoddart used this approach to append a [2]catenane into the interlayer region of a 2D coordination framework **MOF-1011**, using the linear dicarboxylate shown in Figure 5.14b and Cu_2 paddlewheel nodes [64]. They demonstrated that the donor–acceptor [2]catenane units could be integrated into a MOF, leading to a high density of molecular machinery modules positioned precisely in well-defined layered structures. Lewis has reported [2]catenanes for which each ring contains a coordinating moiety [105]. Coordination to Ag^{I} resulted in the formation of a CP in which the mechanical bond between the macrocycles is also integral to the structure. Loeb reported the first CP with a rotaxane linker in which the coordinating groups were attached to the macrocyclic wheel [67], and recently, Berna utilized the [2]rotaxane linker depicted in Figure 5.14c with coordinating groups on either side of the macrocycle to create a robust MOF with a **sql** net [68].

Alternately, a creative strategy for inserting a MIM into a MOF that does not require any of the above strut/linker designs was devised by Stoddart and Farha. They demonstrated that a suitably derivatized MIM, such as that shown in Figure 5.15a, could simply be exchanged for a more labile ligand in the highly porous MOF **NU-1000,** (*N*orthwestern *U*niversity) making it part of the interior pore scaffold [69, 70]. The Zr_6 clusters of **NU-1000** feature coordination sites, which are occupied by monocarboxylate modulators from the MOF synthesis. These modulators are not part of the structural framework of the lattice and can be readily displaced by the addition of a new carboxylate ligand. Therefore, [2]rotaxanes [69] or [2]catenanes [70] featuring a carboxylate moiety could be post synthetically tethered into the large pores of the MOF using this exchange process.

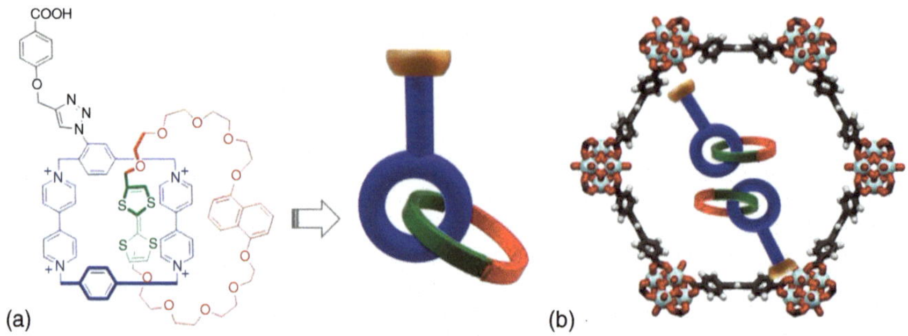

Figure 5.15 Exchanging the MIM linker for a labile group in the large cavities of **NU-1000**. (a) A [2]catenane linker with an electron-poor pyridinium macrocycle (blue) tethered with a coordinating carboxylate group (black) and an interlocked macrocycle containing two recognition sites, an electron-rich (red) naphthyl group and a redox-active tetrathiafulvalene site (green), (b) a cartoon version of the redox-active [2]catenane linker, (c) a structure showing the [2]catenane linkers coordinated to Zr-atoms of the SBU and located inside of the pores of **NU-1000**.

The advantages of this approach are that this ligand exchange is facile for **NU-1000**, and inclusion of the MIM does not affect any lattice structural components; see Figure 5.15b. The disadvantages are that this approach is only adaptable to a very small number of MOFs, and undesirable defects can occur as inclusion of the MIM is often limited by its bulk and diffusion of the linker into the pores.

5.8 Controlling Dynamics and Switching of MIMs in MOFs

When a MIM is imbedded in a MOF, there are two ways that the dynamics can be altered. Either the MIM itself can be altered, or the microenvironment surrounding it can be changed, both of which will influence any dynamic processes. Simply changing the ring size of the macrocycle in the **UWDM-1** linker, as shown in Figure 5.16a, allowed a degree of control of the rate of rotation by strengthening or weakening the hydrogen bonding interactions between the aniline recognition site and the crown ether macrocycle [39]. Amide macrocycles binding a furamide axle, as shown in Figure 5.14c, have been incorporated into MOF linkers by Berna. In the solid-state, ^{2}H SSNMR revealed a pirouette motion of the axle occurs inside the macrocycle [68], however, it is noticeably slower than the rate of rotation observed for the crown ether macrocycles in MIM containing MOFs reported by Loeb and Schurko [39]. This was rationalized as being due to the stronger hydrogen bonds between the macrocycle and axle. Again, stronger interactions between the MIM components provide a larger barrier to their relative motion and this results in the observation of a slower rotation.

The effect of confining solvent molecules in the pores of the MOF can have a marked effect on the dynamics of the MIM components by directly hindering the motion of a MIM component. This effect is more significant for certain types of

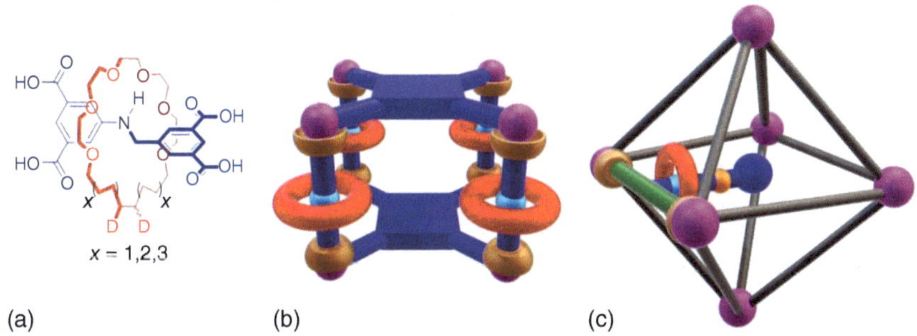

Figure 5.16 Ring size, phase, and solvent effects on dynamics. (a) The linker used to make **UWDM-1** with three different ring sizes, (b) cartoon depiction of a porphyrinic MOF with MIM pillars, (c) a cartoon of a molecular shuttle directed into the center of an octahedral cavity of **PCN-57**.

motion, such as translation in molecular shuttles. For example, Loeb and Schurko found that the nature of the solvent had a minimal effect on the pillared porphyrinic MOF system, with a [2]rotaxane undergoing rotation shown in Figure 5.16c [33]. However, in **UWDM-11**, in which a nondegenerate molecular shuttle is directed via a T-shaped appendage into the octahedral cavity of a **fcu** framework, as depicted in Figure 5.16c, the addition of mesitylene solvent molecules resulted in an increased energy barrier for translational motion when compared to the solvent-free activated form [41]. There is ample evidence [93] that the increased viscosity associated with confining the solvent molecules in the pores is the cause for this increased energy barrier. However, the effect of solvation is much less significant in MOF systems with [2]rotaxanes exhibiting molecular rotation since the free volume occupied by the macrocycle does not change. A rotational motion occupies a constant volume, so the effect of increased viscosity is minimized. Whereas the large amplitude translation of a molecular shuttle requires a great deal more rearrangement of solvent molecules and is, thus, quite dependent on the solvent viscosity.

MOFs are 3D crystalline networks, and therefore, the relative orientation of different components can be controlled with high precision. In the context of MIMs, this provides an avenue for the interaction of MIMs with other components in the MOF. This is particularly important for the advancement of molecular machines, as machines are made up of many interacting components. An early and simple example of this was the phase dependence of the rotational barrier observed for **UWDM-2** and **UWDM-3** [36]. Both materials showed more mobility of the macrocycle after activation and a concomitant phase change. Since the phase change experienced upon activation involved the removal of the solvent and could be reversed by re-introducing the original solvent, this provided a way to turn on and then off the mobility of the ring [36]. The difference in mobility for the two different phases is presumably the result of differing interactions between the macrocycle and the framework in the two different phases.

A more sophisticated example of this was observed for the series of MOFs **UWDM-8**, **-9**, and **-10** [46, 62]. These MOFs are zirconium-based frameworks with

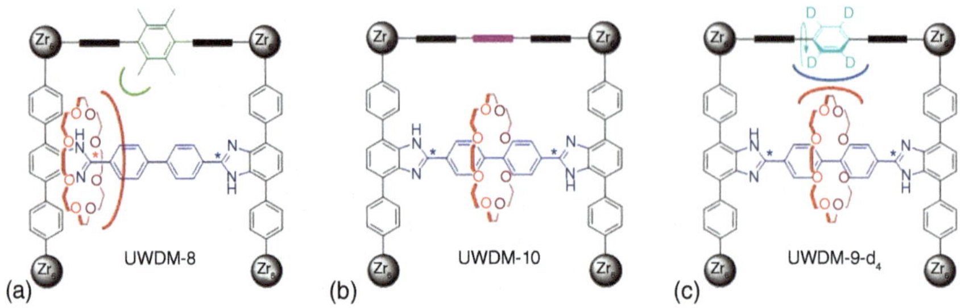

Figure 5.17 The interaction of two different mobile components within a MOF. (a) In **UWDM-8**, the methyl groups of the primary strut block the macrocycle from moving; (b) for **UWDM-10**, the tetrazine linker allows the macrocycle to reside in the center of the cavity since there are no steric constraints; (c) in **UWDM-9**-d_4, the central (deuterated) ring of the strut and the macrocycle interact so the motion of one component affects the dynamics of the other. Source: Reproduced from Wilson et al. [46] with permission from Elsevier.

fcu topology featuring tetrahedral and octahedral cavities, the former of which is cross-linked by a [2]rotaxane molecular shuttle linker as exhibited in Figure 5.17. The remaining four edges of the tetrahedron are made up of a terphenyl dicarboxylate derivative co-linker with either a 2,3,5,6-tetramethylphenyl (**UWDM8**), phenyl (**UWDM-9**), or tetrazine (**UWDM-10**) central spacer. In the case of **UWDM-8**, the steric bulk of the methyl groups on the central spacer of the co-linker locks the macrocycle of the molecular shuttle at one end of the axle. This prevents any translational motion from occurring at any temperature, and the interaction of the co-linker with the macrocycle acts as a molecular brake; see Figure 5.17a. On the other hand, for **UWDM-9** and **UWDM-10**, this steric bulk is not present, and the macrocycle occupies the mid-point of the axle. This position is sterically favorable as it displaces the ring away from the sterically constricted corners of the tetrahedron. For **UWDM-10**, the tetrazine spacer is oriented away from the macrocycle, but for **UWDM-9**, the central phenyl spacer of the primary linker is able to undergo its own molecular motion and rotate about the axis of the linker; see Figure 5.17b,c respectively. This brings the phenyl rotor into close proximity with the macrocycle of the rotaxane, which influences the rotational dynamics of the co-linker. In order to study the rotational dynamics of the system, ^{2}H SSNMR measurements were employed by making an analog of the framework in which the central phenyl ring of the co-linker was deuterated (d_4). Comparison of the VT ^{2}H SSNMR spectra with that of the framework without the cross-linking molecular shuttle showed some significant differences. These differences were attributed to an additional energy barrier in the rotational dynamics of the deuterated phenyl ring when the crosslinking molecular shuttle was present in the MOF, i.e. two different mobile components in a MOF were influencing each other's motions.

In the preceding examples, the dynamics described for MIM systems in MOFs have been continuous thermally induced molecular motions. This occurs when there are two or more conformations that are relatively close in energy with a lower barrier for interconversion. The alternative – significantly different energies and/or a high energy barrier for interconversion – forms the basis of MIM molecular switches. For a MIM system that has one conformation that is significantly favored over another, an external stimulus can invert the thermodynamic conformational preference, resulting in the MIM behaving as a molecular switch. Common external stimuli for MIM molecular switches include light or a change in redox potential or pH. While there are a large number of MIM molecular switches that work in solution, there are only a few that operate while incorporated into MOFs. Stoddart and Farha used the post-synthetic tethering approach, described above, to mount electrochemically active rotaxanes, shown in Figure 5.18a, inside the large pores of **NU-1000**. Ultraviolet–visible–near-infrared (UV–Vis–NIR) spectra shown in Figure 5.18c, as well as electron paramagnetic resonance (EPR) and ^{1}H NMR spectroscopies, all confirmed the capture of redox-active rotaxanes within the mesoscale hexagonal channels of the MOF [69]. Cyclic voltammetry measurements performed on electroactive thin films of the resulting material indicated that redox-active viologen subunits located on the rotaxane components could be accessed electrochemically in the solid state [69]. As the system contained only a single recognition site and the length of the linker was limited in order for it to fit inside the cavity, changing the redox state of the rotaxane did not result in an appreciable change in the ring position. On the other hand, the [2]catenane, shown in Figure 5.14c, displayed a reversible change in ring conformation mediated by the change in the redox state when similarly mounted inside **NU-1000** [70]. The [2]catenane is composed of two rings, one of which is a tetracationic paraquat-like macrocycle that is anchored to the zirconium node, while the other is a polyether macrocycle with naphthyl and tetrathiafulvalene moieties. In the reduced form of the catenane, the free macrocycle adopts a conformation such that the tetrathiafulvalene moiety is bound in the recognition pocket of the other macrocycle. Following a two-electron oxidation of the tetrathiafulvalene to the dication, charge repulsion results in the ring rotating such that the tetrathiafulvalene dication is free from the positively charged binding pocket, which is now occupied by the electron-rich naphthyl ring. Thus demonstrating the operation of a redox-driven molecular switch embedded in a MOF material – true robust dynamics.

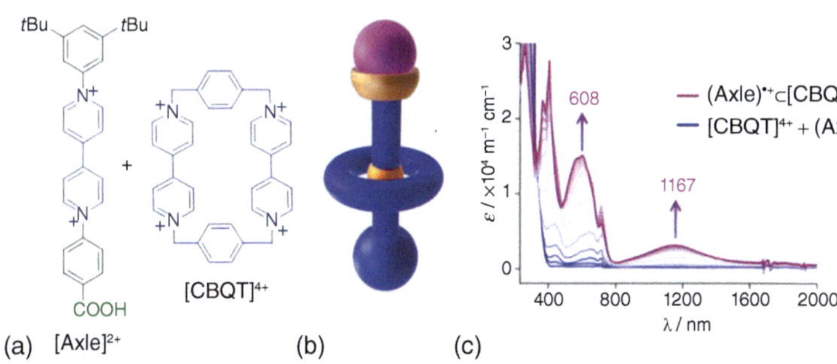

Figure 5.18 Redox addressable MIMs in MOFs. (a) The charged macrocycle $[\textbf{CBQT}]^{4+}$ and axle used to make a semi-rotaxane linker by forming a tris radical, (b) a cartoon version of the intact [2]rotaxane made from the two components in (a), (c) the UV–Vis–NIR spectra showing that the redox activity of the MIM linker can be addressed inside the MOF. Source: Adapted from McGonigal et al. [69].

Berna utilized the [2]rotaxane shown in Figure 5.14c as a MIM linker for which the coordinating sites are on either side of the macrocycle. The advantage of this design is that, as the axle of the rotaxane is not rigidly held in place via coordinate bonds from the framework, it is relatively free to move and undergoes molecular rotations as verified by VT ^{2}H SSNMR. Irradiation of the MOF with 312 nm light results in partial *E* to *Z* (and reverse) bond isomerization of the axle, highlighting how photoactive MIM molecular switches can be incorporated into a MOF [68]. The *Z* isomer exhibits a slightly faster rotational motion of the axle compared to the *E* isomer. Therefore, light irradiation provides an avenue for controlling the rate of the rotational motion by an external stimulus, even in the solid state.

5.9 MIMs to Construct Poly-Threaded MOF Lattices

Although most MIM linkers in MOFs have featured the coordinating groups on either the axle or the macrocyclic wheel of a [2]rotaxane, there is no reason not to use linkers with coordinating sites on both of the MIM components. The incentive for including coordinating groups on both the axle and wheel is that the mechanical bond between them can then form an integral part of the MOF structural makeup. This means that the dynamics of the MIM linker can potentially be harnessed to make a flexible MOF framework. Flexible MOFs are known to have the capability to exhibit novel and useful gas sorption behaviors. For example, negative gas adsorption [104, 106, 107] – a decrease in adsorption of gas upon an increase of pressure – could be used to make a pressure-amplifying material and gated gas sorption [108–112] – a sudden increase in sorption capacity resulting from a structural transition triggered by the entry of gas – could be advantageous for sensing of gas uptake.

The first example of a MIM with coordinating groups on both the axle and wheel of a rotaxane was reported by Loeb and involved a [2]rotaxane with six thioether donors, two on the stoppers of the axle and four attached to a crown ether macrocycle [58]. Coordination of this novel linker to Ag^{I} yielded a CP with a complex interwoven structure combining the mechanical bond of the MIM linker and interpretation of the resulting lattices, demonstrating the potential to create new topologies with woven lattices in which the mechanical bond is more of a structural component and less of an independent dynamic one. Loeb has also utilized molecular shuttle-based linkers like that shown in Figure 5.19a to synthesize materials in which the mechanical bond is integral to the structure of the framework referred to as "poly-threaded MOFs" [113, 114]. The linkers were composed of tetracarboxylate H-shaped axles and crown ether macrocycles featuring either two pyridyl [114] or two carboxy coordinating groups [113]. The linker containing pyridyl moieties on the macrocycle forms frameworks upon coordination with both Zn^{II} and Cu^{II} [114]. The Zn^{II} MOF featured all four carboxylate moieties of the axle and the two pyridyl moieties of the macrocycle coordinating to Zn_2 nodes. All coordinating moieties on each linker coordinated to the same 2D lattice. However, when Cu^{II} was used in place of Zn^{II}, the macrocycle and axle coordinating moieties on each rotaxane were

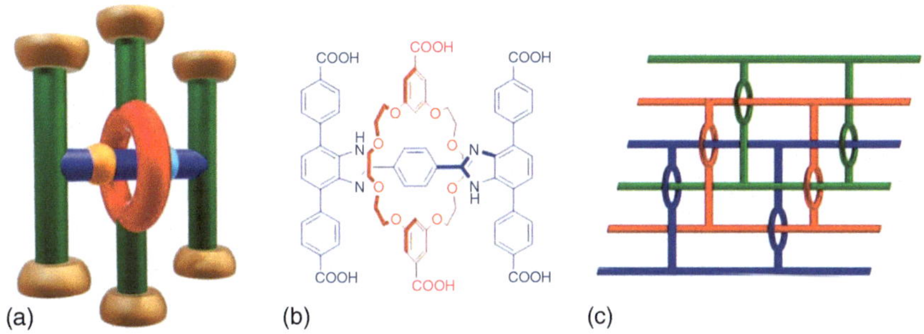

Figure 5.19 Lattice linkers for poly-threaded MOFs. (a) Cartoon of an H-shaped linker with coordinating groups on both the axle and the macrocycle wheel, (b) an example of this type of linker with six carboxylate groups, (c) the type of unique poly-threaded lattice that results from having both interlocked linkers and interpenetrated lattices. Source: Reproduced from Gholami et al. [113] with permission from John Wiley and Sons.

bonded to a different lattice [114]. This resulted in two interpenetrated lattices, which were linked via mechanical bonds between the macrocycle and axle. The complexity of these frameworks was increased upon switching the pyridyl moieties on the macrocycle to carboxylates. Upon coordination with Zn^{II} nodes, three independent 1D CPs formed [113]. Each independent lattice was composed of alternating axles and macrocycles (not mechanically interlocked). These lattices were then interpenetrated such that the axles and macrocycles of one lattice were mechanically interlocked with the axles and macrocycles of the other two independent lattices. Despite these novel structures linking independent lattices via mechanical bonds of molecular shuttle linkers, no translational motion between the frameworks has yet been observed.

5.10 Applications and Future Perspectives

A fundamental feature of pseudorotaxanes and rotaxanes is the binding of the macrocyclic wheel around the recognition site of an axle. Stoddart and Yagi took advantage of the host–guest chemistry between axles and wheels to synthesize MOFs that bind specific analytes. Linear dicarboxylate linkers with appended crown ether macrocycles were used to synthesize the **pcu** frameworks **MOF-1001A**, **MOF-1001**, and **MOF-1002** [115]. The former is triply interpenetrated, while the latter two are non-interpenetrated. The appended crown ether macrocycles have well-established host–guest chemistry and are particularly good at binding electron-deficient guest molecules. The addition of a paraquat derivative to the MOFs resulted in the formation of a pseudorotaxane between the appended macrocycle host and the paraquat guest. When a MOF analog with no appended macrocycle was used, less than 0.06 molecules per MOF linker were incorporated into the framework. Therefore, by exploiting the host–guest chemistry of pseudorotaxanes in solution, the selectivity of MOFs toward various guests can be drastically altered. MOFs with affinities for specific guests have potential application in environmental remediation by selectively removing toxic species from solution, as shown by **MOF-1001** and **MOF-1002**, which bind paraquat, a toxic herbicide. This approach was extended to linkers using pillar[5]arene macrocycles in **P5A-MOF-1**. This macrocycle host demonstrated binding of both cationic and neutral electron-poor guests such as paraquat derivatives and 1,4-dinitrobenzene [116].

The solution chemistry of MIMs has been used to demonstrate their applicability in a plethora of applications. In comparison, the challenges of engineering MIMs into the cavities of MOFs means that the practical application of these hybrid materials is currently limited. However, there are several examples described herein that demonstrate the potential of this field by combining the benefits of MIMs and MOFs into a single material. The partial photo-switching of the alkene that occurs inside the MOF reported by Berna has implications for the solvent-accessible void of the MOF. This was investigated in terms of the ability of the material to release cargo, in this case, quinone. The partial switching of the axle from *E* to *Z* resulted in a more rapid release of quinone than the purely *E* isomer framework [68]. It was theorized that this was due to photoisomerization causing variations in the size and shape of the MOF pore – a MOF behavior known as breathing.

Many applications of MOFs are related to their ability to reversibly adsorb gas molecules. In general, there are two types of adsorption, physisorption and chemisorption, with MOFs usually exhibiting the former. It is this weaker physisorption that allows MOFs to have lower regeneration energy than other gas-sequestration materials that rely on physisorption, such as amine carbon dioxide scrubbers. One of the challenges with gas storage in MOFs is the working capacity (amount of guests that can be adsorbed/desorbed). Typically, to achieve high working capacities, large changes in temperature or pressure are required. Alternativity, flexible MOFs like **MIL-53** [117] and **Co(BDP)** [118] can exhibit stepped or gated adsorption isotherms, resulting in large changes in the adsorbed guest over a very small pressure window. However, flexible MOFs such as these present practical engineering challenges due to the change in volume of the material. The inclusion of a MIM in a MOF could provide a moiety that results in the desired gated gas sorption without the requirement of a macroscopic structural change in the material. This could be potentially implemented by using a molecular shuttle rotaxane linker, which contains a recognition site that also functions as an adsorption site for the gas molecule. This concept was described by Simon, who proposed a statistical mechanical model for this behavior [119]. Upon comparing model Langmuir isotherms, it was found that for the MIM containing MOF, if the adsorption of the gas at the molecular shuttle recognition site resulted in a decrease in the entropy of the macrocycle, the gas adsorption is more temperature sensitive and more heat is released when the gas molecule adsorbs. While the potential for frameworks that exhibit novel gas sorption behaviors has been demonstrated theoretically, there are yet to be any materials that put this into practice. This is very much an area to be further explored in the coming years, and the true impact of MIMs on the gas sorption properties of MOFs is yet to be realized.

Most importantly, the porous nature of MOFs has provided opportunities to explore the dynamics of MIMs in the solid state and understand phenomena that are unique to their organization in a crystalline lattice. It is now possible to control the rate of rotational [33, 36, 39] or translational motion [40, 41] of a MIM in a MOF or operate electrochemically and photochemically actuated molecular switches that are organized in a highly dense and organized format [68–70]. There is even an example of how to precisely arrange mobile components in MOF so that they interact and influence each other's motion [46]. In a recent review, Krause and Feringa eloquently outlined that "Controlling cooperativity, synchronization, amplification, and translation of intermolecular and intramolecular dynamics over different length scales for applications in stimuli-responsive robust solids are considered key challenges in materials sciences" [120]. It is, therefore, reasonable to speculate that the success of MIMs in MOFs, as summarized in this chapter, maybe the start of an exciting future using MOFs to compartmentalize molecular machines so that their actions can be made cooperative and assembled into "molecular factories" designed for specific functions and properties at the nanoscale.

References

1 Amabilino, D.B. and Stoddart, J.F. (1995). Interlocked and intertwined structures and superstructures. *Chem. Rev.* 95: 2725–2828.
2 Bruns, C.J. and Stoddart, J.F. (2016). *Nature of the Mechanical Bond: From Molecules to Machines*, 786. Wiley.
3 Sluysmans, D. and Stoddart, J.F. (2019). The burgeoning of mechanically interlocked molecules in chemistry. *Trends Chem.* 1: 185–197.
4 Stoddart, J.F. (2009). The chemistry of the mechanical bond. *Chem. Soc. Rev.* 38: 1802–1820.
5 Zhou, H.-C., Long, J.R., and Yaghi, O.M. (2012). Introduction to metal–organic frameworks. *Chem. Rev.* 112: 673–674.
6 James, S.L. (2003). Metal-organic frameworks. *Chem. Soc. Rev.* 32: 276–288.
7 Rowsell, J.L.C. and Yaghi, O.M. (2004). Metal–organic frameworks: a new class of porous materials. *Microporous Mesoporous Mater.* 73: 3–14.
8 Yaghi, O.M., O'Keeffe, M., Ockwig, N.W. et al. (2003). Reticular synthesis and the design of new materials. *Nature* 423: 705–714.
9 Eddaoudi, M., Kim, J., Rosi, N. et al. (2002). Systematic design of pore size and functionality in isoreticular MOFs and their application in methane storage. *Science* 295: 469–472.
10 Li, J.-R., Kuppler, R.J., and Zhou, H.-C. (2009). Selective gas adsorption and separation in metal–organic frameworks. *Chem. Soc. Rev.* 38: 1477–1504.
11 Sumida, K., Rogow, D.L., Mason, J.A. et al. (2012). Carbon dioxide capture in metal–organic frameworks. *Chem. Rev.* 112: 724–781.
12 Murray, L.J., Dinca, M., and Long, J.R. (2009). Hydrogen storage in metal–organic frameworks. *Chem. Soc. Rev.* 38: 1294–1314.
13 Khraisheh, M., Mukherjee, S., Kumar, A. et al. (2020). An overview on trace CO_2 removal by advanced physisorbent materials. *J. Environ. Manag.* 255: 109874.

14 Mukherjee, S., Kumar, A., and Zaworotko, M.J. (2019). 2 - Metal-organic framework based carbon capture and purification technologies for clean environment. In: *Metal-Organic Frameworks (MOFs) for Environmental Applications* (ed. S.K. Ghosh), 5–61. Elsevier.

15 Madden, D.G., Scott, H.S., Kumar, A. et al. (2017). Flue-gas and direct-air capture of CO_2 by porous metal–organic materials. *Philos. Trans. A Math. Phys. Eng. Sci.* 375: 20160025.

16 Nugent, P., Belmabkhout, Y., Burd, S.D. et al. (2013). Porous materials with optimal adsorption thermodynamics and kinetics for CO_2 separation. *Nature* 495: 80–84.

17 Kuyuldar, S., Genna, D.T., and Burda, C.J. (2019). On the potential for nanoscale metal–organic frameworks for energy applications. *J. Mater. Chem. C* 7: 21545–21576.

18 Li, S.-L. and Xu, Q. (2013). Metal–organic frameworks as platforms for clean energy. *Energy Environ. Sci.* 6: 1656–1683.

19 Wang, H., Zhu, Q.-L., Zou, R. et al. (2017). Metal-organic frameworks for energy applications. *Chem* 2: 52–80.

20 Qin, J.-S., Du, D.-Y., Li, W.-L. et al. (2012). Ultrastable polymolybdate-based metal–organic frameworks as highly active electrocatalysts for hydrogen generation from water. *Chem. Sci.* 3: 2114–2118.

21 Sun, C.-Y., Qin, C., Wang, X.-L. et al. (2013). Metal-organic frameworks as potential drug delivery systems. *Expert Opin. Drug Deliv.* 10: 89–101.

22 Wu, M.-X. and Yang, Y.-W. (2017). Metal–organic framework (MOF)-based drug/cargo delivery and cancer therapy. *Adv. Mater.* 29: 1606134.

23 Zhu, X., Gu, J., Wang, Y. et al. (2014). Inherent anchorages in UiO-66 nanoparticles for efficient capture of alendronate and its mediated release. *Chem. Commun.* 50: 8779–8782.

24 Eddaoudi, M., Moler, D.B., and Li, H. (2001). Modular chemistry: secondary building units as a basis for the design of highly porous and robust metal–organic carboxylate frameworks. *Acc. Chem. Res.* 34: 319–330.

25 Stock, N. and Biswas, S. (2012). Synthesis of metal-organic frameworks (MOFs): routes to various MOF topologies, morphologies, and composites. *Chem. Rev.* 112: 933–969.

26 Martinez-Diaz, M.V., Spencer, N., and Stoddart, J.F. (1997). The self-assembly of a switchable [2]rotaxane. *Angew. Chem. Int. Ed. Engl.* 36: 1904–1907.

27 Zhu, K., Vukotic, V.N., and Loeb, S.J. (2012). Molecular shuttling of a compact and rigid H-shaped [2]rotaxane. *Angew. Chem. Int. Ed.* 51: 2168–2172.

28 Choi, J.W., Flood, A.H., Steuerman, D.W. et al. (2005). Ground-state equilibrium thermodynamics and switching kinetics of bistable [2]rotaxanes switched in solution, polymer gels, and molecular electronic devices. *Chem. Eur. J.* 12: 261–279.

29 Raehm, L., Kern, J.-M., and Sauvage, J.-P. (1999). A transition metal containing rotaxane in motion: electrochemically induced pirouetting of the ring on the threaded dumbbell. *Chem. Eur. J.* 5: 3310–3317.

30 Blanco, M.-J., Consuelo Jimenez, M., Chambron, J.-C. et al. (1999). Rotaxanes as new architectures for photoinduced electron transfer and molecular motions. *Chem. Soc. Rev.* 28: 293–305.

31 Gholami, G., Zhu, K., Baggi, G. et al. (2017). Influence of axle length on the rate and mechanism of shuttling in rigid H-shaped [2]rotaxanes. *Chem. Sci.* 8: 7718–7723.

32 Farahani, N., Zhu, K., O'Keefe, C.A. et al. (2016). Thermally driven dynamics of a rotaxane wheel about an imidazolium axle inside a metal–organic framework. *ChemPlusChem* 81: 836–841.

33 Martinez-Bulit, P., O'Keefe, C.A., Zhu, K. et al. (2019). Solvent and steric influences on rotational dynamics in porphyrinic metal–organic frameworks with mechanically interlocked pillars. *Cryst. Growth Des.* 19: 5679–5685.

34 Anelli, P.L., Spencer, N., and Stoddart, J.F. (1991). A molecular shuttle. *J. Am. Chem. Soc.* 113: 5131–5133.

35 Vignon, S.A., Jarrosson, T., Iijima, T. et al. (2004). Switchable neutral bistable rotaxanes. *J. Am. Chem. Soc.* 126: 9884–9885.

36 Zhu, K., Vukotic, V.N., O'Keefe, C.A. et al. (2014). Metal–organic frameworks with mechanically interlocked pillars: controlling ring dynamics in the solid-state via a reversible phase change. *J. Am. Chem. Soc.* 136: 7403–7409.

37 Bermudez, V., Capron, N., Gase, T. et al. (2000). Influencing intramolecular motion with an alternating electric field. *Nature* 406: 608–611.

38 Deng, H., Olson, M.A., Stoddart, J.F. et al. (2010). Robust dynamics. *Nat. Chem.* 2: 439–443.

39 Vukotic, V.N., Harris, K.J., Zhu, K. et al. (2012). Metal–organic frameworks with dynamic interlocked components. *Nat. Chem.* 4: 456–460.

40 Zhu, K., O'Keefe, C.A., Vukotic, V.N. et al. (2015). A molecular shuttle that operates inside a metal–organic framework. *Nat. Chem.* 7: 514–519.

41 Wilson, B.H., Abdulla, L.M., Schurko, R.W. et al. (2021). Translational dynamics of a non-degenerate molecular shuttle imbedded in a zirconium metal–organic framework. *Chem. Sci.* 12: 3944–3951.

42 Young, P.G., Hirose, K., and Tobe, Y. (2014). Axle length does not affect switching dynamics in degenerate molecular shuttles with rigid spacers. *J. Am. Chem. Soc.* 136: 7899–7906.

43 Zhu, K., Vukotic, V.N., Noujeim, N. et al. (2012). Bis(benzimidazolium) axles and crown ether wheels: a versatile templating pair for the formation of [2]rotaxane molecular shuttles. *Chem. Sci.* 3: 3265–3271.

44 Zhu, K., Baggi, G., and Loeb, S.J. (2018). Ring-through-ring molecular shuttling in a saturated [3]rotaxane. *Nat. Chem.* 10: 625–630.

45 Stirk, A.J., Wilson, B.H., O'Keefe, C.A. et al. (2021). Applying reticular synthesis to the design of Cu-based MOFs with mechanically interlocked linkers. *Nano Res.* 14: 417–422.

46 Wilson, B.H., Vojvodin, C.S., Gholami, G. et al. (2021). Precise spatial arrangement and interaction between two different mobile components in a metal-organic framework. *Chem* 7: 202–211.

47 Martinez-Bulit, P., Stirk, A.J., and Loeb, S.J. (2019). Rotors, motors, and machines inside metal–organic frameworks. *Trends Chem.* 1: 588–600.

48 Wilson, B.H. and Loeb, S.J. (2020). Integrating the mechanical bond into metal-organic frameworks. *Chem* 6: 1604–1612.

49 Moulton, B. and Zaworotko, M.J. (2001). From molecules to crystal engineering: supramolecular isomerism and polymorphism in network solids. *Chem. Rev.* 101: 1629–1658.

50 Leong, W.L. and Vittal, J.J. (2011). One-dimensional coordination polymers: complexity and diversity in structures, properties, and applications. *Chem. Rev.* 111: 688–764.

51 Kitagawa, S. and Matsuda, R. (2007). Chemistry of coordination space of porous coordination polymers. *Coord. Chem. Rev.* 251: 2490–2509.

52 Kitagawa, S., Kitaura, R., and Noro, S.-I. (2004). Functional porous coordination polymers. *Angew. Chem. Int. Ed.* 43: 2334–2375.

53 Li, B., Wen, H.-M., Cui, Y. et al. (2016). Emerging multifunctional metal–organic framework materials. *Adv. Mater.* 28: 8819–8860.

54 Davidson, G.J.E. and Loeb, S.J. (2003). Channels and cavities lined with interlocked components: metal-based polyrotaxanes that utilize pyridinium axles and crown ether wheels as ligands. *Angew. Chem. Int. Ed.* 42: 74–77.

55 Hoffart, D.J. and Loeb, S.J. (2005). Metal–organic rotaxane frameworks: three-dimensional polyrotaxanes from lanthanide-ion nodes, pyridinium N-oxide axles, and crown-ether wheels. *Angew. Chem. Int. Ed.* 44: 901–904.

56 Loeb, S.J. (2005). Metal–organic rotaxane frameworks; MORFs. *Chem. Commun.* 41: 1511–1518.

57 Loeb, S.J. (2007). Rotaxanes as ligands: from molecules to materials. *Chem. Soc. Rev.* 36: 226–235.

58 Frank, N.C., Mercer, D.J., and Loeb, S.J. (2013). An interwoven metal-organic framework combining mechanically interlocked linkers and interpenetrated networks. *Chem. Eur. J.* 19: 14076–14080.

59 Vukotic, V.N. and Loeb, S.J. (2012). Coordination polymers containing rotaxane linkers. *Chem. Soc. Rev.* 41: 5896–5906.

60 Vukotic, V.N., O'Keefe, C.A., Zhu, K. et al. (2015). Mechanically interlocked linkers inside metal–organic frameworks: effect of ring size on rotational dynamics. *J. Am. Chem. Soc.* 137: 9643–9651.

61 Zhu, K. and Loeb, S.J. (2014). Organizing mechanically interlocked molecules to function inside metal-organic frameworks. In: *Molecular Machines and Motors: Recent Advances and Perspectives* (ed. A. Credi, S. Silvi, and M. Venturi), 213–251. Cham: Springer International Publishing.

62 Gholami, G., Wilson, B.H., Zhu, K. et al. (2021). Exploring the dynamics of Zr-based metal–organic frameworks containing mechanically interlocked molecular shuttles. *Faraday Discuss.* 225: 358–370.

63 Coskun, A., Hmadeh, M., Barin, G. et al. (2012). Metal–organic frameworks incorporating copper-complexed rotaxanes. *Angew. Chem. Int. Ed.* 51: 2160–2163.

64 Li, Q., Zhang, W., Miljanić, O.S. et al. (2010). A metal–organic framework replete with ordered donor–acceptor catenanes. *Chem. Commun.* 46: 380–382.

65 Zhao, Y.-L., Liu, L., Zhang, W. et al. (2009). Rigid-strut-containing crown ethers and [2] catenanes for incorporation into metal–organic frameworks. *Chem. Eur. J.* 15: 13356–13380.

66 Robin, A.Y. and Fromm, K.M. (2006). Coordination polymer networks with O-and N-donors: what they are, why and how they are made. *Coord. Chem. Rev.* 250: 2127–2157.

67 Mercer, D.J., Vukotic, V.N., and Loeb, S.J. (2011). Linking [2]rotaxane wheels to create a new type of metal organic rotaxane framework. *Chem. Commun.* 47: 896–898.

68 Saura-Sanmartin, A., Martinez-Cuezva, A., Bautista, D. et al. (2020). Copper-linked rotaxanes for the building of photoresponsive metal organic frameworks with controlled cargo delivery. *J. Am. Chem. Soc.* 142: 13442–13449.

69 McGonigal, P.R., Deria, P., Hod, I. et al. (2015). Electrochemically addressable trisradical rotaxanes organized within a metal–organic framework. *Proc. Natl. Acad. Sci.* 112: 11161.

70 Chen, Q., Sun, J., Li, P. et al. (2016). A redox-active bistable molecular switch mounted inside a metal–organic framework. *J. Am. Chem. Soc.* 138: 14242–14245.

71 Fernandez, A., Moreno Pineda, E., Ferrando-Soria, J. et al. (2015). A hybrid organic–inorganic molecular daisy chain. *Chem. Commun.* 51: 11126–11129.

72 Lee, E., Heo, J., and Kim, K. (2000). A three-dimensional polyrotaxane network. *Angew. Chem. Int. Ed.* 39: 2699–2701.

73 Lee, E., Kim, J., Heo, J. et al. (2001). A two-dimensional polyrotaxane with large cavities and channels: a novel approach to metal–organic open-frameworks by using supramolecular building blocks. *Angew. Chem. Int. Ed.* 40: 399–402.

74 Whang, D. and Kim, K. (1997). Pseudopolyrotaxanes made to order: cucurbituril threaded on polyviologen. *Chem. Commun.* 33: 2361–2362.

75 Kim, K. (2002). Mechanically interlocked molecules incorporating cucurbituril and their supramolecular assemblies. *Chem. Soc. Rev.* 31: 96–107.

76 Whang, D., Jeon, Y.-M., Heo, J. et al. (1996). Self-assembly of a polyrotaxane containing a cyclic "Bead" in every structural unit in the solid state: cucurbituril molecules threaded on a one-dimensional coordination polymer. *J. Am. Chem. Soc.* 118: 11333–11334.

77 Whang, D. and Kim, K. (1997). Polycatenated two-dimensional polyrotaxane net. *J. Am. Chem. Soc.* 119: 451–452.

78 Knight, L.K., Vukotic, V.N., Viljoen, E. et al. (2009). Eliminating the need for independent counterions in the construction of metal–organic rotaxane frameworks (MORFs). *Chem. Commun.* 45: 5585–5587.

79 Hoffart, D.J. and Loeb, S.J. (2007). The missing link: a 2D metal-organic rotaxane framework (MORF) with one rotaxane linker and one naked linker. *Supramol. Chem.* 19: 89–93.

80 Mercer, D.J., Yacoub, J., Zhu, K. et al. (2012). [2]Pseudorotaxanes,[2]rotaxanes and metal–organic rotaxane frameworks containing tetra-substituted dibenzo [24]crown-8 wheels. *Org. Biomol. Chem.* 10: 6094–6104.

81 Rambo, B.M., Gong, H.-Y., Oh, M. et al. (2012). The "texas-sized" molecular box: a versatile building block for the construction of anion-directed mechanically interlocked structures. *Acc. Chem. Res.* 45: 1390–1401.

82 Gong, H.-Y., Rambo, B.M., Cho, W. et al. (2011). Anion-directed assembly of a three-dimensional metal–organic rotaxane framework. *Chem. Commun.* 47: 5973–5975.

83 Gong, H.-Y., Rambo, B.M., Nelson, C.A. et al. (2012). Rare-earth cation effects on three-dimensional metal–organic rotaxane framework (MORF) self assembly. *Chem. Commun.* 48: 10186–10188.

84 Gong, H.-Y., Rambo, B.M., Nelson, C.A. et al. (2012). Multi component self-assembly: supramolecular organic frameworks containing metal–rotaxane subunits (RSOFs). *Dalton Trans.* 41: 1134–1137.

85 Vukotic, V.N. and Loeb, S.J. (2014). Metal-organic frameworks: polyrotaxane frameworks. In: *Encyclopedia of Inorganic and Bioinorganic Chemistry* (ed. L.R. MacGillivray and C.M. Lukehart), 115–134. Chichester: Wiley.

86 Gładysiak, A., Nguyen, T.N., Bounds, R. et al. (2019). Temperature-dependent interchromophoric interaction in a fluorescent pyrene-based metal–organic framework. *Chem. Sci.* 10: 6140–6148.

87 Shrivastava, A. and Das, D. (2019). Axial positive, negative, and zero thermal expansion in a mixed-metal mixed-linker coordination compound: role of 2D layer in the thermal expansion property. *Cryst. Growth Des.* 19: 4908–4913.

88 Tahier, T. and Oliver, C.L. (2015). *In situ* variable-temperature single crystal X-ray diffraction studies of the single-crystal-to-single-crystal dehydration and rehydration of a mixed-ligand 2D zinc metal–organic framework using trimesate and 4,4′-bipyridine-*N*,*N*′-dioxide as ligands. *CrystEngComm* 17: 8946–8956.

89 Meirovitch, E., Liang, Z., and Freed, J.H. (2015). Protein dynamics in the solid State from ^{2}H NMR line shape analysis: a consistent perspective. *J. Phys. Chem. B* 119: 2857–2868.

90 Meirovitch, E., Liang, Z., and Freed, J.H. (2020). Structural dynamics by NMR in the solid state: the unified MOMD perspective applied to organic frameworks with interlocked molecules. *J. Phys. Chem. B* 124: 6225–6235.

91 Jiang, X., Duan, H.-B., Khan, S.I. et al. (2016). Diffusion-controlled rotation of triptycene in a metal–organic framework (MOF) sheds light on the viscosity of MOF-confined solvent. *ACS Cent. Sci.* 2: 608–613.

92 Furukawa, H., Cordova, K.E., O'Keeffe, M. et al. (2013). The chemistry and applications of metal-organic frameworks. *Science* 341: 974.

93 Barman, S., Khutia, A., Koitz, R. et al. (2014). Synthesis and hydrogen adsorption properties of internally polarized 2,6-azulenedicarboxylate based metal–organic frameworks. *J. Mater. Chem. C* 2: 18823–18830.

94 Burd, S.D., Nugent, P.S., Mohamed, M.H. et al. (2013). Square grid and pillared square grid coordination polymers–fertile ground for crystal engineering of structure and function. *Chimia* 67: 372–378.

95 Burtch, N.C. and Walton, K.S. (2015). Modulating adsorption and stability properties in pillared metal–organic frameworks: a model system for understanding ligand effects. *Acc. Chem. Res.* 48: 2850–2857.

96 Dybtsev, D.N., Chun, H., and Kim, K. (2004). Rigid and flexible: a highly porous metal–organic framework with unusual guest-dependent dynamic behavior. *Angew. Chem. Int. Ed.* 43: 5033–5036.

97 Henke, S., Schneemann, A., Wütscher, A. et al. (2012). Directing the breathing behavior of pillared-layered metal–organic frameworks via a systematic library of functionalized linkers bearing flexible substituents. *J. Am. Chem. Soc.* 134: 9464–9474.

98 Klein, N., Herzog, C., Sabo, M. et al. (2010). Monitoring adsorption-induced switching by ^{129}Xe NMR spectroscopy in a new metal–organic framework $Ni_2(2,6\text{-ndc})_2(dabco)$. *Phys. Chem. Chem. Phys.* 12: 11778–11784.

99 Schneemann, A., Vervoorts, P., Hante, I. et al. (2018). Different breathing mechanisms in flexible pillared-layered metal–organic frameworks: Impact of the metal center. *Chem. Mater.* 30: 1667–1676.

100 Uemura, K., Yamasak, Y., Komagawa, Y. et al. (2007). Two-step adsorption/desorption on a jungle-gym-type porous coordination polymer. *Angew. Chem. Int. Ed.* 46: 6662–6665.

101 Eddaoudi, M., Kim, J., Rosi, N. et al. (2002). Systematic design of pore size and functionality in isoreticular MOFs and their application in methane storage. *Science* 295: 469.

102 Lin, X., Telepeni, I., Blake, A.J. et al. (2009). High capacity hydrogen adsorption in Cu(II) tetracarboxylate framework materials: the role of pore size, ligand functionalization, and exposed metal sites. *J. Am. Chem. Soc.* 131: 2159–2171.

103 Gole, B., Bar, A.K., Mallick, A. et al. (2013). An electron rich porous extended framework as a heterogeneous catalyst for Diels–Alder reactions. *Chem. Commun.* 49: 7439–7441.

104 Krause, S., Evans, J.D., Bon, V. et al. (2018). Adsorption contraction mechanics: understanding breathing energetics in isoreticular metal–organic frameworks. *J. Phys. Chem. C* 122: 19171–19179.

105 Lewis, J.E.M. (2019). Self-templated synthesis of amide catenanes and formation of a catenane coordination polymer. *Org. Biomol. Chem.* 17: 2442–2447.

106 Krause, S., Bon, V., Senkovska, I. et al. (2018). The effect of crystallite size on pressure amplification in switchable porous solids. *Nat. Commun.* 9: 1–8.

107 Krause, S., Bon, V., Senkovska, I. et al. (2016). A pressure-amplifying framework material with negative gas adsorption transitions. *Nature* 532: 348–352.

108 Zhu, A.-X., Yang, Q.-Y., Mukherjee, S. et al. (2019). Tuning the Gate-Opening Pressure in a Switching pcu Coordination Network, X-pcu-5-Zn, by Pillar-Ligand Substitution. *Angew. Chem. Int. Ed.* 58: 18212–18217.

109 Wang, S.-Q., Yang, Q.-Y., Mukherjee, S. et al. (2018). Recyclable switching between nonporous and porous phases of a square lattice (sql) topology coordination network. *Chem. Commun.* 54: 7042–7045.

110 Schneemann, A., Bon, V., Schwedler, I. et al. (2014). Flexible metal–organic frameworks. *Chem. Soc. Rev.* 43: 6062–6096.

111 Wang, S.-Q., Mukherjee, S., and Zaworotko, M.J. (2021). Spiers memorial lecture: Coordination networks that switch between nonporous and porous structures: an emerging class of soft porous crystals. *Faraday Discuss.* 231: 9–50. https://doi.org/10.1039/d1fd00037c

112 Uemura, K., Matsuda, R., and Kitagawa, S. (2005). Flexible microporous coordination polymers. *J. Solid-state Chem.* 178: 2420–2429.

113 Gholami, G., Zhu, K., Ward, J.S. et al. (2016). Formation of a polythreaded, metal–organic framework utilizing an interlocked hexadentate, carboxylate linker. *Eur. J. Inorg. Chem.* 2016: 4524–4529.

114 Gholami, G., Baggi, G., Zhu, K. et al. (2017). Metal–organic frameworks utilising an interlocked, hexadentate linker containing a tetra-carboxylate axle and a bis(pyridine) wheel. *Dalton Trans.* 46: 2462–2470.

115 Li, Q., Zhang, W., Miljanić, O.S. et al. (2009). New tools for chemical biology from main group elements. *Science* 325: 855–861.

116 Strutt, N.L., Fairen-Jimenez, D., Iehl, J. et al. (2012). Incorporation of an A1/A2-difunctionalized pillar[5] arene into a metal–organic framework. *J. Am. Chem. Soc.* 134: 17436–17439.

117 Loiseau, T., Serre, C., Huguenard, C. et al. (2004). A rationale for the large breathing of the porous aluminum terephthalate (MIL-53) upon hydration. *Chem. Eur. J.* 10: 1373–1382.

118 Taylor, M.K., Runčevski, T., Oktawiec, J. et al. (2016). Tuning the adsorption-induced phase change in the flexible metal–organic framework Co(bdp). *J. Am. Chem. Soc.* 138: 15019–15026.

119 Carney, J., Roundy, D., and Simon, C.M. (2020). Statistical mechanical model of gas adsorption in a metal–organic framework harboring a rotaxane molecular shuttle. *Langmuir* 36: 13112–13123.

120 Krause, S. and Feringa, B.L. (2020). Towards artificial molecular factories from framework-embedded molecular machines. *Nat. Rev. Chem.* 4: 550–562.

6

Mechanically Interlocked Proteins

Yu-Xiang Wang, Wen-Hao Wu, and Wen-Bin Zhang

Peking University, Center for Soft Matter Science and Engineering, Beijing National Laboratory for Molecular Sciences, Key Laboratory of Polymer Chemistry & Physics of Ministry of Education, College of Chemistry and Molecular Engineering, 292 Chengfu Rd, Haidian District, Beijing 100871, P. R. China

6.1 Introduction

Since ancient times, mechanically interlocked structures (such as knots and links) have been well known to humankind and widely used in diverse occasions, including art design, climbing, sailing, and camping. However, making their molecular counterparts in chemistry is only a recent endeavor, mainly in the field of molecular machines [1]. As defined by Bruns and Stoddart [2], these mechanically interlocked molecules (MIMs) are "molecules containing mechanical or topological entanglements." A mechanical bond is "an entanglement in space between two or more molecular entities such that they can't be separated without breaking or distorting chemical bond," and a mechanomolecule is "a molecule possessing one or more mechanical bonds" [2]. Hence, MIMs include not only mechanomolecules but also those deeply entangled molecules, such as molecular knots. The latter does possess an entanglement in space, but no mechanical bond can be identified since the entities in the knot can usually slide and become essentially indistinguishable. Common types of MIMs thus include knots, rotaxanes, and catenanes [3]. The concept of MIMs is closely related to chemical topology, an intriguing topic in chemistry dealing with the invariant spatial properties of a molecule under continuous deformation without breaking covalent bonds. Despite some overlap, each of them has its own focus. Whereas bending or stretching chemical bonds are usually considered when distinguishing chemical topology, such operations might disintegrate some MIMs like rotaxanes. In other words, some MIMs possess complex chemical topology, while others may be topologically trivial.

Synthesis of mechanically interlocked small molecules has achieved considerable success over the past decades via the combination of assembly and reaction. Not only have some extremely complex MIMs, such as Solomon links [4], Borromean rings [5], and a Star of David catenane [6], been successfully prepared, but also various applications of MIMs have been demonstrated from molecular shuttles [7], switches [8], and machines [9], to allosteric catalysts [10], molecular assemblers [11],

Mechanically Interlocked Materials: Polymers, Nanomaterials, MOFs, and more, First Edition.
Edited by Emilio M. Pérez.

and synthesizers [12]. Nevertheless, the synthesis of mechanically interlocked macromolecules remains a formidable challenge. As the molecular size increases, the number of possible reactive sites and molecular conformations increases dramatically [13]. The complexity of both intra- or intermolecular interactions further complicated their synthesis. Although it has been shown in limited cases that mechanically interlocked structures indeed strongly affect the macromolecular properties in either static or dynamic ways [14], their challenging synthesis plagues a thorough understanding of their structure–property relationships.

As one of the most common natural biopolymers, proteins play a vital but versatile role in living systems. The nascent protein chain synthesized by ribosome has a linear configuration with specific amino acid sequences that define their final 3D geometry and biological functions upon folding. When combined with site-specific ligation, folding can help manipulate the polypeptides' conformation and achieve a selective synthesis of mechanically interlocked proteins (MIPs). Elegant examples exist in Nature. Among all the protein structures deposited in the Protein Data Bank (PDB), ~6% of them are entangled and can be classified into various mechanically interlocked structures such as knots, slipknots, lassos, and links [15]. Such structures are often associated with certain functional benefits like enhanced mechanical, chemical, and thermodynamic stability [16]. It was speculated that these MIPs may result from long-term evolution under the pressure to survive under harsh circumstances [17]. Following this line, artificial MIPs have also been designed and synthesized to exploit mechanical interlocking for protein stabilization. In this chapter, we will summarize current findings about both natural and artificial MIPs with respect to their classification, preparation, biological significance, and functional benefits.

6.2 Classification of Mechanically Interlocked Proteins

Based on the conventional classification of MIMs, MIPs can also be generally divided into two categories, as shown in Figure 6.1. The first type is MIPs that do not contain mechanical bonds, which is mainly composed of various protein knots and slipknots such as the MJ0366 protein with 3_1 knot from *Methanocaldococcus jannaschii* and the α-haloacid dehalogenase protein with 6_1 knot [18]. Due to the translation mechanism of ribosomes, those knotted proteins always have separated N- and C-termini. Strictly speaking, they are not knots in the mathematical sense, but the entanglement and the resulting mechanical interlocking are real under native state. Assuming an extra segment connecting the N-terminus to the C-terminus, one can identify the type of a knotted protein by Alexander–Briggs notation [19]. The second type is mechanoproteins with mechanical bonds. Consistent with that in MIMs, the mechanoproteins can be further divided into sub-categories such as lassos, pseudo-rotaxanes, rotaxanes, and catenanes. A lasso structure can be considered as tadpole-like chain with its tail threading through the loop. For lasso peptide, the loop area tends to be small. The side-chain steric hindrances could fix the threading chain, leading to a mechanical bond [20]. For lasso proteins with a large-sized ring, the threading structure is usually stabilized

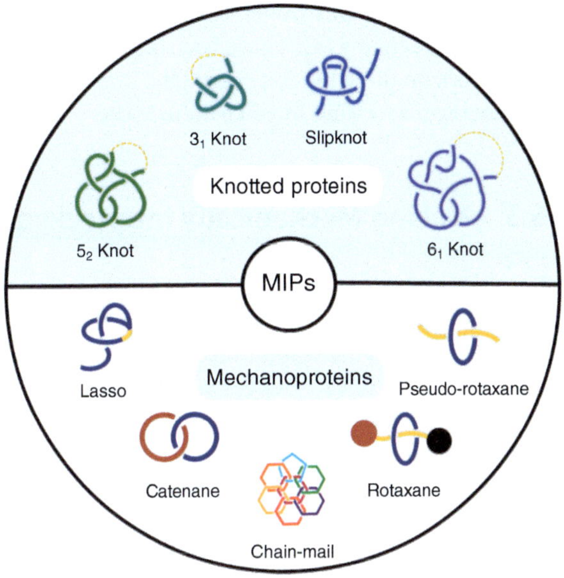

Figure 6.1 Two categories of MIPs: knotted proteins and mechanoproteins.

through intramolecular interactions [21]. When the covalent bond between the threading chain and the loop is broken, the lasso structure is converted into a pseudo-rotaxane where the entanglement is due to secondary interactions. When pulled, the threading chain can slide through the ring. Adding two bulky domains on each side of the threading chain may restrict such movements, making it more like a rotaxane [22]. A catenane is formed if the two ends of the threading chain are connected by covalent bond(s) [23]. Similarly, multiple rings can be interlocked together to form a chain-mail structure [24]. The molecular mobility is further restrained in these mechanically interlocked rings. There are MIPs with even more complex structures, such as the Solomon link and Brunnian link [25]. Sulkowska's group has developed some computation methods to search for the proteins with entangled structures and built some databases such as KnotProt [26], LassoProt [27], and LinkProt [28] to categorize these proteins in PDB.

A close examination of mechanoproteins reveals that it is actually hard to define the mechanical bond in lasso proteins or protein rotaxanes according to the classical definition. In these cases, the entanglement is due to protein folding and the entangled structure is held by collective secondary interactions. It would disappear when denatured. Note that this happens without "breaking or distorting chemical bonds." In line with our recent discussion on identifying macromolecular topology [29], we feel that it is necessary to consider the bonding of interest (as set by the bonding threshold) in defining mechanical bonds in the context of macromolecules. We propose that a mechanical bond exists when an entanglement in space exists between two or more molecular entities such that they cannot be separated without breaking or distorting "the bonds as set by the bonding threshold of interest." In this case, the folding-induced entanglement leads to an "effective" mechanical bond in lasso proteins or protein rotaxanes as long as the bonds of interest exist to hold the folding in place. We feel that this would also apply to other biomacromolecules like DNA and RNA. To date, many MIPs have been discovered in nature, and there has been quite some work on the synthesis of artificial MIPs in protein engineering. Below, we will discuss the major ways to make these MIPs.

6.3 Making Mechanically Interlocked Proteins

Despite tremendous progress in the chemical synthesis of proteins [30], most proteins are synthesized by the cellular machinery – the ribosome. Limited by the template polymerization mechanism, the nascent proteins prepared this way are intrinsically linear. Hence, making MIPs depends on posttranslational events like folding, cleavage, and reaction. There are basically two pathways to make MIPs, one by folding alone and the other by folding and reaction (Figure 6.2). The principle holds for both natural and artificial MIPs. Knotted proteins are formed solely upon folding with four possible folding mechanisms. The first is the "native twisted loop" pathway for most middle-sized proteins [31]. A native twisted loop is formed to serve as a nucleation site for folding, followed by the threading of the tail through plugging directly or a slipknot intermediate state [18a, 32]. This folding process has been proved in some cases such as the smallest knotted protein MJ0366 known to date [18a]. The same situation was also observed in an artificial knotted protein designed by Yeates' group [33]. The knotted structure was obtained by making a tandem dimer of the protein HP0242 with an entangled dimer structure through genetic fusion, connecting the N-terminus of one monomer to the C-terminus of the other via a flexible linker. However, the model is not applicable for some more

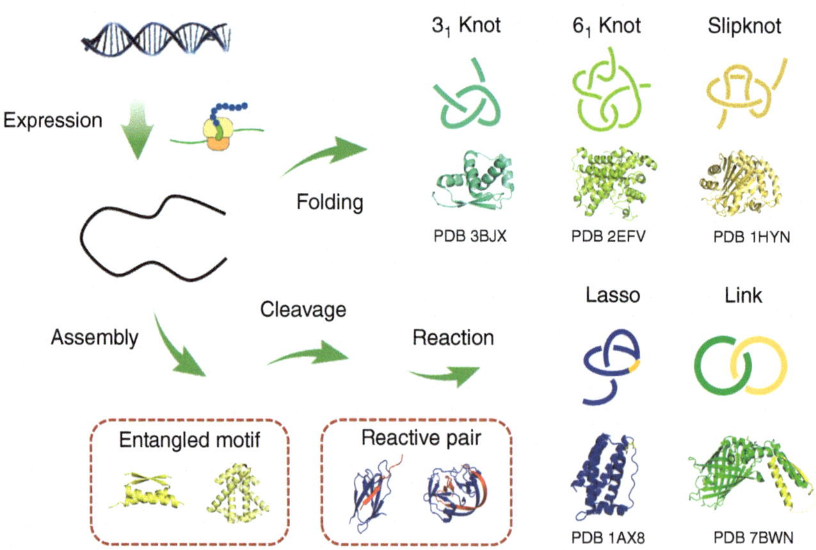

Figure 6.2 Two pathways to make MIPs: one is by folding only, while the other combines both folding and reaction.

complex knots like 6_1 knots in α-haloacid dehalogenase. The "loop-flipping" model has been proposed to account for its knotting [18b]. A twisted loop forms when the protein is almost folded, and the knotting occurs when the disordered big-loop flips over the small-loop followed by the threading of the terminus through the small-loop, leading to a 6_1 knot. As the most complex knotted protein known to date, the folding pathway is expected to possess many kinetic traps and is mostly slow in these cases [31, 34]. The third mechanism is called the "ribosome-based" mechanism. It is suggested to be found on extremely deep knots such as that found in the Treponema pallidum protein, which is composed of a knotted middle domain and two large terminal domains. The model proposes that ribosome plays an active role in directing the folding of knotted proteins. When the protein chain leaves the ribosome channel after translation, it forms a twisted loop around the channel port and the chain is continuously pushed by the ribosome through the loop as it is translated further [34b, 35]. The fourth mechanism is based on the chaperon. A chaperonin can attach the protein chains and provide a relatively narrow, confined environment [36]. The process called encapsulation increases the knotting probability under confinement. As a result, the folding energy landscape is smoothed by the encapsulation. With the help of the chaperonin, one of the ubiquitin C-terminal hydrolase (UCH) family members can form a complex 5_2 knot [37]. Some knotted proteins, such as VirC2 and DndE protein, can only self-tie and untie in the presence of chaperonin, which indicates that the limitation of conformational entropy may increase the possibility of overcoming thc folding barrier [38].

While folding can lead to entanglement in space, additional chemical reactions are often necessary to form mechanically interlocked structures. The "assembly-reaction" synergy has been demonstrated as an effective strategy in making MIMs in supramolecular chemistry. It is also the strategy adopted by Nature in making MIPs. Natural lasso proteins can be divided into two groups based on their different closing bond. The first is connected by the amide bond called lasso peptide, which is formed posttranslationally [20]. The second is disulfide-based lassos [39]. For example, leptin contains a lasso structure locked by an intramolecular disulfide formed under oxidizing conditions. Natural proteins with Hopf and Solomon links in a single chain have also been identified [25]. Two possible folding pathways have been proposed to explain the formation of the Hopf link. First, if one disulfide bond is formed, the remaining long chain will thread the loop by "slip-knotting" process, and as the lasso-like structure is formed, another disulfide bond will be formed to lock the two loops. Second, a crossover structure may form between two β-sheets as stabilized by hydrogen bonding, and the Hopf link is formed by subsequent formation of two disulfide bonds. Barwin-like endoglucanases superfamily is one of the typical protein superfamily with Hopf links [25]. Some proteins, such as families of flocculation and epithelial adhesion proteins, can form a Solomon link. However, their folding mechanisms remain elusive. Intriguingly, many chains can assemble to form oligomers with entwining features. Covalent bonding could lead to large-sized MIPs in this case, as shown by the famous example of the capsid of bacteriophage HK97 with a "chain mail" structure. The 60 hexamer rings and 12 pentamer rings are mutually interlocked and further stabilized by 420 intermolecular isopeptide bonds to form a catenane cage. Its formation takes multistep processing, including folding and assembly of monomers into the pro-head structure, packaging of DNA into the cage and subsequent maturation via isopeptide bond formation [24, 40]. The entire approach is a true masterpiece resulting from an orchestra among multiple components and processes.

Complementary to natural MIPs, artificial MIPs can help understand the mechanisms of MIP formation and the structure–property relationships of MIPs. Earlier works in this direction mainly involve reengineering of natural MIPs. Yeates' group synthesized a series of knotted protein polymers using the knotted carbonic anhydrase II protein as a monomer [41]. Depending on the location of the cysteines, they can form either serially or superficially knotted protein polymers. Link's group obtained pseudo-rotaxane by digesting natural lasso peptides, which self-assembled through disulfide bond formation in solution to form [*n*]catenanes [42]. Dawson's group successfully prepared homocatenanes and heterocatenanes based on an entwining dimer motif from the tetramerization domain of the tumor suppressor protein p53 using solid-phase peptide synthesis and native chemical ligation [23, 43]. However, these methods lack generality. One challenge is the site-selective ligation. Disulfide bonds may form between any sulfuryl of cysteine residues, leading to side-products. Therefore, it is very important to develop protein ligation reactions that are entirely based on natural amino acids (thus genetically encoded), occur spontaneously under mild physical conditions, and are highly selective and efficient.

The emergence of genetically encoded protein reactions provides a possible solution to programming posttranslational processing, such as cyclization in cells [44]. Common tools for backbone ligation include split-intein-mediated ligation [45] and enzyme (e.g. sortase, butelase)-mediated ligation [46]. They are useful for synthesizing cyclic proteins with high efficiency [47]. By splitting the CnaB2 domain of *Streptococcus pyogenes*, the Howarth group developed the well-known SpyTag-SpyCatcher pair for side-chain coupling via isopeptide formation [48]. The isopeptide bond is spontaneously formed between Lys31 and Asp117 catalyzed by Glu77 on the CnaB domain [49]. This seminal work leads to a family of genetically encoded peptide-protein reactive pairs, such as SnoopTag-SnoopCatcher [50] and SdyTag-SdyCatcher [51]. Through further splitting, three-component ligation systems were also obtained with much smaller reactive sequences, such as the SpyLigase-KTag-SpyTag [52], SnoopLigase-SnoopTagJr-DogTag [53], and SpyStapler-BDTag-SpyTag coupling systems [54]. These tools facilitate the design and synthesis of various MIPs. Two approaches have been established for topological protein synthesis: the passive [55] and the active template methods [56]. The former uses natural existing entanglement motifs such as p53dim to direct folding and uses genetically encoded protein coupling tools such as SpyTag-SpyCatcher reaction to fix the MIPs. For example, the SpyX modules with both p53dim and SpyTag/SpyCatcher motifs facilitate the synthesis of [2]catenanes and branched [2]catenanes. Here, assembly and reaction are two individual processes, and a synergy between the two is required to form the desired structures [55b, 57]. By contrast, the latter combines assembly and reaction together in one motif [56].

For example, a reactive linear protein with BDTag at N-terminus and SpyTag at C-terminus can thread through a cyclic protein containing SpyStapler to reconstitute the folded domain, which auto-catalyzes the isopeptide bond formation between BDTag and SpyTag for ring closure and catenane synthesis [54, 58]. Finally, the synthesis of topological proteins may also be achieved via topological transformation. By combining natural entangled dimers, orthogonal protein reactive pairs and proteolytic digestion, we successfully demonstrated the cellular synthesis of various MIPs, such as protein rotaxanes and heterocatenanes, through a lasso intermediate [59]. Despite these works, the richness of MIPs has been largely untapped, especially compared to that of small molecules. With the recent development in computer-assisted protein design and biochemical tools, we anticipate that novel MIPs with unique topology and functions will be continuously uncovered.

6.4 Biological Significance of Natural MIPs

Ever since the discovery of MIPs in Nature (mostly in enzymes), the biological significance of mechanical interlocking has been a puzzling conundrum in the community. Compared to their linear counterparts, MIPs are more difficult to form due to kinetic traps. It remains unclear whether mechanical interlocking was merely a random structural feature that arose during evolution or preferentially kept due to better fitness. Indeed, scattered examples, both computational and experimental, suggest that stability enhancement may be a major advantage of mechanical interlocking (Figure 6.3) [17, 60]. Reduction of conformational entropy via mechanical

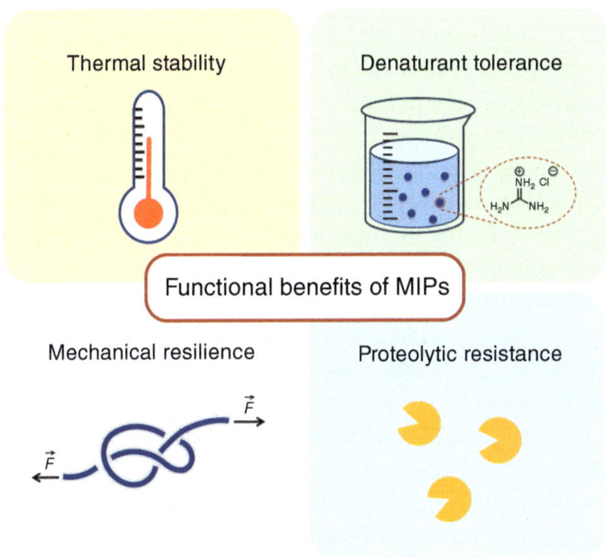

Figure 6.3 Functional benefits of MIPs include thermal stability, denaturant tolerance, proteolytic resistance, and mechanical resilience.

interlocking is more apparent in the unfolded state, leading to increased stability of the folded state. The rigid entangled structure may also render the active site tolerant to mutations during evolution [17]. For example, *N*-acetylornithine transcarbamylase, which shows a trefoil knot structure, has much higher mechanical and thermal stability compared to its linear counterpart, ornithine transcarbamylase. This enhanced stability is vital for the arginine synthesis in several eubacteria [61]. It was found that the knotted structure of the active site possesses strong local interactions and a small solvent-accessible surface, leading to a strong shielding effect and thus becoming resistant to external effects. Similar phenomena were also observed in the other two types of enzymes, the SPOUT superfamily for rRNA or tRNA-modifying methyltransferases and the splicing factor Rds3p [62]. TrmD is a member of the SPOUT superfamily and has a trefoil knot structure. The active sites can bind to the sugar ring of *S*-adenosyl methionine (SAM) and allow the enzyme to complete the methyl-transferring function. The knotted structure brings local stiffness, making it easier to capture the free energies of substrate binding for subsequent catalysis [63]. By circular permutation, Hsu's group removed the knot feature from the protein and found that unknotted mutants show higher folding kinetics and lower stability with the loss of ligand-binding capacity [64]. Hence, the conformational restriction by knotting makes the active site suitable for binding to SAM and for catalyzing the methylation reaction [65]. The 5_2 knotted UCHs participate in the hydrolysis of ubiquitin [17]. Interestingly, the knotted structure prevents itself from being degraded when it works.

Lasso peptides are an important type of ribosomally synthesized and posttranslationally modified peptides possessing a wide range of biological activities. Some of them can inhibit the uptake channels of bacterial nucleoside triphosphate (NTP) [66]. Complex lasso proteins are different. leptin is a typical complex lasso protein. The formation of disulfide bond under oxidizing conditions leads to the rigid lasso structure, but the part distal to the mechanical interlocking becomes more dynamic. Thus, the lasso structure influences the global dynamics, which in turn affects the subsequent cellular pathways [67]. The proteins with deterministic links, such as the endoglucanases superfamily, are stable even under very harsh conditions. For example, fungal endoglucanase is stable under 60 °C in pH from 3.0 to 11.0. Even after heating at 95 °C for five minutes, it still maintains 45% of its initial activity [68]. Other animal endoglucanases behave similarly with well-restored activity [69]. The mechanically interlocked rings can hardly disassociate without breaking the disulfide bonds. The embedding of these disulfide bridges inside the hydrophobic core of the proteins further stabilizes the structure [25]. Impressively, the chain-mail structure brings extraordinary stability to HK97, which could even resist denaturation in 5 M guanidine hydrochloride [24]. Therefore, complex mechanical interlocking may indeed enhance protein stability in different ways. The underlying rationale is the restricted or reshaped conformational space of the proteins. The energy barriers brought by mechanical interlocking increase the difficulty both for folding and unfolding. In view of the natural MIPs with exceptional properties, one would be optimistic about the future and promise of utilizing mechanical interlocking as a unique way of protein engineering.

6.5 Cultivating Mechanically Interlocked Proteins

The biological significance of natural MIPs offers a strong incentive for people to cultivate artificial MIPs for diverse applications such as protein therapeutics and materials. In principle, it is anticipated that natural and artificial MIPs share similar structure-activity relationships. While the former often exhibits moderate functional advantages with mere fitness, the latter has the potential to fully unleash the power of mechanical interlocking through molecular engineering. The p53dim protein catenane shows higher ability to resist thermal and chemical denaturation [23, 43]. At a concentration of 50 μM, its apparent melting temperature (T_m) is raised by 59 °C and its denaturation midpoint concentration (C_m of guanidine hydrochloride) is increased by 4.5 M compared to the open-ended dimer. In order to rationalize the stability enhancement, Zhou theoretically analyzed entropy restriction in the model system and attributed the effects to two key factors: backbone cyclization and constrained relative motion due to chain interlocking [70]. A 10 to 1000-fold more powerful stabilizing effect than simple covalent linking is predicted. Experimentally, similar stability improvement was also observed in artificial protein catenanes [57]. Notably, for a catenane with dihydrofolate reductase (DHFR), not only was the T_m increased, but a mild enhancement in catalytic activity (~27%) was also observed, probably due to the multivalent effects (Figure 6.4a). A mechanically interlocked 4-arm star protein based on SpyX module becomes highly tolerant to trypsin digestion as compared to the linear control [55b]. The heterocatenane containing green fluorescent protein (GFP) and DHFR synthesized by the active template method exhibits great improvement in its resistance to proteolysis, thermal denaturation, and freeze/thaw-induced mechanical denaturation [56]. Theoretically, mechanical interlocking is even tighter in knotted proteins with a higher number of crossings. For example, a serially knotted protein polymer of the trefoil knot carbonic anhydrase II did show increased T_m [41]. These improvements are of great significance to the practical applications of proteins.

The dynamic features of MIPs are also very intriguing. To date, for most protein rotaxanes, the ring components are mostly protein nanopores penetrating the membranes. Under different voltages, the threading complex can be stuck in the nanopores or pass through the channels. As a result, protein rotaxanes can be used as a macromolecular switch to regulate the transport of proteins inside and outside the membrane [22b]. In addition, this kind of transmembrane protein rotaxanes also provides a platform for single-molecule analysis and the study of refolding mechanism of translocated substrates [72]. Besides, by replacing the linker between the two bulky domains with a single-strand DNA of a specific sequence, this assembly can also be used for transporting DNA [73]. With more cell signaling pathways being studied, the protein rotaxane transport tool may have broader application prospects in the future. Lasso proteins are another type of dynamic MIP. Lasso peptide benenodin-1 exhibits a temperature-dependent switching behavior between different threaded states [74]. For lasso proteins with p53dim motif, they can be transformed into a tadpole-like structure upon heating and restore the

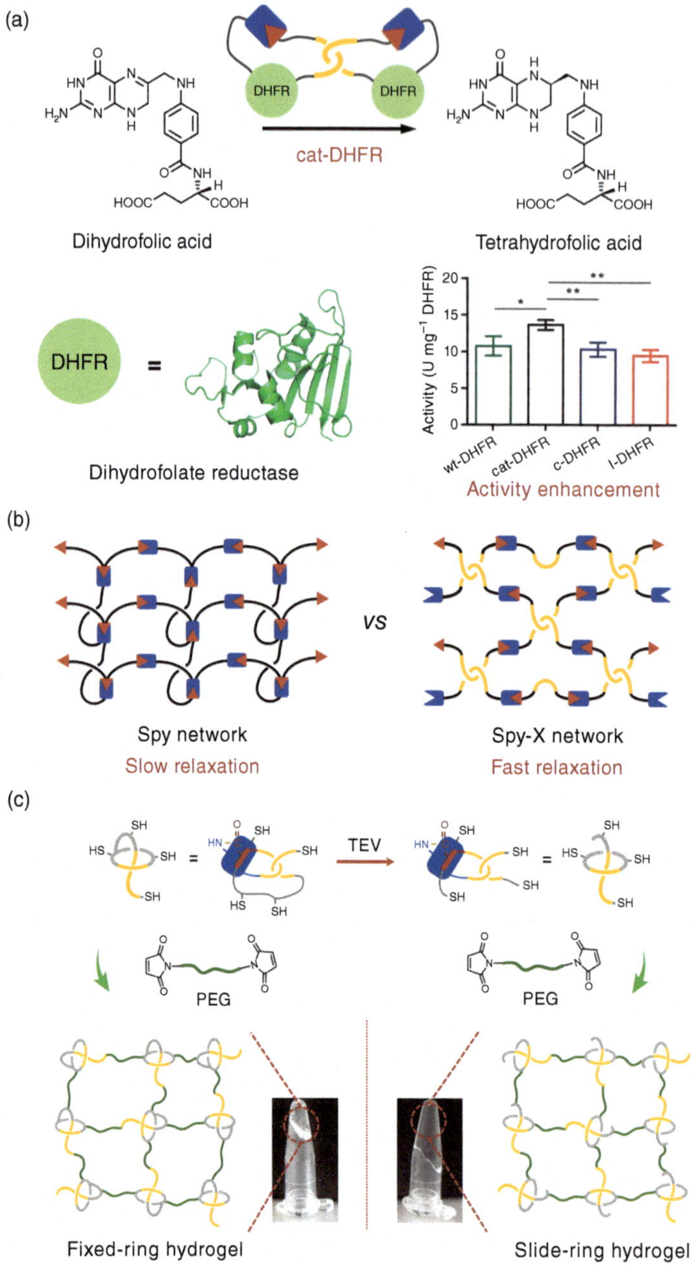

Figure 6.4 Cultivating artificial MIPs. (a) The homocatenane containing two DHFR domains demonstrates higher catalytic activity than the wild type (wt-DHFR). Source: Adapted from Wang and Zhang [57]. (b) The Spy-X network with designed entanglements exhibits faster stress-relaxation behavior than the control Spy network with only chemical crosslinking. Source: Adapted from Yang et al. [71]. (c) Lasso protein may be converted to protein pseudo-rotaxane by digestion, and the hydrogels obtained by these two isomeric motifs possess distinct properties in terms of swelling ratio and relaxation behaviors. Source: Adapted from Liu et al. [59a].

lasso structure upon cooling [59a]. Lasso proteins may also be converted to protein rotaxanes by enzymatic digestion at the loop region.

Assembling these dynamic MIPs into networks shall lead to interesting macroscopic hydrogel materials. For example, the cross-entwined p53dim motif may be introduced into the network as a rationally designed entanglement, which is also a physical crosslinking point. It was found that in the Spy network assembled by SpyTag/SpyCatcher reactions, the presence of p53dim motif significantly accelerates the relaxation behaviors of the hydrogel and helps promote the proliferation of the cells encapsulated in the cells (Figure 6.4b) [71]. Slide-ring and fixed-ring hydrogels could be prepared from the corresponding rotaxane and lasso precursors [59a]. Interestingly, the former exhibits much higher swelling ratio and distinct rheological behavior at the terminal region, consistent with the presence of sliding crosslinking points (Figure 6.4c). Other chemically crosslinked hydrogels may also be prepared whose properties may be programmed through embedding specific functional sequences (such as Arg-Gly-Asp peptide (RGD) for cell adhesion or p53dim for controlled relaxation) for various biomedical applications [71]. These bioactive hydrogels are promising matrixes for directing cell fate [75] by controlling the mechanical properties (such as elasticity [76], degradation rate [77], and stress relaxation [78]) and the bioactive signals (such as those for cell adhesion, proliferation, and differentiation) [79]. They are particularly useful as model systems to elucidate the influence of certain factors since their properties could be drastically changed even just by one mutation [80]. Nevertheless, the exploration of MIPs' application is still in its infancy, and the field is plagued by the unknown structure–property relationship of MIPs and the ultimate protein complexity. There are equally grand challenges and opportunities in cultivating artificial MIPs in protein engineering.

6.6 Conclusion and Future Perspective

In summary, MIPs mainly include protein knots, slipknots, lassos, rotaxanes, and links. Compared with their linear and cyclic counterparts, MIPs do possess certain functional benefits, particularly in protein stability enhancement. Enzymes account for a significant portion of natural MIPs discovered to date. Their active sites are often closely related to the entangled structure, whose rigidity enhances their ability to tolerate mutations as well as environmental changes. The past decade has witnessed an increasing number of artificial MIPs, mostly knots and catenanes, developed in the hope of enhancing protein stability and activity for biomedical applications such as synergistic therapeutics or bioactive materials.

Nonetheless, there are still many unresolved issues. First of all, we lack a thorough understanding of the structure–property relationship of MIPs. The study is plagued by the lack of MIP samples and also by the protein complexity to certain extent. Apparently, we need new chemical tools for a clean design and synthesis of MIPs without affecting their native structure and function. The most desirable candidates, in our opinion, are the genetically encoded protein ligation tools with traceless features, such as split-intein technology, the three-component protein/peptide reactive pairs mentioned in Section 6.3, and the use of noncanonical amino acid for preprogrammed reactivity. Another issue lies in the determination of MIPs' topology. The structure analysis techniques for proteins such as cryo-EM and X-ray crystal diffraction are quite advanced and no doubt will be among the first choices to unambiguously determine the structure and topology, but they are usually time and resource consuming and may fail in certain cases, especially when the entangled regions are intrinsically flexible and disordered. New physical tools are thus urgently needed to help facilely characterize the mechanically interlocked structures, such as ion mobility spectrometry–mass spectrometry, nanopore-based technology, or structure ensemble characterization by NMR. A plethora of artificial designer MIPs is a prerequisite for this study. It is highly desirable to have a fast and convenient screening method for selecting samples potentially with unconventional topologies before devoting a great deal of time and effort to identifying unambiguously their topology. Moreover, MIPs should be placed in a specific context to demonstrate certain "killer" applications to justify the significance of such research efforts. The most promising areas are perhaps therapeutic proteins and industrial enzymes. Introducing mechanical interlocking into these proteins is excepted to improve their practical performance via optimizing their biophysical properties, including thermal/chemical stability, organic solvent tolerance, *in vivo* circulation time, and so on. There is certainly a long way to go for MIPs. However, we are confident that mechanical interlocking provides a unique way for protein stabilization, and MIPs hold great promise for diverse applications in biomedicine and biomaterials.

Acknowledgments

We are grateful for the financial support from the National Natural Science Foundation of China (No. 21925102, 21991132, and 92056118), the National Key R&D Program of China (No. 2020YFA0908100), Beijing Municipal Natural Science Foundation (No. L182003), Beijing National Laboratory for Molecular Sciences (BNLMS-CXXM-202006), and Clinical Medicine Plus X project of Peking University, Fundamental Research Funds for the Central Universities.

References

1 Heard, A.W. and Goldup, S.M. (2020). Simplicity in the design, operation, and applications of mechanically interlocked molecular machines. *ACS Cent. Sci.* 6: 117.

2 Bruns, C.J. and Stoddart, J.F. (2016). *The Nature of the Mechanical Bond: From Molecules to Machines*. Hoboken, NJ: Wiley.

3 Anelli, P.L., Ashton, P.R., Ballardini, R. et al. (1992). Molecular meccano. 1. [2]rotaxanes and a [2]catenane made to order. *J. Am. Chem. Soc.* 114: 193.

4 Nierengarten, J.F., Dietrich-Buchecker, C.O., and Sauvage, J.P. (1994). Synthesis of a doubly interlocked [2]-catenane. *J. Am. Chem. Soc.* 116: 375.

5 Chichak, K.S., Cantrill, S.J., Pease, A.R. et al. (2004). Molecular borromean rings. *Science* 304: 1308.

6 Leigh, D.A., Pritchard, R.G., and Stephens, A.J. (2014). A star of David catenane. *Nat. Chem.* 6: 978.

7 Anelli, P.L., Spencer, N., and Stoddart, J.F. (1991). A molecular shuttle. *J. Am. Chem. Soc.* 113: 5131.

8 Asakawa, M., Ashton, P.R., Balzani, V. et al. (1998). A chemically and electrochemically switchable [2]catenane incorporating a tetrathiafulvalene unit. *Angew. Chem. Int. Ed.* 37: 333.

9 Balzani, V., Credi, A., Raymo, F.M., and Stoddart, J.F. (2000). Artificial molecular machines. *Angew. Chem. Int. Ed.* 39: 3348.

10 Dommaschk, M., Echavarren, J., Leigh, D.A. et al. (2019). Dynamic control of chiral space through local symmetry breaking in a rotaxane organocatalyst. *Angew. Chem. Int. Ed.* 58: 14955.

11 Sell, H., Gehl, A., Plaul, D. et al. (2019). Towards a light driven molecular assembler. *Commun. Chem.* 2: 62.

12 De Bo, G., Gall, M.A.Y., Kuschel, S. et al. (2018). An artificial molecular machine that builds an asymmetric catalyst. *Nat. Nanotechnol.* 13: 381.

13 Zhang, W.-B. and Cheng, S.Z.D. (2020). Giant is different: size effects and the nature of macromolecules. *Giant* 1: 100011.

14 (a) Cao, P.-F., Rong, L.-H., Mangadlao, J.D., and Advincula, R.C. (2017). Synthesizing a trefoil knotted block copolymer via ring-expansion strategy. *Macromolecules* 50: 1473. (b) Wu, Q., Rauscher, P.M., Lang, X. et al. (2017). Poly[n]catenanes: synthesis of molecular interlocked chains. *Science* 358: 1434.

15 (a) Sulkowska, J.I. (2020). On folding of entangled proteins: knots, lassos, links and theta-curves. *Curr. Opin. Struct. Biol.* 60: 131. (b) King, N.P., Yeates, E.O., and Yeates, T.O. (2007). Identification of rare slipknots in proteins and their implications for stability and folding. *J. Mol. Biol.* 373: 153. (c) Taylor, W.R. (2007). Protein knots and fold complexity: some new twists. *Comput. Biol. Chem.* 31: 151.

16 Yeates, T.O., Norcross, T.S., and King, N.P. (2007). Knotted and topologically complex proteins as models for studying folding and stability. *Curr. Opin. Chem. Biol.* 11: 595.

17 Dabrowski-Tumanski, P. and Sulkowska, J.I. (2017). To tie or not to tie? *That is the question. Polymers* 9: 454.

18 (a) Noel, J.K., Sulkowska, J.I., and Onuchic, J.N. (2010). Slipknotting upon native-like loop formation in a trefoil knot protein. *Proc. Natl. Acad. Sci. U. S. A.* 107: 15403. (b) Bolinger, D., Sulkowska, J.I., Hsu, H.P. et al. (2010). A Stevedore's protein knot. *PLoS Comput. Biol.* 6: e1000731.

19 (a) Alexander, J.W. and Briggs, G.B. (1926). On types of knotted curves. *Ann. Math.* 28: 562. (b) Sulkowska, J.I., Sulkowski, P., Szymczak, P., and Cieplak, M. (2008). Tightening of knots in proteins. *Phys. Rev. Lett.* 100: 058106.

20 Hegemann, J.D., Zimmermann, M., Xie, X., and Marahiel, M.A. (1909). Lasso peptides: an intriguing class of bacterial natural products. *Acc. Chem. Res.* 2015: 48.

21 Haglund, E., Sulkowska, J.I., He, Z. et al. (2012). The unique cysteine knot regulates the pleotropic hormone leptin. *PLoS One* 7: e45654.

22 (a) Franceschini, L., Soskine, M., Biesemans, A., and Maglia, G. (2013). A nanopore machine promotes the vectorial transport of DNA across membranes. *Nat. Commun.* 4: 2415. (b) Biesemans, A., Soskine, M., and Maglia, G. (2015). A protein rotaxane controls the translocation of proteins across a ClyA nanopore. *Nano Lett.* 15: 6076.

23 Blankenship, J.W. and Dawson, P.E. (2003). Thermodynamics of a designed protein catenane. *J. Mol. Biol.* 327: 537.

24 Wikoff, W.R., Liljas, L., Duda, R.L. et al. (2000). Topologically linked protein rings in the bacteriophage HK97 capsid. *Science* 289: 2129.

25 Dabrowski-Tumanski, P. and Sulkowska, J.I. (2017). Topological knots and links in proteins. *Proc. Natl. Acad. Sci. U. S. A.* 114: 3415.

26 (a) Jamroz, M., Niemyska, W., Rawdon, E.J. et al. (2015). KnotProt: a database of proteins with knots and slipknots. *Nucleic Acids Res.* 43: D306. (b) Dabrowski-Tumanski, P., Rubach, P., Goundaroulis, D. et al. (2019). KnotProt 2.0: a database of proteins with knots and other entangled structures. *Nucleic Acids Res.* 47: D367.

27 (a) Dabrowski-Tumanski, P., Niemyska, W., Pasznik, P., and Sulkowska, J.I. (2016). LassoProt: server to analyze biopolymers with lassos. *Nucleic Acids Res.* 44: W383. (b) Gierut, A.M., Niemyska, W., Dabrowski-Tumanski, P. et al. (2017). PyLasso: a PyMOL plugin to identify lassos. *BMC Bioinform.* 33: 3819.

28 (a) Dabrowski-Tumanski, P., Jarmolinska, A.I., Niemyska, W. et al. (2017). LinkProt: a database collecting information about biological links. *Nucleic Acids Res.* 45: D243. (b) Gierut, A.M., Dabrowski-Tumanski, P., Niemyska, W. et al. (2019). PyLink: a PyMOL plugin to identify links. *BMC Bioinform.* 35: 3166.

29 Qu, Z., Cheng, S.Z.D., and Zhang, W.-B. (2021). *Macromolecular topology engineering. Trends Chem.* 3: 402.

30 Zhang, B., Li, Y., Shi, W. et al. (2020). Chemical synthesis of proteins containing 300 amino acids. *Chem. Res. Chin. Univ.* 36: 733.

31 Sułkowska, J.I., Sułkowski, P., and Onuchic, J. (2009). Dodging the crisis of folding proteins with knots. *Proc. Natl. Acad. Sci. U. S. A.* 106: 3119.

32 Noel, J.K., Onuchic, J.N., and Sulkowska, J.I. (2013). Knotting a protein in explicit solvent. *J. Phys. Chem. Lett.* 4: 3570.

33 (a) King, N.P., Jacobitz, A.W., Sawaya, M.R. et al. (2010). Structure and folding of a designed knotted protein. *Proc. Natl. Acad. Sci. U. S. A.* 107: 20732. (b) Sulkowska, J.I., Noel, J.K., and Onuchic, J.N. (2012). Energy landscape of knotted protein folding. *Proc. Natl. Acad. Sci. U. S. A.* 109: 17783.

34 (a) Bryngelson, J.D., Onuchic, J.N., Socci, N.D., and Wolynes, P.G. (1995). Funnels, pathways, and the energy landscape of protein folding: a synthesis. *Proteins Struct. Funct. Bioinform.* 21: 167. (b) Mallam, A.L., Onuoha, S.C., Grossmann, J.G., and Jackson, S.E. (2008). Knotted fusion proteins reveal unexpected possibilities in protein folding. *Mol. Cell* 30: 642. (c) Gosavi, S., Chavez, L.L., Jennings, P.A., and Onuchic, J.N. (2006). Topological frustration and the folding of interleukin-1 beta. *J. Mol. Biol.* 357: 986.

35 Dabrowski-Tumanski, P., Piejko, M., Niewieczerzal, S. et al. (2018). Protein knotting by active threading of nascent polypeptide chain exiting from the ribosome exit channel. *J. Phys. Chem. B* 122: 11616.

36 Brinker, A., Pfeifer, G., Kerner, M.J. et al. (2001). Dual function of protein confinement in chaperonin-assisted protein folding. *Cell* 107: 223.

37 (a) Zhao, Y., Dabrowski-Tumanski, P., Niewieczerzal, S., and Sulkowska, J.I. (2018). The exclusive effects of chaperonin on the behavior of proteins with 5_2 knot. *PLoS Comput. Biol.* 14: e1005970. (b) Mallam, A.L. and Jackson, S.E. (2011). Knot formation in newly translated proteins is spontaneous and accelerated by chaperonins. *Nat. Chem. Biol.* 8: 147.

38 Niewieczerzal, S. and Sulkowska, J.I. (2017). Knotting and unknotting proteins in the chaperonin cage: effects of the excluded volume. *PLoS One* 12: e0176744.

39 Niemyska, W., Dabrowski-Tumanski, P., Kadlof, M. et al. (2016). Complex lasso: new entangled motifs in proteins. *Sci. Rep.* 6: 36895.

40 Helgstrand, C., Wikoff, W.R., Duda, R.L. et al. (2003). The refined structure of a protein catenane: the HK97 bacteriophage capsid at 3.44 A resolution. *J. Mol. Biol.* 334: 885.

41 Sayre, T.C., Lee, T.M., King, N.P., and Yeates, T.O. (2011). Protein stabilization in a highly knotted protein polymer. *Protein Eng. Des. Sel.* 24: 627.

42 Allen, C.D. and Link, A.J. (2016). Self-assembly of catenanes from lasso peptides. *J. Am. Chem. Soc.* 138: 14214.

43 Blankenship, J.W. and Dawson, P.E. (2007). Threading a peptide through a peptide: protein loops, rotaxanes, and knots. *Protein Sci.* 16: 1249.

44 Zhang, F. and Zhang, W.B. (2020). Encrypting chemical reactivity in protein sequences toward information-coded reactions. *Chinese J. Chem.* 38: 864.

45 Panne, D., Liu, X.-Q., Evans, T.C. et al. (2000). Protein trans-splicing and cyclization by a naturally split intein from the dnaE Gene of synechocystis species PCC6803. *J. Biol. Chem.* 275: 9091.

46 Antos, J.M., Popp, M.W., Ernst, R. et al. (2009). Stabilization of proteins by covalent cyclization. *J. Biol. Chem.* 284: 16028.

47 Purkayastha, A. and Kang, T.J. (2019). Spontaneous intermolecular amide bond formation between side chains for irreversible peptide targeting. *Biotechnol. Bioproc. E.* 24: 702.

48 Zakeri, B. and Howarth, M. (2010). Peptide tag forming a rapid covalent bond to a protein, through engineering a bacterial adhesin. *J. Am. Chem. Soc.* 132: 4526.

49 Zakeri, B., Fierer, J.O., Celik, E. et al. (2012). Programmable polyproteams built using twin peptide superglues. *Proc. Natl. Acad. Sci. U. S. A.* 109: E690.

50 Veggiani, G., Nakamura, T., Brenner, M.D. et al. (2016). Kinetic controlled tag-catcher interactions for directed covalent protein assembly. *Proc. Natl. Acad. Sci. U. S. A.* 113: 1202.

51 Tan, L.L., Hoon, S.S., and Wong, F.T. (2016). SpyLigase peptide-peptide ligation polymerizes affibodies to enhance magnetic cancer cell capture. *PLoS One* 11: e0165074.

52 Fierer, J.O., Veggiani, G., and Howarth, M. (2014). SpyLigase peptide-peptide ligation polymerizes affibodies to enhance magnetic cancer cell capture. *Proc. Natl. Acad. Sci. U. S. A.* 111: E1176.

53 Buldun, C.M., Jean, J.X., Bedford, M.R., and Howarth, M. (2018). SnoopLigase catalyzes peptide-peptide locking and enables solid-phase conjugate isolation. *J. Am. Chem. Soc.* 140: 3008.

54 Wu, X.L., Liu, Y., Liu, D. et al. (2018). An intrinsically disordered peptide-peptide stapler for highly efficient protein ligation both in vivo and in vitro. *J. Am. Chem. Soc.* 140: 17474.

55 (a) Wang, X.W. and Zhang, W.B. (2016). Cellular synthesis of protein catenanes. *Angew. Chem. Int. Ed.* 55: 3442. (b) Liu, D., Wu, W.H., Liu, Y.J. et al. (2017). Topology engineering of proteins in vivo using genetically encoded, mechanically interlocking SpyX modules for enhanced stability. *ACS Cent. Sci.* 3: 473.

56 Da, X.D. and Zhang, W.B. (2019). Active template synthesis of protein heterocatenanes. *Angew. Chem. Int. Ed.* 58: 11097.

57 Wang, X.W. and Zhang, W.B. (2017). Protein catenation enhances both the stability and activity of folded structural domains. *Angew. Chem. Int. Ed.* 56: 13985.

58 Zhang, X.J., Wu, X.L., Wang, X.W. et al. (2018). SpyCatcher-N(TEV): a circularly permuted, disordered SpyCatcher variant for less trace ligation. *Bioconjug. Chem.* 29: 1622.

59 (a) Liu, Y., Wu, W.H., Hong, S. et al. (2020). Lasso Proteins: modular design, cellular synthesis, and topological transformation. *Angew. Chem. Int. Ed.* 59: 19153. (b) Liu, Y., Duan, Z., Fang, J. et al. (2020). Cellular synthesis and X-ray crystal structure of a designed protein heterocatenane. *Angew. Chem. Int. Ed.* 59: 16122.

60 Virnau, P., Mirny, L.A., and Kardar, M. (2006). Intricate knots in proteins: function and evolution. *PLoS Comput. Biol.* 2: e122.

61 Sułkowska, J.I., Sułkowski, P., Szymczak, P., and Cieplak, M. (2008). Stabilizing effect of knots on proteins. *Proc. Natl. Acad. Sci. U. S. A.* 105: 19714.

62 Dabrowski-Tumanski, P., Stasiak, A., and Sulkowska, J.I. (2016). In Search of functional advantages of knots in proteins. *PLoS One* 11: e0165986.

63 (a) Christian, T., Sakaguchi, R., Perlinska, A.P. et al. (2016). Methyl transfer by substrate signaling from a knotted protein fold. *Nat. Struct. Mol. Biol.* 23: 941. (b) Hou, Y.M., Matsubara, R., Takase, R. et al. (2017). TrmD: a methyl transferase for tRNA methylation with m^1G37. *Enzymes* 41: 89.

64 Chuang, Y.C., Hu, I.C., Lyu, P.C., and Hsu, S.D. (2019). Untying a protein knot by circular permutation. *J. Mol. Biol.* 431: 857.

65 Perlinska, A.P., Stasiulewicz, A., Nawrocka, E.K. et al. (2020). Restriction of S-adenosylmethionine conformational freedom by knotted protein binding sites. *PLoS Comput. Biol.* 16: e1007904.

66 (a) Yuzenkova, J., Delgado, M., Nechaev, S. et al. (2002). Mutations of bacterial RNA polymerase leading to resistance to microcin J25. *J. Biol. Chem.* 277: 50867. (b) Mukhopadhyay, J., Sineva, E., Knight, J. et al. (2004). Antibacterial peptide microcin J25 inhibits transcription by binding within and obstructing the RNA polymerase secondary channel. *Mol. Cell* 14: 739. (c) Adelman, K., Yuzenkova,

J., La Porta, A. et al. (2004). Molecular mechanism of transcription inhibition by peptide antibiotic microcin J25. *Mol. Cell* 14: 753.

67 (a) Haglund, E., Sulkowska, J.I., Noel, J.K. et al. (2014). Pierced lasso bundles are a new class of knot-like motifs. *PLoS Comput. Biol.* 10: e1003613. (b) Haglund, E., Pilko, A., Wollman, R. et al. (2017). Pierced lasso topology controls function in leptin. *J. Phys. Chem. B* 121: 706.

68 Hayashida, S., Ohta, K., and Mo, K. (1988). Cellulases of Humicola insolens and Humicola grisea. *Method Enzymol.* 160: 323.

69 Xu, B., Hellman, U., Ersson, B., and Janson, J.-C. (2000). Purification, characterization and amino-acid sequence analysis of a thermostable, low molecular mass endo-β-1,4-glucanase from blue mussel, Mytilus edulis. *Eur. J. Biochem.* 267: 4970.

70 (a) Zhou, H.-X. (2003). Effect of catenation on protein folding stability. *J. Am. Chem. Soc.* 125: 9280. (b) Zhou, H.-X. (2004). Loops, linkages, rings, catenanes, cages, and crowders: entropy-based strategies for stabilizing proteins. *Acc. Chem. Res.* 37: 123.

71 Yang, Z., Kou, S., Wei, X. et al. (2018). Genetically programming stress-relaxation behavior in entirely protein-based molecular networks. *ACS Macro Lett.* 7: 1468.

72 Feng, J., Martin-Baniandres, P., Booth, M.J. et al. (2020). Transmembrane protein rotaxanes reveal kinetic traps in the refolding of translocated substrates. *Commun. Biol.* 3: 159.

73 Bayoumi, M., Nomidis, S.K., Willems, K. et al. (2021). Autonomous and active transport operated by an entropic DNA piston. *Nano Lett.* 21: 762.

74 Zong, C., Wu, M.J., Qin, J.Z., and Link, A.J. (2017). Lasso peptide benenodin-1 is a thermally actuated [1]rotaxane switch. *J. Am. Chem. Soc.* 139: 10403.

75 Wong, S.W., Lenzini, S., Bargi, R. et al. (2020). Controlled deposition of 3D matrices to direct single cell functions. *Adv. Sci.* 7: 2001066.

76 Engler, A.J., Sen, S., Sweeney, H.L., and Discher, D.E. (2006). Matrix elasticity directs stem cell lineage specification. *Cell* 126: 677.

77 Khetan, S., Guvendiren, M., Legant, W.R. et al. (2013). Degradation-mediated cellular traction directs stem cell fate in covalently crosslinked three-dimensional hydrogels. *Nat. Mater.* 12: 458.

78 Chaudhuri, O., Gu, L., Klumpers, D. et al. (2016). Hydrogels with tunable stress relaxation regulate stem cell fate and activity. *Nat. Mater.* 15: 326.

79 DeForest, C.A. and Anseth, K.S. (2012). Advances in bioactive hydrogels to probe and direct cell fate. *Annu. Rev. Chem. Biomol. Eng.* 3: 421.

80 Dooling, L.J. and Tirrell, D.A. (2016). Engineering the dynamic properties of protein networks through sequence variation. *ACS Cent. Sci.* 2: 812.

7

Recent Advances on Catenanes and Rotaxanes Made of DNA

Yinzhou Ma[1], *Ze Yu*[2], *and Julián Valero*[3]

[1]*Wenzhou Institute University of Chinese Academy of Sciences, Wenzhou, Zhejiang 325001, China*
[2]*University of Bonn, Chemical Biology and Medicinal Chemistry Unit, Life and Medical Sciences (LIMES) Institute, Bonn, Germany*
[3]*Aarhus University, Interdisciplinary Nanoscience Center (iNANO) and Department of Molecular Biology and Genetics (MBG), Gustav Wieds Vej, 14, 8000 Aarhus, Denmark*

7.1 Introduction

Mimicking the structure and function of proteins is a longstanding goal in chemistry and synthetic biology. In particular, biological molecular motors are exceptional examples of complex proteins essential for living organisms that display key functions in cells. Biological motors operate out-of-equilibrium and are driven by physical inputs (e.g. mechano- and heat sensory transduction proteins such as the mechanosensitive ion channel protein PIEZO [1] and transient receptor potential cation channel (TRPV1) [2] protein channels that selectively open and activate for ion transport upon mechanical stress or heat response, respectively) or highly efficient catalytic reactions in which chemical energy is converted into work – a textbook example is kinesin that uses adenosine triphosphate (ATP) hydrolysis to move along microtubules [3, 4]. An essential feature shared in all protein motors is their ability to move and switch conformations in a controlled fashion with (sub-)nanometer precision. Due to these unique features and biological relevance, over the last decades, there has been an increasing interest in developing molecules that can mimic the features and functions of protein motors in order to better understand their mechanisms and precise nanoscale-directed movement.

One of the most prominent examples of synthetic systems used as building blocks for mimicking biological motors are mechanically interlocked molecules (MIMs) [5, 6]. MIMs are molecules consisting of two or more components joined together through mechanical bonds. These topologically complex architectures allow specific intramolecular control of the movement and conformational rearrangement of their mechanically interlocked components. These features are crucial to building molecular systems that can switch between two or more different states or harness the controlled motion of the interlocked components to design molecular motors. Most relevant examples of MIMs are catenanes – two or more rings mechanically

Mechanically Interlocked Materials: Polymers, Nanomaterials, MOFs, and more, First Edition.
Edited by Emilio M. Pérez.

interlocked – and rotaxanes – one or more rings threaded into a rod-like molecule, although more exotic and topologically complex structures such as molecular Borromean rings and knots [7] have also been described. The first examples of synthetic MIMs were reported in the 1960s by Wasserman [8] and Schill and Lüttringhaus [9] for catenanes and Harrison and Harrison [10] for rotaxanes. However, the field was revolutionized decades later by Sauvage [11] and Stoddart, among others, by introducing a template-directed synthesis that considerably improved the yields of these constructs. Since then, MIMs have found a wide range of applications in material science, optics, electronics, biosensors, and biomedicine [12, 13].

Despite the progress in the field [14], synthetic organic MIMs have drawbacks such as complex multistep synthesis or poor biocompatibility and water solubility, which limits their broad applicability. To overcome these issues, in recent years, scientists have alternatively exploited the use of DNA and DNA nanotechnology to easily design and assemble a plethora of mechanically interlocked structures [15].

In particular, DNA nanotechnology has become a powerful toolbox that uses the programmability of DNA to build complex 2D and 3D structures with high precision in the nanoscale [16–18]. Using simple base-pairing rules and structural principles, DNA has been employed as building block to develop nanomechanical devices, chemically and physically driven switches, membrane nanopores, molecular motors and multivalent recognition elements, among others [19–21].

DNA is highly programmable and easy to self-assemble. Moreover, it is biocompatible and water-soluble, thus enabling a straightforward implementation in biomedical and diagnosis applications. A key feature of nucleic acids is the possibility to establish sequence-specific interactions with different proteins. These binding properties can be explored to design new biohybrid systems that use DNA as a supportive architecture to scaffold enzymes and proteins with specific catalytic and molecular recognition properties.

Herein, we present the most relevant examples and recent advances in DNA-based interlocked nanostructures, current applications, and future perspectives in the field.

7.2 DNA Catenanes

Catenanes represent one of the most fundamental classes of MIMs. They consist of two or more ring structures that are connected through interlocked bonds instead of covalent bonds (Figure 7.1). In nature, DNA catenanes often serve as the intermediates during the replication of circular DNA [22, 23] and can be created by making a transient break on a double-stranded DNA (dsDNA) ring, which threads to another dsDNA ring in the presence of specific topoisomerases [24, 25]. However, there have been only a few attempts to synthesize DNA catenanes with DNA rings and topoisomerases *in vitro* [24–26], given the fact that topoisomerases can also disentangle and resolve topologically linked DNA catenanes into individual rings [24, 25].

The general strategy for synthesizing catenane structures in DNA nanotechnology involves three steps: first, threading and hybridization of a linear DNA strand to a

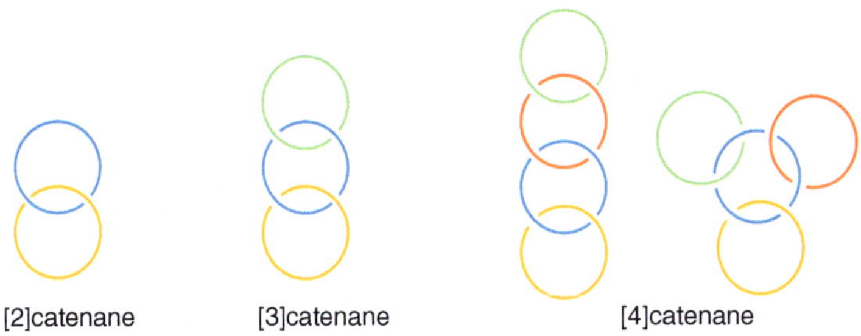

Figure 7.1 Representation of a [2]catenane, [3]catenane, and [4]catenanes.

DNA ring; then, circularizing the threading DNA strand to form a DNA structure where the interlocked ring remains immobile; and finally, releasing the newly made DNA ring to form a mechanically interlocked DNA catenane. One of the earliest examples of this strategy was reported by Liang et al. in 2006 [27]. In this report, the authors presented two different approaches for synthesizing a [2]catenane by allowing different parts of the threading single-stranded DNA (ssDNA) strand to hybridize with a ssDNA ring (Figure 7.2a). However, due to the helical structure of dsDNA, the length of the hybridization region between the ssDNA ring and the ssDNA strand also affects the mobility of the resulting catenane. According to Liang' s study, when the length of the hybridization region is 12 base pairs (bp), the main product was a catenane with the linking number of 1 (LK1, see Figure 7.2a), while when the length of the hybridization region is 15 bp or more, catenanes with larger linking number values were predominantly formed (Figure 7.2a). Based on a similar concept, other methods have also been developed. For example, Liu et al. synthesized a tailed DNA/organic catenane by creating a PEG-amide-PEG linkage across one full turn of the threading DNA (Figure 7.2b) [28]. Instead of relying on DNA hybridization to fix the threading DNA to the DNA ring, Heckel and coworkers assembled a dsDNA catenane by employing polyamides that bind specific DNA sequences to guide the assembly of two DNA half-circles containing repetitive AT-tracts forming an interlocked structure and pointing their ends into opposite directions (Figure 7.2c) [29].

Following similar strategies, complex catenanes with three or more DNA rings have also been described. Valero et al. reported a method of synthesizing dsDNA [3]catenane by ligating two dsDNA rings which have pre-hybridized with half of the middle ring in the [3]catenane respectively (Figure 7.2d) [30]. ssDNA catenanes consisting of five rings (Figure 7.2e) [31] and seven rings (Figure 7.2f) [32] were successfully synthesized by Lu et al.

An alternative approach for catenane assembly was reported by Wu et al. [33]. In order to avoid strong duplex DNA interactions for catenane formation (especially in ssDNA catenanes), which hampers the mobility of the interlocked rings, they developed an *in vitro* selection strategy to engineer ssDNA [2]catenanes without a linking duplex. Using such weak physical interactions, the rings are capable of freely operating in an independent manner, which is especially useful for designing molecular machines.

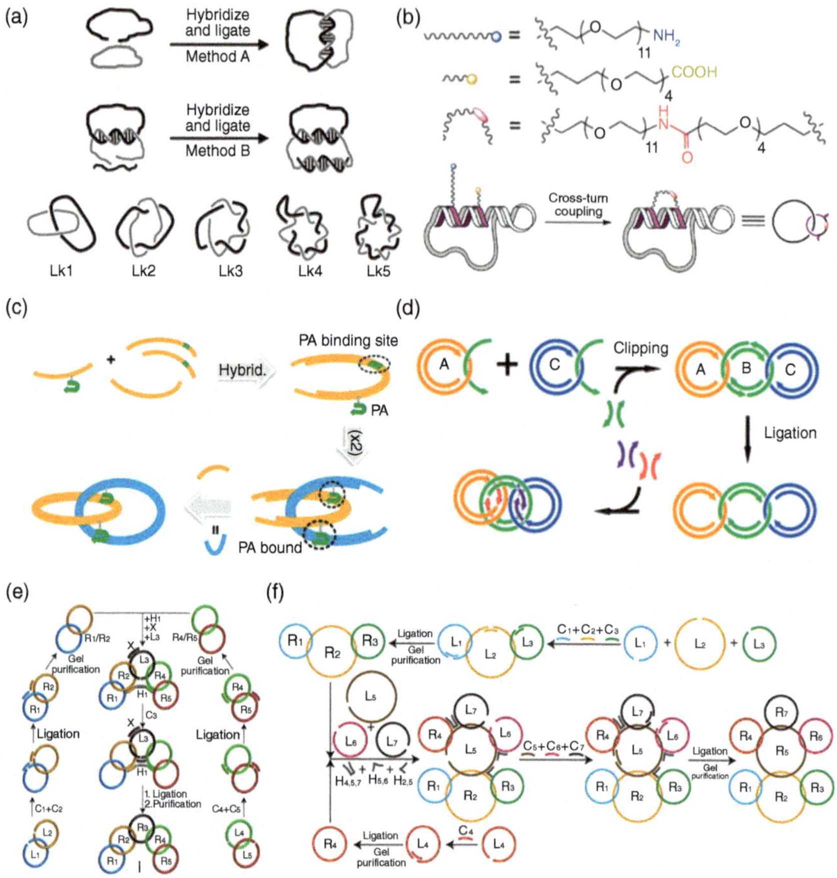

Figure 7.2 (a) Two methods to form [2]DNA catenanes and the topology of DNA catenanes with a different linking number (LK). In method A, the two termini of the linear ssDNA strand hybridize first to the ssDNA ring, and then the nick is sealed by DNA ligase. In method B, the central part of the linear ssDNA strand is with the ssDNA ring first, then two ends of the linear ssDNA strand are ligated after hybridizing with a short splint ssDNA. Source: Reproduced with permission from Liang et al. [27]/Elsevier. (b) Formation of a [2]DNA/organic catenane by creating a PEG-amide-PEG linkage with modified ends of the threading ssDNA. Source: Reproduced with permission from Liu et al. [28]/American Chemical Society. (c) Synthesis of a dsDNA [2]catenane by employing polyamides as anchors to bind two DNA half-circles in an interlocked structure. Source: Reproduced with permission from Schmidt and Heckel [29]/American Chemical Society. (d) Synthesis of a dsDNA [3]catenane by clipping two previously assembled and threaded rings. Source: Reproduced with permission from Valero et al. [30]/Springer Nature. (e) Schematic synthesis of a five-ring catenane. Source: Reproduced with permission from Lu et al. [31]/John Wiley & Sons. (f) Schematic synthesis of a seven-ring interlocked catenane. Source: Reproduced with permission from Lu et al. [32]/American Chemical Society.

DNA origami technology has also been employed for the formation of larger catenated structures. DNA origami are self-assembled structures consisting of a long DNA "template" strand that is folded into a designed 2 or 3D shape using small "staple" strands. Peil et al. recently described the formation of DNA origami catenanes templated by gold nanoparticles (AuNPs). In particular, curve-shaped DNA origami was decorated in specific positions with thiol-functionalized strands for capturing AuNPs, which were used to drive the assembly and hold together different interlocked catenated DNA nanostructures [34]. A recent example of using topoisomerases for developing complex DNA origami catenated nanostructures was reported by the Heddle lab. Starting from a designed double-stranded plasmid, the addition of Tn3 resolvase with topoisomerase activity and subsequent nicking and exonuclease treatment yielded two catenated ssDNA rings that serve as scaffolds for the assembly of a new class of topologically linked origami, coined as toporigami [35].

Compared to other "static" DNA nanostructures, the key feature of a DNA catenane relies on its mechanically interlocked structure. This structure can form topological bonds to hold multiple independent DNA nanostructures, such as DNA rings, together as a single "supra-structure." Although those topological bonds in the catenane structure restrict the degrees of freedom of movement of the components relative to each other, they still provide much greater mobility than any other type of chemical bond can offer, allowing all components to move freely within the interlocked space without structural disassembly.

Based on these features, Lu et al. first reported a three-station rotor system composed of a DNA [2]catenane and triggered by Hg^{2+}/cysteine and H^+/OH^- stimuli in 2013 [36]. In this study, two interlocked ssDNA rings (Ring α and Ring β) were stabilized via inter-ring hybridization (Figure 7.3a). Ring α was hybridized with an ssDNA strand (5), which was modified with a BHQ-2 quencher used for indicating the different states of [2]catenane. Ring β was designed with three different domains (brown, red, and blue) that can hybridize with Ring α in different conditions. A ssDNA strand (number 6 in Figure 7.3a) with a cytosine-rich sequence was hybridized with one of the domains (blue) and served as a blocker. In its initial state (State I), Ring α hybridizes with the brown domain of Ring β, resulting in a fluorescent signal of Cy3 and quenching of Cy5. In the presence of Hg^{2+}, the [2]catenane will switch from state I to state II, in which the two rings are stabilized by an energetically favored duplex domain characterized by cooperative T-Hg^{2+}-T bridges that assist base-pairing. In this state, the fluorophore Cy3 is quenched, resulting in a fluorescent signal of Cy5. The addition of cysteine to the catenane in state II will release the Hg^{2+} ions from the duplex and regenerate state I.

On the other hand, due to its cytosine-rich sequence, strand (6) can form an i-motif structure at pH 5.2, consequently exposing another complementary domain (blue) for Ring α on Ring β. Since the blue domain has a longer complementary sequence with Ring α than the brown domain, this process will cause the [2]catenane to switch from state I to the energetically favored state III, resulting in the fluorescent signals of both Cy3 and Cy5. By adjusting the pH of the system to 7.2, the i-motif structure of strand (6) will again be disassembled. Consequently, strand (6) will displace Ring

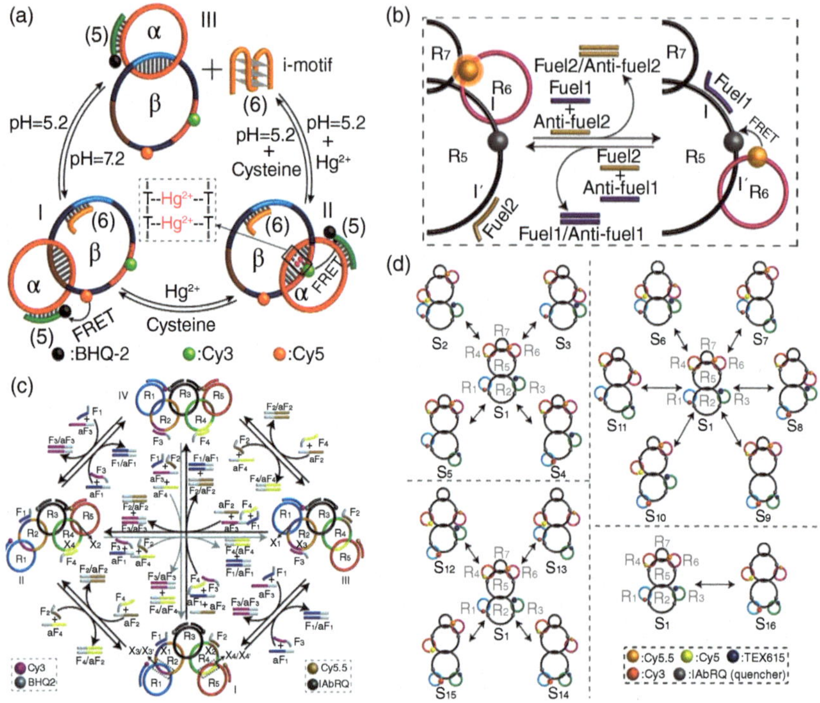

Figure 7.3 (a) Reversible switch of the rotor Ring α between three states (I, II, and III) controlled by Hg^{2+}/cysteine and H^+/OH^-. Source: Reproduced with permission from Lu et al. [36]/American Chemical Society. (b) Using strand displacement reaction to switch a DNA ring between different positions on a DNA catenane with "fuel" and "anti-fuel" strands. Source: Reproduced with permission from Lu et al. [32]/American Chemical Society. (c) Cyclic reconfiguration of the system across states I–IV of the [5] catenane system by using appropriate "fuel" and "anti-fuel" strands. Source: Reproduced with permission from Lu et al. [31]/John Wiley & Sons. (d) Reversible reconfiguration between 16 different states of the seven-ring catenane with strand-displacement reaction. Source: Reproduced with permission from Lu et al. [32]/American Chemical Society.

α from the blue domain and reset the [2]catenane to state I. Hence, the authors claimed that by using Hg^{2+}/cysteine and H^+/OH^- as stimuli, this DNA [2]catenane can perform a stepwise reversible rotation. However, one of the limitations of these systems is that the transient formation of the different states is driven by the Brownian, nondirectional motion of the interlocked components.

Based on a similar concept, Lu et al. reported two examples of switchable DNA catenanes controlled by DNA strand displacement [31, 32]. In comparison with the use of ions and pH as stimuli, DNA strand displacement is easier to design and implement and can be applied orthogonally by using sequence-specific targeting to control the DNA catenanes. In such systems, the interring hybridization regions on ssDNA rings in the DNA catenane system are controlled by pairs of so-called "fuel" and "anti-fuel" ssDNA strands. The "fuel" strand can displace a DNA ring from the hybridization region, creating a mobile DNA ring in the catenane system. The "anti-fuel" strand can remove the "fuel" strand and expose the interring hybridization region, allowing the mobile DNA ring to hybridize with the DNA catenane system again. Therefore, by combining different "fuel" strands and "anti-fuel" strands, a DNA ring in such catenane system can be easily shifted from one position to another (Figure 7.3b). As shown in Lu' s examples, the [5]catenane system (Figure 7.3c) is controlled by four pairs of "fuel" and "anti-fuel" ssDNA strands and can switched between four different configurations, while the catenane system is composed of seven rings, [7]catenane (Figure 7.3d) can switch between 16 different states, thus allowing for 240 dynamic transitions.

Besides the mere structural switching between different static states of a hybridized catenane, the transitions from static to freely mobile (dynamic) catenane structure can also be used in numerous applications. In 2014, Li et al. presented a dsDNA [3]catenane that was used for creating a reversible logic tandem YES and AND gates switching between different static and dynamic conformations of the catenane [37]. As shown in Figure 7.4a, this dsDNA [3]catenane is controlled by two different inputs: the toehold releasing oligodeoxynucleotides (ROs) and H^+. After the addition of ROs input, the DNA rings L and R become freely movable in the interlocked structure, resulting in the spatial separation of rhodamine green (RG) from BHQ-1 and the increase of the RG fluorescence intensity at 534 nm (Fl_{534}) as an output. Although Ring L and Ring R can move freely without an input of H^+, the overhangs on both rings cannot hold close enough to enable FRET between Cy3 and Cy5 (yellow and red dots, respectively). The input of H^+ alone does not allow the two DNA rings to become freely movable. Only when both inputs, ROs and H^+, are used together the overhangs on both rings can form an i-motif and enable a FRET signal at 663 nm (Fl_{663}). Thus, this system behaves as two tandem logic gates (YES and AND) with two outputs of Fl_{534} and Fl_{663}.

In 2020, Ma et al. reported a reversible DNA nanojoint for thermal sensing based on the transformation between the hybridized and mobile DNA catenane [38]. As shown in Figure 7.4b, the nanojoint is initially assembled in its hybridized catenated form. In this state, the nanojoint remains static; however, when the temperature reaches the melting temperature of the hybridized region between two dsDNA rings, the structure is converted into a freely mobile [2]catenane. Since the switching of the DNA nanojoint is controlled by the melting of the hybridized region between two dsDNA rings, the thermal response range of the DNA nanojoint can be easily tuned by changing the length or the sequence of the hybridized region between two DNA rings. Moreover, because of its interlocked structure, the temperature response range of the DNA nanojoint is largely unaffected by its own concentration. This is a key advantage compared with other thermal sensing systems that consist of separated DNA components.

The mechanical properties of interlocked DNA catenanes can find applications as structural or functional components in different nanostructures. As mentioned, the interlocked structure itself can be seen as a highly stable topological bond between different molecules. Compared to the Watson–Crick base pairing widely used for connecting different DNA nanostructures, the topological bonds formed in the interlocked structure are more robust in different conditions and allow its components to perform motion that otherwise would be restricted by chemical bonds. One example

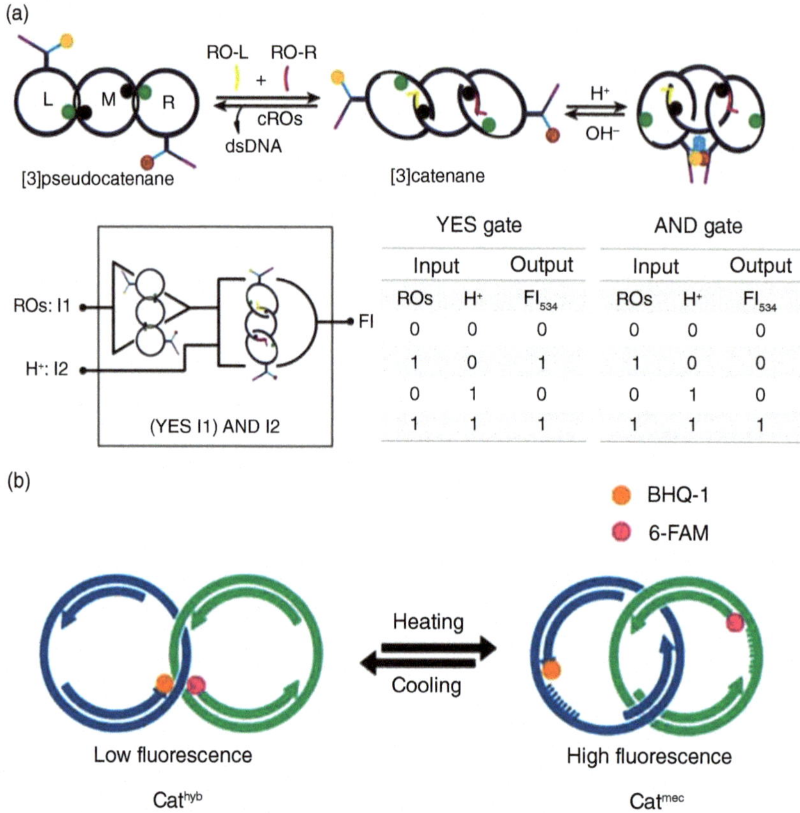

YES gate			AND gate		
Input		Output	Input		Output
ROs	H+	FI_{534}	ROs	H+	FI_{534}
0	0	0	0	0	0
1	0	1	1	0	0
0	1	0	0	1	0
1	1	1	1	1	1

Figure 7.4 (a) Reversible logic circuit based on the transformation of a dsDNA [3]catenane. The logic circuit consists of two tandem gates. A one-input YES gate equivalent to the two-input AND gate is used. Source: Reproduced with permission from Li et al. [37]/Springer Nature/CC BY 4.0. (b) Mechanism of thermal sensing with a hybridized dsDNA [2]catenane. As temperature increases, the hybridized region between two DNA rings begins to melt, and the hybridized [2]catenane is converted into a freely mobile [2]catenane, resulting in an increase in fluorescence intensity. Source: Reproduced with permission from Ma et al. [38]/John Wiley & Sons.

of utilizing the topological bond in a DNA nanostructure for enhanced structural stability was reported by Cassinelli et al. [39]. In this study, the authors designed 6-helix DNA tile tubes consisting of 24 DNA strands carrying alkyne groups on their 3′-ends and azides on their 5′-ends. With a mild click reaction, the two ends of the modified oligonucleotides in the DNA tile tubes were covalently connected to form DNA rings, resulting in a "chain-armor" structure consisting of 24 interlocked DNA rings (Figure 7.5a). Since all the DNA rings are connected through topological bonds, this "chain-armor" structure showed a great resistance to conditions that can normally destabilize DNA nanostructures, such as high temperature, low salt concentration, exonuclease activity and even EtOH precipitation.

Catenanes have also been used for specific and sensitive detection of essential proteins for anticancer therapy, such as topoisomerase II (Topo II). Kristoffersen et al.

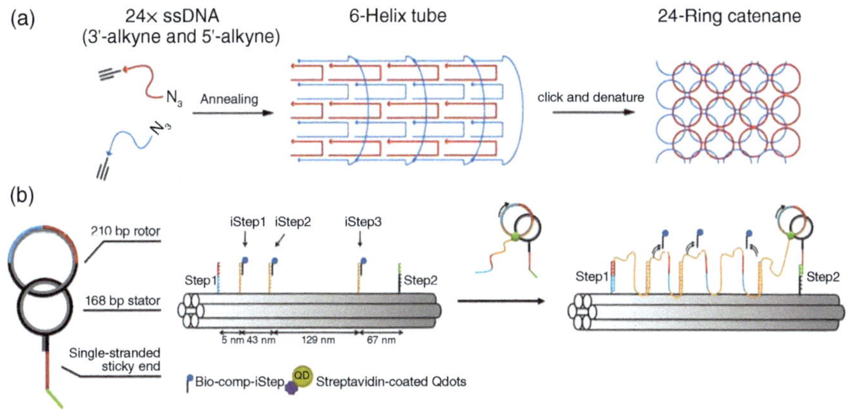

Figure 7.5 (a) Synthesis of a 6-helix tube with 24 single-stranded DNA strands modified with azide and alkyne. The click chemistry covalently circularizes all DNA strands, forming an interlocked "chain-armor" structure. Source: Reproduced with permission from Cassinelli et al. [39]/John Wiley & Sons. (b) Scheme of the DNA rotor–stator nanoengine and the walking mechanism of the DNA nanoengine on a six-helix-bundle DNA origami controlled by RCT. Source: Reproduced with permission from Valero et al. [40]/Springer Nature.

described the use of [2]catenanes for the detection and monitoring of Topo II enzymatic activity at the single-molecule level. The decatenation events upon Topo II treatment of interlocked structures could be followed by combining rolling circle amplification as well as single-molecule fluorescence techniques. This technology paves the way for the application of interlocked DNA nanostructures in the detection of key cancer biomarkers [41].

In 2018, Valero et al. presented a DNA rotor–stator nanoengine based on a dsDNA [2]catenane that utilizes the topological features of this structure to function [40]. As shown in Figure 7.5b, the rotor–stator nanoengine consists of two interlocked dsDNA rings, a rotor ring and a stator ring. The stator ring serves as the stationary portion of the nanoengine, anchoring the whole structure on predefined ssDNA tracks protruding from a DNA origami nanorod. The rotor ring contains a T7 promoter, allowing T7 RNA polymerase to perform rolling circle transcription (RCT). Since stator and rotor rings are connected through a topological bond instead of a chemical bond, the rotor ring can rotate unidirectionally during RCT, producing long, repetitive RNA transcripts that transiently remain attached to the nanoengine. Through a programmed and spatiotemporally controlled series of strand-displacement reactions, the transcribed RNA displaces the stator ring in a stepwise manner from the predefined tracks arranged on the DNA nanorod, guiding the linear motion of the nanoengine along the path.

7.3 DNA Rotaxanes

Rotaxanes are interlocked structures with one, or several macrocycles threaded onto an axle confined by bulky structures at each end of the axle, called stoppers, that avoid macrocycle release. Therefore, the interlocked macrocycles are joined to the rest of the structure through mechanical bonds that allow their rotational and longitudinal motion along the axle. Two main strategies are used for rotaxane assembly. The clipping strategy involves macrocycle closing around an axle with the stoppers already assembled. The capping strategy employs a closed macrocycle that threads onto an axle with at least one free end, which will be subsequently capped with one or two stoppers.

The first example of a dsDNA rotaxane was described by the Famulok group in 2010 [42]. The design of the dsDNA rotaxane was based on the threading of a dsDNA macrocycle onto a DNA axle via hybridization of specific single-stranded "gap" regions and confined in the interlocked space by a capping mechanism in which two bulkier DNA rings or stoppers are bound by three-way junctions to the end-parts of the axle (Figure 7.6a). The assembly of the different dsDNA rings composing the rotaxane was performed using the inherent bending of A-tract segments that can provide the curvature needed to form ring-shaped structures with less tension and good production yields [44, 45]. The addition of releasing oligonucleotides (ROs) allows displacement of the macrocycle from the gap of the axle, therefore switching the pseudorotaxane (hybridized form) into a mechanically interlocked rotaxane. The formation of the different structures was confirmed by gel electrophoresis, fluorescence experiments and atomic force microscopy (AFM). Interestingly, the stability of the system is highly dependent on the size of the macrocycle and the stoppers. For dsDNA ring stoppers, dethreading of the macrocycle occurs even when the ring stoppers are bigger than the macrocycle due to their flexibility. The stability of the interlocked structure increases considerably by replacing the ring stoppers with spherical stoppers that combine two rings with four-way junctions and are more rigid and bulkier, thus preventing the macrocycle from escaping (Figure 7.6a).

Interlocked nanostructures can be employed to perform precise mechanical motions. Rotaxanes are excellent examples where the macrocycle can move translationally along the axis vector as well as rotate around it. To employ rotaxanes for converting mechanical motion to mechanical force in the direction of the motion, their axles are required to be structurally stable. Despite its persistence length, dsDNA is inherently flexible, and thus, it is paramount that the axle is not deformed during the macrocycle movement and operation. To overcome this issue, the axle requires to be mechanically reinforced, for example, employing "paranemic crossover" (PX) structures, where two DNA duplexes are interwoven (Figure 7.6b) [46, 47]. Following this premise, the Famulok group introduced a rotaxane with a paranemic crossover axle (PX100) [43]. The axle design consists of two parallel dsDNA strands that are joined by six double crossovers and incorporates a region spanning 16 bp in the center of the axle without crossovers to introduce an 8 nt ss-gap pointing outward of the entire PX100 axle. This ss-region enables hybridization with the macrocycle (Figure 7.6b). AFM analysis showed a stable rotaxane structure with a reinforced and more rigid axle without observable distortion.

One of the key features of interlocked structures, and rotaxanes in particular, is their ability to reversibly control the shuttling of interlocking macrocycles between

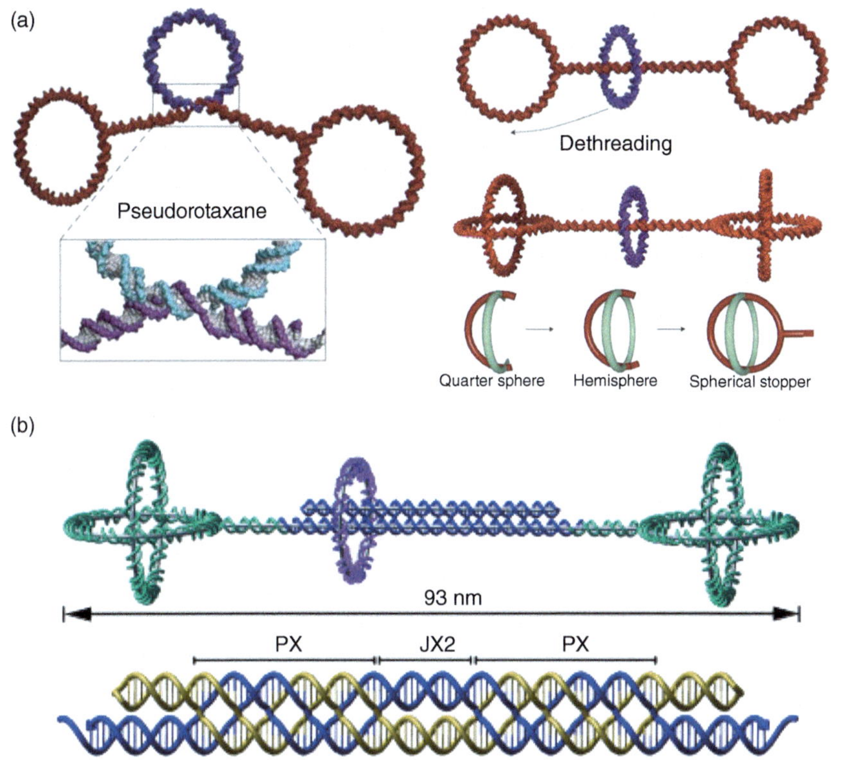

Figure 7.6 (a) Model of a dsDNA (pseudo-) rotaxane. The hybridization between the macrocycle and single-stranded gap on the axle is depicted in the zoomed box. In the presence of ROs, the structure was converted to a dsDNA rotaxane with an interlocked macrocycle, where the macrocycle can dethread due to the flexibility of the ring stoppers. dsDNA rotaxane with two spherical stoppers that combines two 168 bp dsDNA ring will enhance the stability of the structure. Source: Reproduced with permission from Ackermann et al. [42]/Springer Nature. (b) Model of PX100 rotaxane with an axle consisting of paranemic crossover (PX) structures. Source: Reproduced with permission from Ackermann et al. [43]/John Wiley & Sons.

mobile and stalled states. For DNA rotaxanes, the switching between these two states can be performed by a toehold-mediated strand displacement mechanism, where releasing oligonucleotides (ROs) can displace the hybridized macrocycle from the axle of the rotaxane (pseudorotaxane) and induce the structural change to an interlocked rotaxane state in which the macrocycle can freely move along the axle. An alternative approach is to use light as physical stimulus to control the hybridization for switching the position of the macrocycle ring or the rotaxane conformation. Light is non-expensive, it is noninvasive, and it does not require the addition of chemicals and thus avoids waste accumulation that ultimately lowers the switching performance of the system.

The implementation of photoswitching strategies on DNA has been successfully achieved by using modified oligodeoxynucleotides (ODNs) with azobenzene derivatives [48]. Asanuma first reported that the *cis*/*trans* isomerization of azobenzene residues incorporated into ODNs can be used to control the stability of DNA duplexes as they undergo a planar/non-planar conformational change at specific wavelengths of light. Lohmann et al. implemented this photo-switching strategy in a dsDNA rotaxane by using a RO that contains 2′,6′-dimethylazobenzene (DMAB) residues, thermally more stable than unsubstituted azobenzenes, that isomerize to a *cis* conformation at 365 nm and reversibly into *trans*-isomer at 450 nm (Figure 7.7a) [52]. The isomerization of DMABs allowed controlling the hybridization of this RO to the complementary ssDNA gap in the axle and, therefore, regulated the macrocycle switching between hybridization and interlocking states. Several switching cycles were performed by alternating different wavelengths. Light-controlled

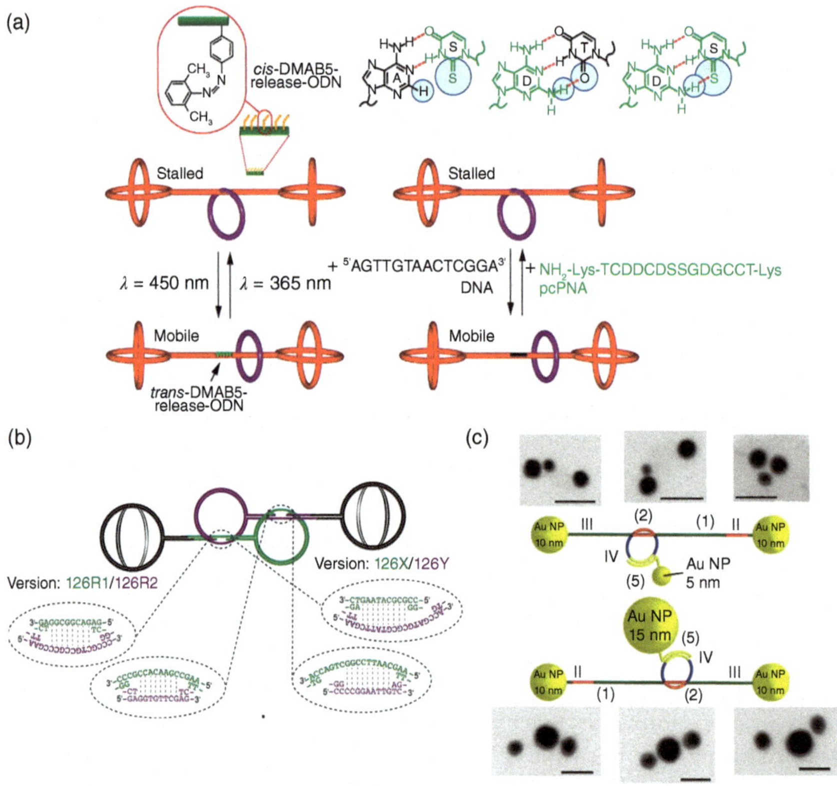

Figure 7.7 (a) Two strategies to control the switching between pseudo-rotaxane and rotaxane: light-induced switching by an RO modified with azobenzene molecules (DMAB5) and reversible switching by a DNA RO and a pcPNA. Source: Reproduced with permission from Jester and Famulok [49]. (b) Schematic representation of a dsDNA daisy chain rotaxane (DCR) and its sequence design in the mutual hybridization regions. Source: Reproduced with permission from Weigandt et al. [50]/John Wiley & Sons. (c) Au-DNA hybrid rotaxanes. Two 10 nm AuNPs are used as bulky stoppers, while a 5 or 15 nm AuNPs modified oligodeoxynucleotide (ODN) complementary to the macrocycle is used for labeling and visualization of the interlocked ring. XHR STEM images (bar = 20 nm) illustrate the correct formation and conformational switching of these rotaxanes. Source: Reproduced with permission from Cecconello et al. [51]/American Chemical Society.

switching offers additional benefits compared to switching strategies that rely on toehold-mediated strand displacement, as the latter requires the addition of RO strands and promotes their accumulation during switching cycles.

An alternative tool for strand displacement of the interlocked macrocycle is to use pseudo-complementary peptide nucleic acids (pcPNAs), which can improve the hybridization efficiency with the target sequence due to the improved binding features of pcPNAs even in the absence of a toehold domain. PcPNAs derive from the introduction of synthetic bases into PNA (a DNA analog where the nucleobases are connected via peptide bonds instead of phosphodiester linkages) [53]. Their synthetic bases allow pairing with their respective natural counterparts but not with each other. Therefore, pcPNA sequences can invade the DNA duplex and replace one of the strands. The resulting pcPNA–DNA hybrid duplexes are also highly sequence-specific and thermodynamically more stable than DNA duplexes. The use of pcPNAs was applied for switching between the static to the mobile interlocked form in a dsDNA rotaxane (Figure 7.7a) [54]. The alternating addition of DNA and PNA strands was used for reversible switching. The use of pcPNA can be orthogonally combined with photo-switching and/or toehold-strand displacement approaches, providing new opportunities to operate DNA nanomachines that require simultaneous triggering in a continuous and reversible fashion.

Besides these relevant DNA rotaxane examples, more topologically complex and exotic DNA interlocked structures have been recently reported. For instance, a daisy chain rotaxane-like structure (DCR containing two macrocycles connected to different axles was developed by Weigandt et al. (Figure 7.7b) [50]. The assembly starts with the synthesis of two different macrocycle-axle complexes, followed by threading and hybridization of one of the axles to the other macrocycle and vice versa, therefore forming a macrocycle-axle dimer. Finally, two bulky stoppers are capping the sticky ends of each axle, leading to the formation of the DCR. The addition of sequence-specific ROs allows the release of the macrocycles from the axles and switches the DCR from the hybridized state to the interlocked state. Sliding of the macrocycles along the axles in the interlocked state leads to the expansion and contraction of the nanostructure and, therefore, holds great potential for developing molecular muscles controlling or guiding linear movement.

There are fewer but interesting examples of ssDNA-based rotaxanes reported in the literature. These DNA interlocked structures are more flexible, mechanically less robust, more difficult to characterize (especially by atomic force microscopy (AFM) and transmission electron microscopy (TEM)) and difficult to control, as there are more single-stranded regions that can fold and establish unwanted interactions, which will affect the overall structure and performance. On the other hand, due to their smaller size and lower structural complexity, they are cheaper and easier to synthesize. Pioneering work in this area was reported by the Willner lab, which developed a rotaxane in which an ss-macrocycle was threaded onto an axle containing at least two ss sites (or stations) complementary to the macrocycle (Figure 7.7c) [51]. Two 10 nm Au-NP were clipped to the thiolated termini of the axle and used as bulky stoppers to avoid macrocycle disassembly. Two types of Au-NP (of 5 and 15 nm diameter, respectively) were used to functionalize an

ODN that specifically hybridizes with the ss-macrocycle in order to unequivocally characterize the assembly. Macrocycle shuttling between the two stations in the rotaxane axle was controlled by the addition of ROs via strand displacement mechanism. This hybrid Au-DNA rotaxane structure can be used to control the assembly, nanoscale position and scaffolding of different Au NPs, thus leading to novel and dynamic plasmonic nanodevices.

More structurally rigid interlocked DNA structures have been reported using DNA tiles and DNA origami (Figure 7.8). Valero et al. reported a DNA rotaxane with single-stranded tile (SST) stoppers that consisted of 97 short ODN strands assembled into a square-shaped structure (Figure 7.8a) [55]. The SST stoppers were used as bulkier alternatives to the ring or spherical stoppers previously described. The size of the SST stoppers was large enough to maintain a 14 nm diameter macrocycle threaded to the rotaxane axle. Additionally, these SST stoppers can be easily visualized by AFM and used to accurately determine the position of the macrocycle. The incorporation of SSTs allows the introduction of additional functionalities such as proteins, small molecules, aptamers and even nanoparticles. The assembly of SSTs is also straightforward and versatile. As a downside, the system requires relatively high salt concentrations to prevent disassembly.

A more structurally complex DNA origami rotaxane with the ability to perform long-range movements was reported by List et al. [56]. The rotaxane is based on a dumbbell-shaped DNA origami structure, which is connected to a flexible DNA origami hinge that is circularized and threaded onto the dumbbell by adding closing strands following a "clipping" strategy and then detached by the addition of releasing fuel strands (Figure 7.8b). The authors showed that the annealing temperature ramp was a key optimization parameter for the stepwise self-assembly of the interlocked structure. The switching between the mobile and hybridized states was successfully achieved by strand displacement and unequivocally characterized by TEM. DNA origami-based designs open new avenues for controlling the stoichiometry and position of modifications and cargo molecules with nanometer precision, as well as having the potential to deliver these molecules over hundreds of nm distances.

Powell et al. described a different assembly approach towards DNA origami-based rotaxanes [57]. In this example, the rotaxane is composed of three DNA origami parts, two half-dumbbells consisting of a square-shaped stopper and an eight-helix bundle axle, and a macrocycle with a square shape (Figure 7.8c). For the assembly of the nanostructure, the macrocycle was first connected to the stopper in one of the half-dumbbells with four ssDNA connectors termed as "tendons." Subsequently, threading of the macrocycle onto the short end of the axle in half-dumbbell occurs when the notched surface of the macrocycle stacks to the surface of the stopper and hybridizes to the ssDNA tendons, thus yielding an intermediate half-dumbbell/macrocycle complex. The assembly was completed by connecting the long end of the axle with another half-dumbbell. Strand displacement by ROs was used to release the DNA macrocycle and switch the rotaxane into the mobile interlocked state. The average movement range of the macrocycle was assessed by TEM. Interestingly, the authors incorporated a small trigger to expand the

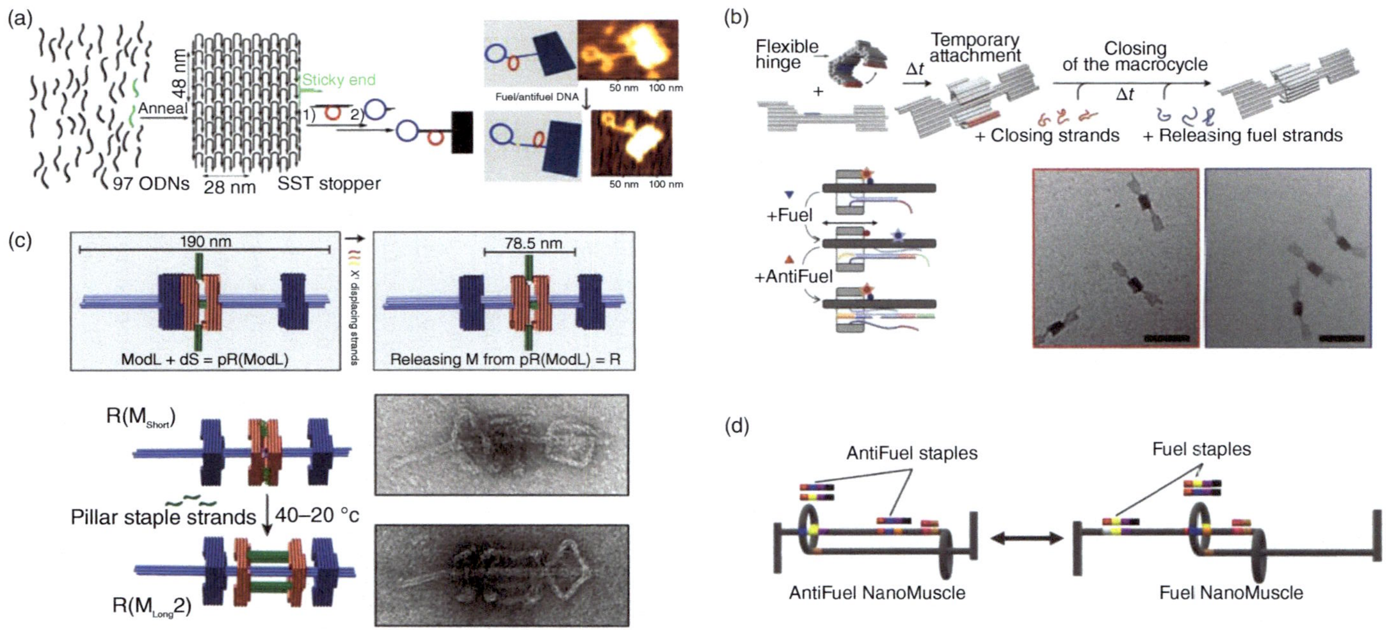

Figure 7.8 (a) Assembly of a single-stranded tile rotaxane. The SST stopper was assembled by annealing 97 different ODNs. The AFM images showed the correct formation of the SST rotaxane. Source: Reproduced with permission from Valero et al. [55]/John Wiley & Sons. (b) Assembly and switching of a DNA origami-based rotaxane. The macrocycle was first assembled as a flexible hinge and then closed and released by different fuel strands. The switching mechanism is designed as a fuel/anti-fuel strategy. TEM images showed the switching in the absence (left) or presence (right) of releasing ODNs. Source: Reproduced with permission from List et al. [56]/Springer Nature/CC BY 4.0. (c) Schematic representation of the structural design and switching mechanism of a DNA origami-based rotaxane with a tubular-shaped macrocycle. The system allowed the expansion of the macrocycle to a tubular-shaped structure by pillar staple strands after assembly, and this mechanism was corroborated by TEM imaging. Source: Reproduced with permission from Powell et al. [57]/John Wiley & Sons. (d) Schematic representation of a DNA origami-based DCR and its mechanism contraction and expansion by fuel/anti-fuel strands. Source: Reproduced with permission from Chao et al. [58]/Royal Society of Chemistry.

macrocycle to a tubular cage after rotaxane assembly. This sequential assembly design enables the cage-shaped macrocycle to be threaded onto the axle, efficiently assisting the tubular macrocycle docking as the latest structure is difficult to interlock during the initial rotaxane assembly.

An interesting and structurally more complex DNA origami interlocked system capable of contracting and expanding was described by Chao et al. [58]. The authors showed the design and assembly of a DNA origami-based DCR structure, which can be used as a molecular muscle with expansion and contraction movements in the range of 310–160 nm upon fuel/antifuel ODN addition. The system was successfully characterized by TEM (Figure 7.8d). Switching between expanded and contracted configurations on these DNA origami muscles provides an opportunity to develop robust nanomachines that can perform continuous and relatively long nanometer-range mechanical movements.

Rotaxanes containing different switching stations are optimal scaffolds for designing dynamic nanodevices (Figure 7.9). An illustrative example is the [3]rotaxane-based shuttle system developed by Lohmann et al. [59], in which two DNA macrocycles of different sizes can be independently or simultaneously switched between two different "stations" (gaps) located on the axle (Figure 7.9a). The migration of macrocycles can be triggered by using ROs to block the ss-gap regions and release the macrocycles by strand displacement or alternatively using azobenzene-modified photo-responsible ROs that can be controlled at different light irradiation wavelengths. Since the design allows the macrocycles to dethread from the axle when both stations are blocked, a dethreading cascade can be implemented depending on the sequence in which the different ROs are added as input. The resulting rotaxane structures (outputs) can be defined as different states of specific logic gates, thus allowing the use of the interlocked system for complex DNA computing and logic operations.

Structural switching of the interlocked macrocycle in a DNA rotaxane has also been explored to control catalytic chemical reactions. In that regard, Centola et al. designed a split DNAzyme-based nanostructure in which a dsDNA rotaxane with two docking stations was used as a scaffold decorated with the different subunits of a split DNAzyme in the mobile macrocycle and one of the stoppers, respectively (Figure 7.9b) [60]. Therefore, the assembly of the DNAzyme – a horseradish peroxidase mimicking DNAzyme – can be controlled depending on the position of the macrocycle threaded on the rotaxane axle. In particular, the functional DNAzyme can only be assembled and perform oxidative catalysis when the macrocycle and the stopper containing one of the halves of the DNAzyme are in close proximity. Moreover, this system can be switched on and off over multiple cycles, making it advantageous with respect to other systems with restricted reversibility.

Some mechanical devices contain linear actuators that move directionally in a straight line in order to generate force or work. They can also be used to convert rotatory motion to a linear motion. In that regard, stiff DNA origami structures have been employed to design a linear rotaxane actuator that is composed of a rail with a locked slider (Figure 7.9c) [61]. For the assembly of such nanostructures, the authors compared two different strategies: one-pot and two-pot assembly, in which

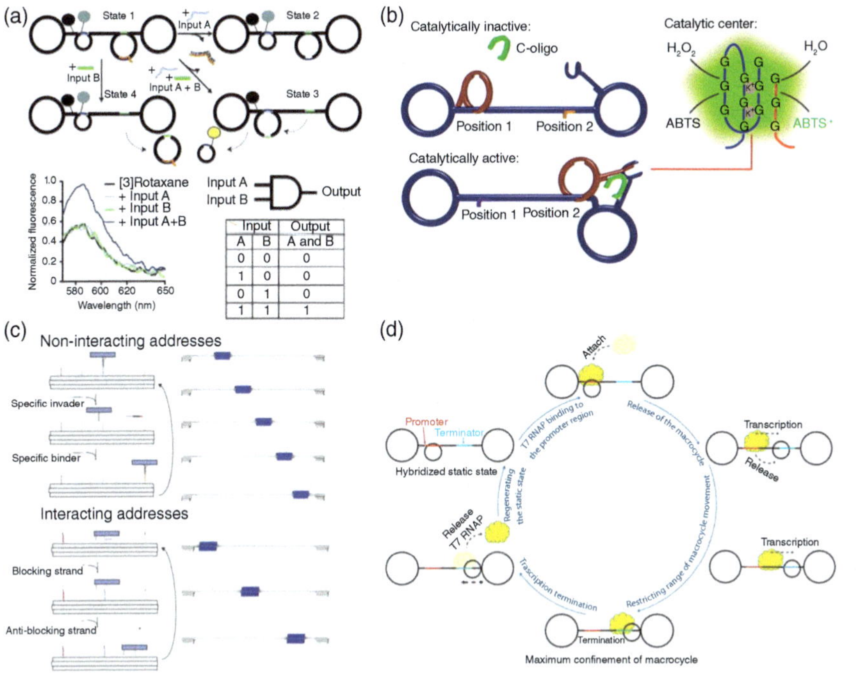

Figure 7.9 (a) Switching of a DNA [3]rotaxane in response to input strands. The fluorescence spectra of [3]rotaxane in the presence of input strands A, B, or A + B, where the fluorescence signal increases with the simultaneous addition of A + B. The AND logic gate and its truth table (with 0.6 normalized fluorescence signal as a threshold). Source: Reproduced with permission from Lohmann et al. [59]/John Wiley & Sons. (b) Design and switching mechanism of a DNA rotaxane with a split DNAzyme distributed over the macrocycle and one of the stoppers. The operation of the rotaxane allows the reversible formation of a DNAzyme consisting of a DNA G-quadruplex structure, which works in the presence of potassium ions, hemin and H_2O_2 to achieve oxidative catalysis. Source: Reproduced with permission from Centola et al. [60]/American Chemical Society. (c) Positioning control of an interlocked cylinder in a DNA origami rotaxane-based linear actuator and their respective molecular models. Source: Reproduced with permission from Benson et al. [61]/John Wiley & Sons/CC BY 4.0. (d) Scheme of an operation cycle in a T7 RNAP-driven rotaxane linear actuator, which starts with the interaction of a T7 RNAP to a rotaxane with a hybridized macrocycle. Upon transcription initiation, the T7 RNAP continuously limits the movement range of the macrocycle during transcription, eventually pushing it to the end of the axle. Finally, a new cycle can start after transcription termination and resetting of the macrocycle to the original anchor position. Source: Reproduced with permission from Yu et al. [62]/American Chemical Society.

the slider and rail are assembled by one or two DNA origami, respectively. The precise and reversible control of the relative position of the slider on the rail is achieved by exchanging ssDNA tethers between both components, and the switching can be tracked by TEM and DNA-PAINT. Such linear nanoactuators provide positioning accuracy over long distances of approximately 200 nm, allowing the development of nanomachines and their potential applications in nanoscale tasks such as the manufacturing of semiconductors or the control of chemical reactions.

In addition to the programmable features of DNA as a building block for nanostructure assembly, DNA can also interact with proteins in a sequence-specific manner, such as motor-like enzymes, which can be used to generate directional forces. A remarkable example of such biohybrid nanodevices is an RNA polymerase-DNA rotaxane complex that can perform linear motion in an automated and repetitive fashion. Such autonomous DNA nanostructures powered by an RNA polymerase protein (T7RNAP) represent optimal systems for developing linear actuators that can control T7RNAP transcription (Figure 7.9d) [62]. In the design reported by Yu et al., the axle and threading macrocycle of a DNA rotaxane encoded parts of the transcription promoter. When hybridizing together, the full promoter sequence is formed, allowing transcription initiation. During transcription, the macrocycle is released from the ss gap region in the promoter by the T7RNAP and the range of movement of the interlocked macrocycle, smaller in diameter than the T7RNAP, is reduced by the progression of the polymerase, which functions as a movable stopper on the axle. When reaching the transcription terminator sequences encoded at the end of the axle, T7RNAP detaches from the axle, and the system is reset, i.e. the macrocycle shifts back and rehybridizes with the promoter gap to reassemble the double-stranded promoter required for the next transcription cycle. The structural and functional characterization of this rotaxane actuator was performed by AFM and bulk fluorescence experiments. OxDNA simulations were used to investigate the stability of the interlocked structure when applying external forces. The T7RNAP-DNA rotaxane nanodevice was shown to produce continuous linear motion without any intervention and achieve self-controlled transcription without any intermediate, therefore bearing great potential to develop delivery systems and RNA production lines that can function autonomously.

7.4 Conclusions and Outlook

Over the last decade, interlocked DNA nanostructures have shown great promise as building blocks for designing complex molecular devices and motors. The mechanical bonds that hold the interlocked components are key to efficiently driving intramolecular interactions and dynamic processes and to provide an intrinsic stability that allows the continuous and long-term performance of these systems. This has been successfully shown with exceptional examples of DNA catenanes performing unidirectional rotational motion or rotaxanes in which the position of the interlocked macrocycle can be controlled with nanoscopic precision. ssDNA interlocked systems are straightforward to synthesize, but compared to the double-stranded versions, they are structurally less well-defined and more difficult to operate due to unwanted base pairing interactions. DNA origami technology has also been applied for the construction of structurally complex and highly dynamic interlocked nanostructures with enhanced mechanical stability displaying long-range motion. In general, DNA-based MIMs are easy to design and assemble, highly programmable and biocompatible, thus holding potential for therapeutic or biomedical applications.

Moreover, these systems can establish sequence-specific interactions with other nucleic acids (e.g. RNA or XNAs), peptides or proteins, and they can be functionalized with a wide variety of small molecules and ligands, which increases the portfolio of different chemical (e.g. ions, ODNs, and ligands) and physical (e.g. light, temperature, etc.) inputs to control the motion of interlocked components. Despite the enormous progress, the field is still in its infancy, and it has certain limitations that need to be addressed, such as the slow kinetics of motion derived from the strand displacement reactions and the low chemical stability and scalability of DNA interlocked systems compared to traditional MIMs, which hamper their use in material science applications. To overcome that, new inputs to drive faster motion of DNA-based interlocked molecules have been reported and can alternatively be used. Moreover, with the new requirements for sequencing and production of oligonucleotide-based therapies, the synthesis of DNA is now more accessible and affordable. Future directions in the field will explore the incorporation of novel chemistries and functionalities in interlocked DNA nanostructures to reduce nuclease degradation and undesired immune responses and to enhance their performance both *in vivo* and *in vitro*, which will open new opportunities for designing biohybrid oligonucleotide-based catenanes and rotaxanes with novel functions and properties.

References

1 Martinac, B. (2022). 2021 Nobel Prize for mechanosensory transduction. *Biophys. Rev.* 14 (1): 15–20.

2 Caterina, M.J., Schumacher, M.A., Tominaga, M. et al. (1997). The capsaicin receptor: a heat-activated ion channel in the pain pathway. *Nature* 389 (6653): 816–824.

3 Schliwa, M. and Woehlke, G. (2003). Molecular motors. *Nature* 422 (6933): 759–765.

4 van den Heuvel, M.G. and Dekker, C. (2007). Motor proteins at work for nanotechnology. *Science* 317 (5836): 333–336.

5 Stoddart, J.F. (2017). Mechanically interlocked molecules (MIMs)—molecular shuttles, switches, and machines (Nobel Lecture). *Angew. Chem. Int. Ed.* 56 (37): 11094–11125.

6 Sluysmans, D. and Stoddart, J.F. (2019). The burgeoning of mechanically interlocked molecules in chemistry. *Trends Chem.* 1 (2): 185–197.

7 Ashbridge, Z., Fielden, S.D.P., Leigh, D.A. et al. (2022). Knotting matters: orderly molecular entanglements. *Chem. Soc. Rev.* 51 (18): 7779–7809.

8 Wasserman, E. (1960). The preparation of interlocking rings: a catenane. *J. Am. Chem. Soc.* 82 (16): 4433–4434.

9 Schill, G. and Lüttringhaus, A. (1964). The preparation of catena compounds by directed synthesis. *Angew. Chem. Int. Ed.* 3 (8): 546–547.

10 Harrison, I.T. and Harrison, S. (1967). Synthesis of a stable complex of a macrocycle and a threaded chain. *J. Am. Chem. Soc.* 89 (22): 5723–5724.

11 Sauvage, J.-P. (2019). EurJOC – 50 years of rotaxanes. *Eur. J. Org. Chem.* 2019 (21): 3287–3288.

12 Heard, A.W. and Goldup, S.M. (2020). Simplicity in the design, operation, and applications of mechanically interlocked molecular machines. *ACS Central Sci.* 6 (2): 117–128.

13 Riebe, J. and Niemeyer, J. (2021). Mechanically interlocked molecules for biomedical applications. *Eur. J. Org. Chem.* 2021 (37): 5106–5116.

14 Mena-Hernando, S. and Pérez, E.M. (2019). Mechanically interlocked materials. Rotaxanes and catenanes beyond the small molecule. *Chem. Soc. Rev.* 48 (19): 5016–5032.

15 Valero, J., Lohmann, F., and Famulok, M. (2017). Interlocked DNA topologies for nanotechnology. *Curr. Opin. Biotechnol.* 48: 159–167.

16 Seeman, N.C. (2005). Structural DNA nanotechnology: an overview. *Methods Mol. Biol.* 303: 143–166.

17 Zhang, F., Nangreave, J., Liu, Y., and Yan, H. (2014). Structural DNA nanotechnology: state of the art and future perspective. *J. Am. Chem. Soc.* 136 (32): 11198–11211.

18 Nummelin, S., Kommeri, J., Kostiainen, M.A., and Linko, V. (2018). Evolution of structural DNA. *Nanotechnology* 30 (24): 1703721.

19 Keller, A. and Linko, V. (2020). Challenges and perspectives of DNA nanostructures in biomedicine. *Angew. Chem. Int. Ed.* 59 (37): 15818–15833.

20 Dunn, K.E. and Elfick, A. (2023). Harnessing DNA nanotechnology and chemistry for applications in photonics and electronics. *Bioconjug. Chem.* 34 (1): 97–104.

21 Linko, V. (2020). At the dawn of applied DNA nanotechnology. *Molecules* 25 (3): 639.

22 Sundin, O. and Varshavsky, A. (1980). Terminal stages of SV40 DNA replication proceed via multiply intertwined catenated dimers. *Cell* 21 (1): 103–114.

23 Laurie, B., Katritch, V., Sogo, J. et al. (1998). Geometry and physics of catenanes applied to the study of DNA replication. *Biophys. J.* 74 (6): 2815–2822.

24 Kreuzer, K.N. and Cozzarelli, N.R. (1980). Formation and resolution of DNA catenanes by DNA gyrase. *Cell* 20 (1): 245–254.

25 Baldi, M., Benedetti, P., Mattoccia, E., and Tocchini-Valentini, G. (1980). In vitro catenation and decatenation of DNA and a novel eucaryotic ATP-dependent topoisomerase. *Cell* 20 (2): 461–467.

26 Stark, W.M., Parker, C.N., Halford, S.E., and Boocock, M.R. (1994). Stereoselectivity of DNA catenane fusion by resolvase. *Nature* 368 (6466): 76–78.

27 Liang, X., Kuhn, H., and Frank-Kamenetskii, M.D. (2006). Monitoring single-stranded DNA secondary structure formation by determining the topological state of DNA catenanes. *Biophys. J.* 90 (8): 2877–2889.

28 Liu, Y., Kuzuya, A., Sha, R. et al. (2008). Coupling across a DNA helical turn yields a hybrid DNA/organic catenane doubly tailed with functional termini. *J. Am. Chem. Soc.* 130 (33): 10882–10883.

29 Schmidt, T.L. and Heckel, A. (2011). Construction of a structurally defined double-stranded DNA catenane. *Nano Lett.* 11 (4): 1739–1742.

30 Valero, J., Centola, M., Ma, Y. et al. (2019). Design, assembly, characterization, and operation of double-stranded interlocked DNA nanostructures. *Nat. Protoc.* 14 (10): 2818–2855.

31 Lu, C.H., Qi, X.J., Cecconello, A. et al. (2014). Switchable reconfiguration of an interlocked DNA olympiadane nanostructure. *Angew. Chem. Int. Ed.* 53 (29): 7499–7503.

32 Lu, C.-H., Cecconello, A., Qi, X.-J. et al. (2015). Switchable reconfiguration of a seven-ring interlocked DNA catenane nanostructure. *Nano Lett.* 15 (10): 7133–7137.

33 Wu, Z.-S., Shen, Z., Tram, K., and Li, Y. (2014). Engineering interlocking DNA rings with weak physical interactions. *Nat. Commun.* 5 (1): 4279.

34 Peil, A., Zhan, P., and Liu, N. (2020). DNA origami catenanes templated by gold nanoparticles. *Small* 16 (6): 1905987.

35 Sakai, Y., Wilkens, G.D., Wolski, K. et al. (2022). Topogami: topologically linked DNA origami. *ACS Nanosci. Au* 2 (1): 57–63.

36 Lu, C.-H., Cecconello, A., Elbaz, J. et al. (2013). A three-station DNA catenane rotary motor with controlled directionality. *Nano Lett.* 13 (5): 2303–2308.

37 Li, T., Lohmann, F., and Famulok, M. (2014). Interlocked DNA nanostructures controlled by a reversible logic circuit. *Nat. Commun.* 5: 4940.

38 Ma, Y., Centola, M., Keppner, D., and Famulok, M. (2020). Interlocked DNA nanojoints for reversible thermal sensing. *Angew. Chem. Int. Ed. Eng.* 59 (30): 12455–12459.

39 Cassinelli, V., Oberleitner, B., Sobotta, J. et al. (2015). One-step formation of "Chain-Armor"-stabilized DNA nanostructures. *Angew. Chem. Int. Ed.* 54 (27): 7795–7798.

40 Valero, J., Pal, N., Dhakal, S. et al. (2018). A bio-hybrid DNA rotor–stator nanoengine that moves along predefined tracks. *Nat. Nanotechnol.* 13 (6): 496.

41 Kristoffersen, E.L., Givskov, A., Jørgensen, L.A. et al. (2017). Interlinked DNA nano-circles for measuring topoisomerase II activity at the level of single decatenation events. *Nucleic Acids Res.* 45 (13): 7855–7869.

42 Ackermann, D., Schmidt, T.L., Hannam, J.S. et al. (2010). A double-stranded DNA rotaxane. *Nat. Nanotechnol.* 5 (6): 436–442.

43 Ackermann, D., Jester, S.S., and Famulok, M. (2012). Design strategy for DNA rotaxanes with a mechanically reinforced PX100 axle. *Angew. Chem.* 51 (27): 6771–6775.

44 MacDonald, D., Herbert, K., Zhang, X. et al. (2001). Solution structure of an A-tract DNA bend. *J. Mol. Biol.* 306 (5): 1081–1098.

45 Ulanovsky, L., Bodner, M., Trifonov, E.N., and Choder, M. (1986). Curved DNA: design, synthesis, and circularization. *Proc. Natl. Acad. Sci. U. S. A.* 83 (4): 862–866.

46 Sa-Ardyen, P., Vologodskii, A.V., and Seeman, N.C. (2003). The flexibility of DNA double crossover molecules. *Biophys. J.* 84 (6): 3829–3837.

47 Li, X., Yang, X., Qi, J., and Seeman, N.C. (1996). Antiparallel DNA double crossover molecules as components for nanoconstruction. *J. Am. Chem. Soc.* 118 (26): 6131–6140.

48 Nishioka, H., Liang, X., and Asanuma, H. (2010). Effect of the ortho modification of azobenzene on the photoregulatory efficiency of DNA hybridization and the thermal stability of its cis form. *Chemistry* 16 (7): 2054–2062.

49 Jester, S.S. and Famulok, M. (2014). Mechanically interlocked DNA nanostructures for functional devices. *Acc. Chem. Res.* 47 (6): 1700–1709.

50 Weigandt, J., Chung, C.L., Jester, S.S., and Famulok, M. (2016). Daisy chain rotaxanes made from interlocked DNA nanostructures. *Angew. Chem.* 55 (18): 5512–5516.

51 Cecconello, A., Lu, C.H., Elbaz, J., and Willner, I. (2013). Au nanoparticle/DNA rotaxane hybrid nanostructures exhibiting switchable fluorescence properties. *Nano Lett.* 13 (12): 6275–6280.

52 Lohmann, F., Ackermann, D., and Famulok, M. (2012). Reversible light switch for macrocycle mobility in a DNA rotaxane. *J. Am. Chem. Soc.* 134 (29): 11884–11887.

53 Lohse, J., Dahl, O., and Nielsen, P.E. (1999). Double duplex invasion by peptide nucleic acid: a general principle for sequence-specific targeting of double-stranded DNA. *Proc. Natl. Acad. Sci. U. S. A.* 96 (21): 11804–11808.

54 Ackermann, D. and Famulok, M. (2013). Pseudo-complementary PNA actuators as reversible switches in dynamic DNA nanotechnology. *Nucleic Acids Res.* 41 (8): 4729–4739.

55 Valero, J., Lohmann, F., Keppner, D., and Famulok, M. (2016). Single-stranded tile stoppers for interlocked DNA architectures. *ChemBioChem* 17 (12): 1146–1149.

56 List, J., Falgenhauer, E., Kopperger, E. et al. (2016). Long-range movement of large mechanically interlocked DNA nanostructures. *Nat. Commun.* 7: 12414.

57 Powell, J.T., Akhuetie-Oni, B.O., Zhang, Z., and Lin, C. (2016). DNA origami rotaxanes: tailored synthesis and controlled structure switching. *Angew. Chem.* 55 (38): 11412.

58 Chao, Y.C., Hong, Y.J., Lee, C.Y. et al. (2020). NanoMuscle: controllable contraction and extension of mechanically interlocked DNA origami. *Nanoscale* 12 (5): 2992–2998.

59 Lohmann, F., Weigandt, J., Valero, J., and Famulok, M. (2014). Logic gating by macrocycle displacement using a double-stranded DNA [3]rotaxane shuttle. *Angew. Chem.* 53 (39): 10372–10376.

60 Centola, M., Valero, J., and Famulok, M. (2017). Allosteric control of oxidative catalysis by a DNA rotaxane nanostructure. *J. Am. Chem. Soc.* 139 (45): 16044–16047.

61 Benson, E., Carrascosa Marzo, R., Bath, J., and Turberfield, A.J. (2021). Strategies for constructing and operating DNA origami linear actuators. *Small* 17 (20): e2007704.

62 Yu, Z., Centola, M., Valero, J. et al. (2021). A self-regulating DNA rotaxane linear actuator driven by chemical energy. *J. Am. Chem. Soc.* 143 (33): 13292–13298.

8

Oligo- and Poly-catenanes from Molecular and Supramolecular Building Blocks

Sougata Datta[1], Atsushi Isobe[2], and Shiki Yagai[1,2,3]

[1] *Institute for Advanced Academic Research (IAAR), Chiba University, 1-33 Yayoi-cho, Inage-ku, Chiba 263-8522, Japan*
[2] *Division of Advanced Science and Engineering, Graduate School of Science and Engineering, Chiba University, 1-33 Yayoi-cho, Inage-ku, Chiba 263-8522, Japan*
[3] *Department of Applied Chemistry and Biotechnology, Graduate School of Engineering, Chiba University, 1-33 Yayoi-cho, Inage-ku, Chiba 263-8522, Japan*

8.1 Introduction

[*n*]Catenanes represent a major class of mechanically interlocked molecular ensembles comprising two or more mechanically bonded macrocyclic molecules that cannot be separated without rupturing at least one of them. Poly[*n*]catenanes are, accordingly, a polymeric or oligomeric version of [*n*]catenanes composed of a large number of interlocked rings. Such topologically intriguing molecular ensembles not only possess the strength and integrity of covalent bonds [1] but also endow the material with the dynamic properties associated with the high degrees of freedom in the motion of their constituent rings [2]. By virtue of the mobility of the constituent rings, poly[*n*]catenanes exhibit a large loss modulus and a low activation energy for flow and, accordingly, potentially act as energy-damping materials and/or elastomers with excellent toughness and mechanical properties under stimulus control [3, 4]. Moreover, the chain-like structure of poly[*n*]catenanes has the potential ability to repair the damaged chains by replacing only the damaged ring. All of these exceptional characteristics are highly intriguing as next-generation soft polymer materials and also might be useful as nanotechnology materials for molecular machineries and soft robotics.

In 1960, before the rise of supramolecular chemistry, the synthesis of a molecular [2]catenane was pioneered by Wasserman through a statistical interlocking approach [5]. Unfortunately, the initial effort of Wasserman ended up with extremely poor yield (less than 0.0001%) due to the following two conflicting factors: the probability of threading of a linear precursor through another cyclic precursor increases at high concentrations, while intramolecular cyclization of the linear component is favored at lower concentrations [6]. A few years later, Schill and coworkers reported the directed and multistep synthesis of [2]catenanes involving

Mechanically Interlocked Materials: Polymers, Nanomaterials, MOFs, and more, First Edition.
Edited by Emilio M. Pérez.

cleavage of covalent bonds [7, 8]. In 1983, Sauvage and coworkers demonstrated a supramolecular approach, so-called template-directed synthesis, to prepare [2]catenane using metal–ligand complexation [9]. The template-directed synthesis includes preorganization and macrocyclization of precursors as the major steps, wherein the entropic cost associated with the interlocking of the macrocycles is compensated by the preorganization through metal–ligand coordination.

Without any doubt, the ingenious discovery of Sauvage created a spark that ignited the thrust for the synthesis of catenanes through the use of other non-covalent interactions such as hydrogen bonding, π–π interactions, hydrophobic effects, halogen bonding, etc. For example, Hunter [10], Vögtle et al. [11], and Leigh and coworkers [12] individually reported the template-directed synthesis of [2]catenanes by using hydrogen bonds. Beer and coworkers demonstrated by the potential for halogen bonding to template the formation of catenanes. Catenanes were obtained by preassembly of precursors through Br–Br–Br– or iodide–pyridine interactions, which was followed by ring-closure metathesis (RCM) [13, 14]. After about ten years of Sauvage' s seminal report, Stoddart' s group first realized a [5]catenane as an oligo[*n*]catenane comprising five macrocyclic units, named as "Olympiadane," by employing charge–transfer interactions for preorganization of the precursor components [15]. The 2016 Nobel Prize in Chemistry was awarded in the area of molecular machines to two pioneers of mechanically interlocked molecules, Sauvage [16] and Stoddart [7], together with the most distinguished non-mechanically interlocked molecular machinist, Feringa [17].

In this chapter, we describe various synthetic strategies for the synthesis of oligo- and poly[*n*]catenanes. The discussion is divided into various sections depending on the location, connection, and character of the catenane-building blocks (Figure 8.1). First, we discuss about "radial polycatenanes," which are also known as [*n*]molecular necklaces. These are topological isomers of [*n*]catenanes consisting

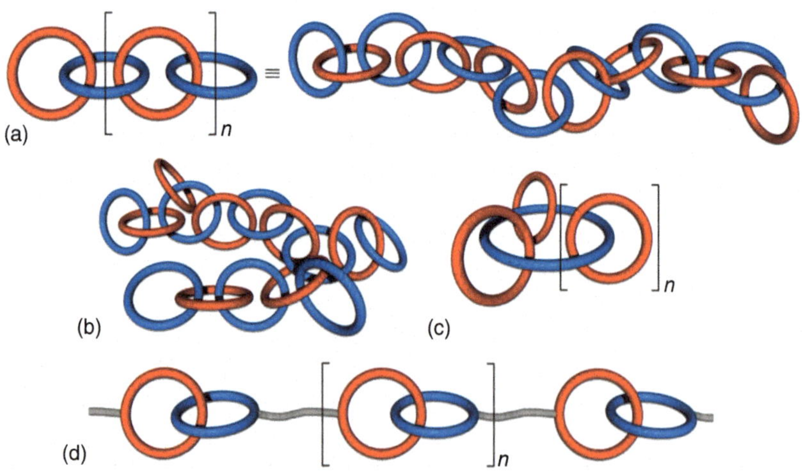

Figure 8.1 Cartoon representations of (a) linear poly[*n*]catenane, (b) branched poly[*n*]catenane, (c) radial poly[*n*]catenane, and (d) poly[2]catenane.

of $(n-1)$ subrings $(n>3)$ mechanically interlocked with a central larger ring. Compared to the other types of polycatenanes, they are much less explored because of the requirement of threading a linear polymer through multiple macrocycles prior to the chain cyclization.

The above is followed by a section on the main chain poly[*n*]catenanes, which are constructed by solely covalent macrocyclic building blocks. The step-wise strategies are not efficient in obtaining long main chain poly[*n*]catenanes, and their efficient preparation could be approached with the supramolecular polymerization of macrocycle precursors. We specifically focus on Rowan' s report of poly[*n*]catenanes, including linear poly[7–27]catenanes, branched poly[13–130]catenanes obtained by metallosupramolecular polymer templating strategies, followed by an efficient ring-closing reaction [18].

We also introduce the synthesis of main chain poly[*n*]catenanes composed of interlocked "non-covalent macrocycles." If we could construct constituent rings by self-assembly and find a mechanism whereby their formation occurs continuously in the vicinity of a preformed ring, we could achieve polycatenane formation entirely by self-assembly alone. In this regard, we describe our recent effort to prepare nano-poly[*n*]catenanes composed of a specific nanosized circular topology consisting of molecular assemblies constructed by a bottom-up approach [19, 20].

Finally, we describe the synthesis of poly[2]catenane by polymerizing a [2]catenane precursor. Suitable functionalization of one of the two macrocycles followed by polymerization leads to the formation of a side chain poly[2]catenane. On the other hand, if both macrocycles are involved in the polymerization process, a main chain poly[2]catenane will be the product.

8.2 [*n*]Molecular Necklaces by Cyclization of Polypseudorotaxanes

After reporting the metal-templated directed synthesis of a [2]catenane [11], Sauvage and coworkers employed the same strategy to obtain a [3]catenane. Instead of using the diiodo linear precursor, the tetrahedral copper(I) phenanthroline complex was subjected to a ring-closure reaction with a shorter α,ω-dibromo linear compound, affording [3]catenane in 6% yield (Figure 8.2a) [21]. After a few years, Sauvage and coworkers prepared a slightly modified tetrahedral copper(I) phenanthroline complex with alkyne termini, which was subjected to undergo oxidative coupling reaction leading to cyclo-oligomerization (Figure 8.2b). A mixture of catenane homologs containing three to seven interlocked macrocycles was formed, as characterized by electrospray mass spectrometry. The [3], [4], and [5]catenane products were isolated in 23, 23, and 10% yields, respectively [22]. Anderson' s group also reported the synthesis of a radial [4]catenane in 62% yield through a palladium-catalyzed oxidative alkyne coupling reaction of a porphyrin [2]rotaxane precursor containing two terminal alkynes in the presence of a radial hexa-pyridyl coordination template [23]. Notably, a "figure-of-eight" shaped [7]catenane also formed as a byproduct in 6% yield, as confirmed by ^{1}H NMR nuclear magnetic

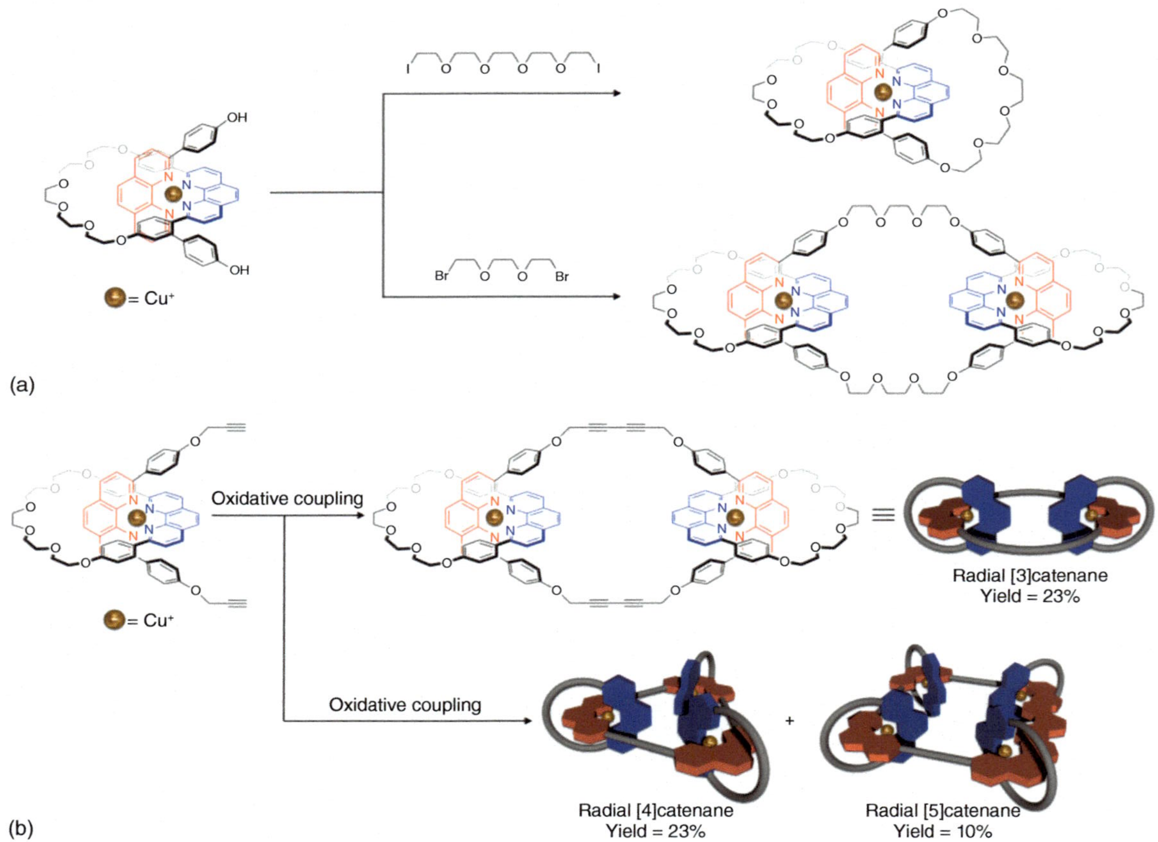

Figure 8.2 Sauvage' s syntheses of (a) [2, 3]catenanes templated by copper(I). (b) A series of [3], [4], and [5]catenanes prepared by oxidative homocoupling cyclization.

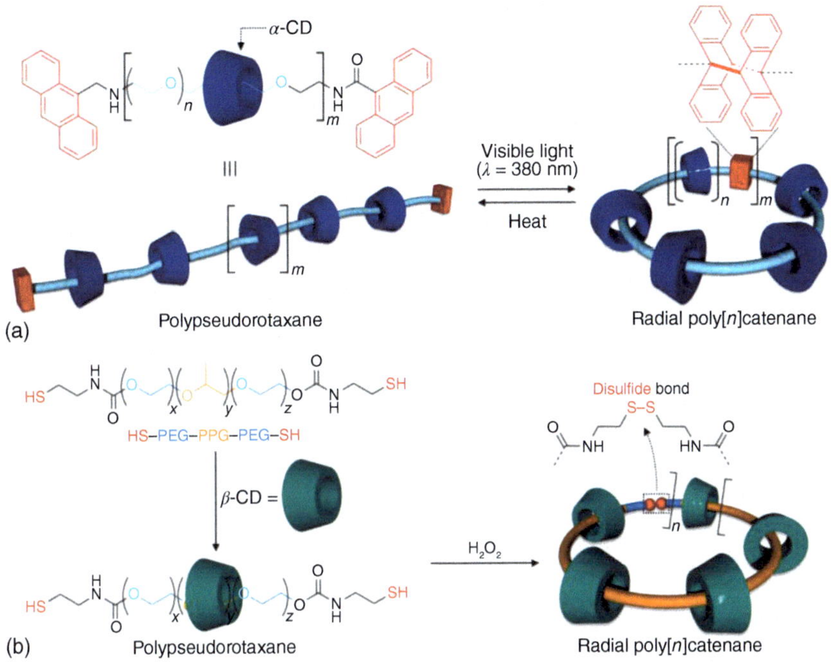

Figure 8.3 (a) Schematic representation of the formation of a radial poly[*n*]catenane by reversible dimerization of Harada' s CD/PEG poly[*n*]rotaxane with anthracene stoppering units. (b) Schematic illustration for the formation of a radial poly[*n*]catenane through oxidative cyclization of a dithiol block co-polymer threaded with CD macrocycles.

resonance (^{1}H NMR), matrix-assisted laser desorption ionization time-of-flight mass spectrometry (MALDI-TOF MS), and gel permeation chromatography (GPC), in addition to ultraviolet-visible (UV–Vis) absorption spectroscopy [23].

In contrast to the large number of studies on the synthesis of cyclodextrin (CD)-based polyrotaxanes, a much smaller number of investigations have been carried out on CD-based polycatenanes. Harada and coworker reported the preparation of poly[*n*]catenane by macrocyclization of a poly[*n*](pseudo)rotaxane (Figure 8.3a) [24]. One end of a poly(ethylene glycol) (PEG) ($M_w = 3350$) chain was functionalized with an anthracene stopper. Nineteen α-CD units were threaded through the modified polymer chain to form a pseudo-polyrotaxane, which was subsequently transformed into a polyrotaxane by coupling the open end with 9-anthracenecarboxylic acid. Photodimerization of the anthracene stoppers by visible light irradiation (λ = 340 nm) led to the formation of a polycatenane in addition to a poly(polyrotaxane). However, the polycatenane could not be separated from the poly(polyrotaxane) because of their similar properties.

Li and coworkers demonstrated a successful one-pot strategy for the synthesis and isolation of CD-based radial poly[*n*]catenanes ($n - 1 > 10$) in which the core ring is composed of PEG–poly(propylene glycol)–PEG copolymer (PEG–PPG–PEG) cyclized through a disulfide bond (Figure 8.3b) [25]. Although a dithiol-PEG was initially chosen for threading with α-CD, the oxidative cyclization of the corresponding pseudopolyrotaxane through disulfide bond formation did not result in a cyclized product owing to the rigidity of the polymer chain. Therefore, PEG–PPG–PEG dithiol (HS–PEG–PPG–PEG–SH) block copolymer was synthesized by activating the terminal hydroxyl groups of PEG–PPG–PEG with 1,1′-carbonyldiimidazole, followed by reaction with cystamine and reduction with 1,4-dithiothreitol (DTT). When the HS–PEG–PPG–PEG–SH block copolymer was treated with aqueous solution of β- and γ-CD macrocycles, polypseudorotaxanes formed as precipitates after 24 h. The resulting polypseudorotaxanes were oxidized by H_2O_2 to obtain the corresponding radial polycatenanes. Notably, β- and γ-CDs encapsulated the PPG moieties of the copolymer chain in their corresponding polypseudorotaxanes, providing the terminal thiol groups with sufficient mobility to undergo cyclization. In contrast, α-CD incorporated polypseudorotaxane did not undergo cyclization because α-CD incorporated PEG moieties of the copolymer chain decreased the flexibility of the terminal thiol groups. The β-CD macrocycles could be reversibly released from the polycatenanes by reducing the disulfide bonds using DTT.

Kim and colleagues exploited the orthogonality between coordination-driven self-assembly and cucurbit[6]uril (CB[6])-based host–guest interaction to achieve a radial [4]-catenane consisting of three small CB[6] rings threaded onto a large metallocycle ring [26, 27]. Refluxing an aqueous solution of a 1 : 1 : 1 mixture of CB[6], *N,N″*-bis(4-pyridylmethyl)-1,4-diaminobutane dihydronitrate and *cis*-Pt (en)-$(NO_3)_2$ (en = ethylenediamine) yielded the radial [4]catenane in 90% yield. The radial [4]-catenane could be obtained in quantitative yield when a preassembled host–guest complex was refluxed with *cis*-Pt(en)-$(NO_3)_2$. Notably, a mixture of radial [4] and [5]catenanes was produced when the precursors were stirred at room temperature for an even longer time (Figure 8.4a). Therefore, it is clearly evident that the entropic contribution upon refluxing could overcome the enthalpic cost associated with the sterically congested [4]catenane structure. Changing the 4-pyridyl group in *N,N″*-bis(4-pyridylmethyl)-1,4-diaminobutane dihydronitrate by a 3-pyridyl group enabled the authors to access [5]catenane as the major product, in which four CB[6] units are threaded on a square metallacycle framework. In another approach, the authors further designed a preorganized pseudorotaxane comprising an L-shaped dipyridyl ligand with phenanthroline corners and two threaded CB[6] units (Figure 8.4b). Complexation reaction with $Cu(NO_3)_2$ or *cis*-Pt(en)-$(NO_3)_2$ joined the ends of the ligand to form [5]catenane structures which were confirmed by X-ray diffraction (XRD) crystallography. When $Cu(NO_3)_2$ was used, the corresponding [5]catenane contained two Cu^{2+} ions coordinated to the phenanthroline corners in addition to the two Cu^{2+} ions at the linkers.

Later, they also reported the synthesis of a radial [6]catenane based on intermolecular charge-transfer complexation within CB[8], of which a larger cavity could accommodate a pair of electron-rich and electron-deficient guest molecules [28]. A bent monomer containing a naphthalene and a dipyridyliumylethylene units connected by a methylene linker self-assembled into a cyclic pentamer supported by the host–guest complexation with five CB[8] units. The charge–transfer complex was characterized by various NMR methods, including ^1H-^1H correlation

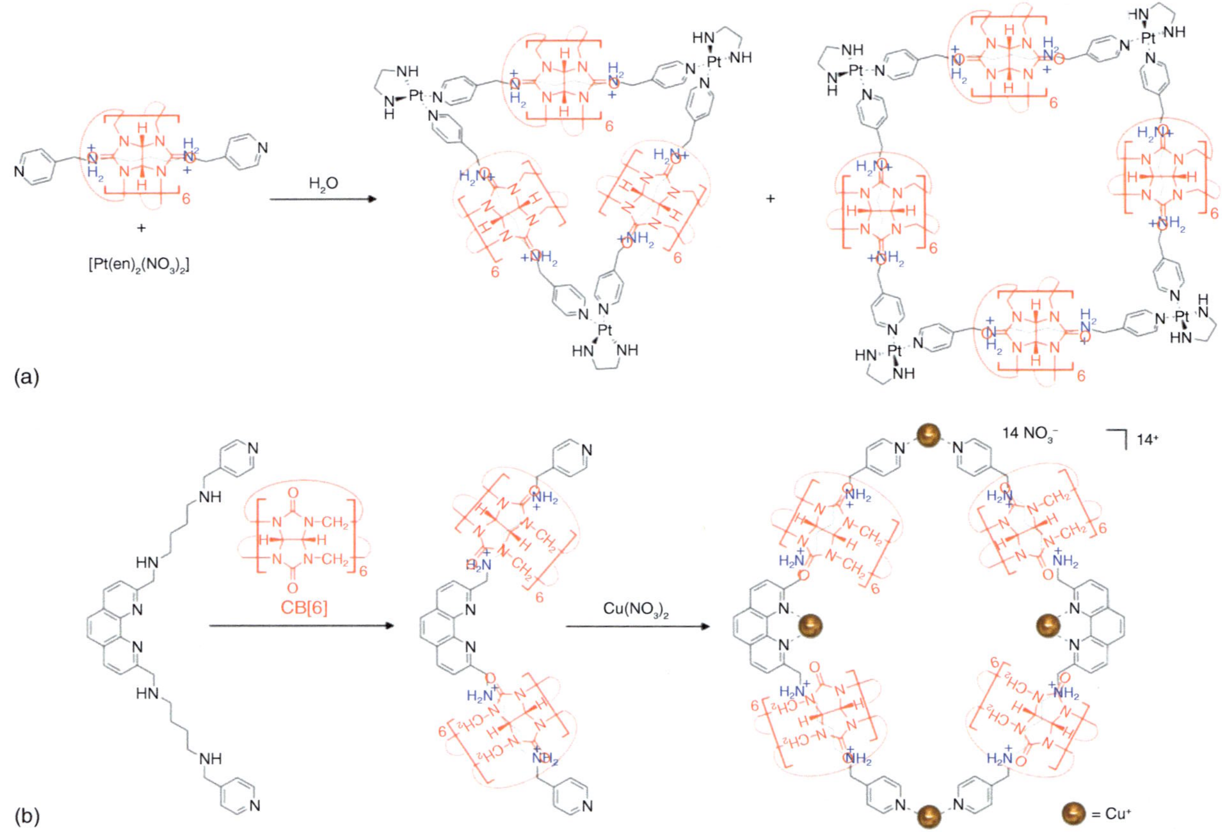

Figure 8.4 Schematic illustration of the preparation of radial (a) [4, 5]catenanes assembled from a linear dipyridyl ligand and (b) a [5]catenane based on a L-shaped dipyridyl ligand reported by Kim' s group.

spectroscopy (COSY), nuclear Overhauser spectroscopy (NOESY), pulsed-field gradient NMR, and electrospray ionization mass spectrometry (ESI MS).

Wu and coworker reported the synthesis of a radial [5]catenane by threading a tetrakis(dialkylammonium)-based precursor through dibenzo-24-crown-8 (DB24C8) followed by a ring-closing metathesis reaction of the terminal alkenes using second generation Grubbs catalyst [29]. They required an excess of DB24C8 (5 equivalent per dialkylammonium motif) to achieve complete threading of the crown on all four dialkylammoniums. Stang and coworkers utilized the combined benefits of dialkylammonium-crown ether molecular recognition and coordination-driven self-assembly to produce a radial [4]catenane and a radial [7]catenane based on a metalacyclic core [30]. Unlike previously described Kim's *cis*-Pt(en)-$(NO_3)_2$, they employed *trans*-bis(phosphene)platinum(II) acceptors to synthesize the radial [*n*]catenanes. Coordination-driven self-assembly of 1,2-bis([4,4′-bipyridin]-1-ium)ethane with 60° or 120° platinum(II) acceptor, followed by anion exchange furnished a triangular and hexagonal macrocycles, respectively, (Figure 8.5).

Subsequently, DB24C8 was threaded into the metallacycles through host–guest recognition with the 1,2-bis(pyridinium)ethane unit to afford [4] and [7]catenanes in quantitative yield. The formation of [*n*]catenanes was accompanied by a color change from pale yellow to bright yellow, attributed to the intermolecular

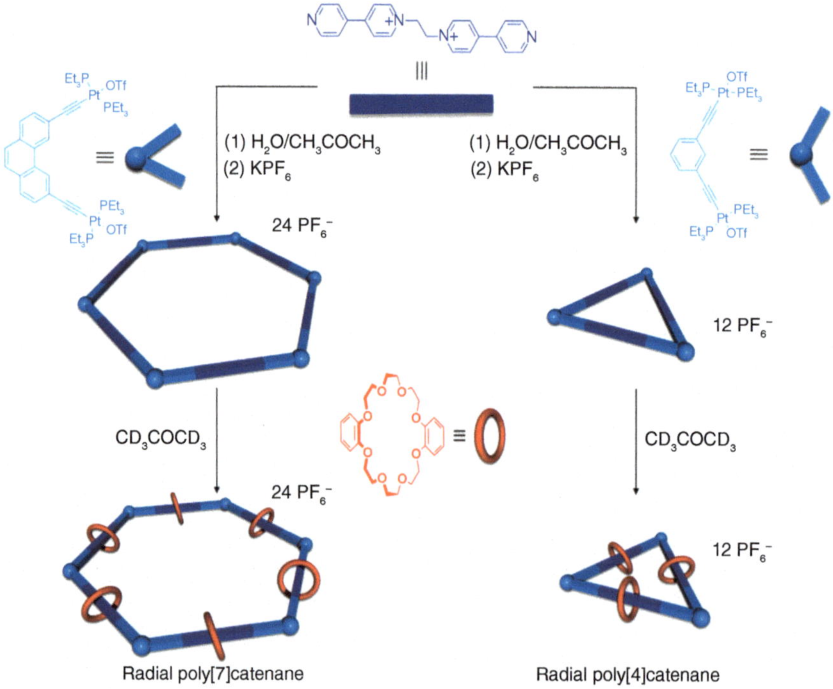

Figure 8.5 Schematic illustration of synthesis of radial (a) [4] and (b) [7]catenanes by Stang and coworkers.

charge–transfer transitions between the host and guest. The dynamics of the threading process were studied using ^{1}H NMR, ^{31}P NMR, 2D NOESY NMR, and ESI-MS. Unlike the threading of CB[6] at the 1: 1 host–guest ratio, excess DB24C8 (4 equivalent hosts per guest motif) was required to drive the formation of catenanes because of the weak ammonium–crown ether binding ($K \approx 200$) [31, 32].

Later, similar radial [*n*]catenanes were reported by using a much less amount (1.2–1.5 equivalent host per guest motif) of a crown ether-based cryptand [33] which possesses much better binding affinities than the corresponding simple crown ether analog [34]. Stirring a mixture of a paraquat-derived carboxylate ligand with a 60° Pt(II) acceptor in H_2O/acetone mixture furnished a triangular metallacycle, while a rectangular metallacycle was obtained from the same ligand by reaction with a dipyridyl ligand and *cis*-$Pt(PEt_3)_2(OTf)_2$. Subsequently, threading of bis(*m*-phenylene)-32-crown-10 (BMP32C10) into the rectangular and triangular metallacycles resulted in radial [4] and [3]catenanes, respectively.

Shi, Yu, Yang, and coworkers reported a facile synthesis of a heterometallic triangular necklace by combining two orthogonal 2,9-disubstituted Cu(I)-bis (phenanthroline)s ($[Cu(phen)_2]^+$) and platinum(II)–pyridine coordination [35] (Figure 8.6a). The coexistence of Cu and Pt metals endowed the molecular necklace

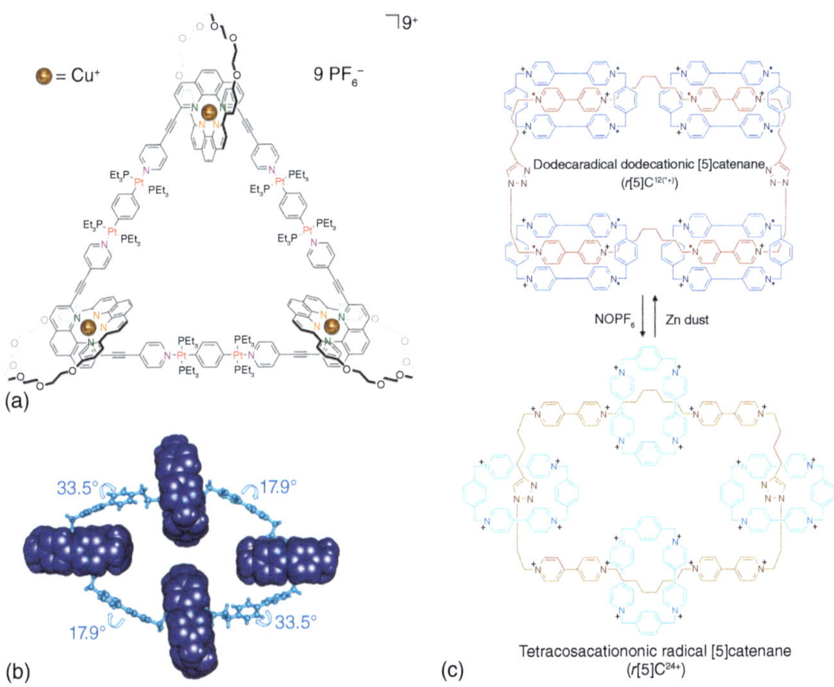

Figure 8.6 (a) Chemical and (b) single-crystal X-ray crystal structures of a heterometallic radial [4]catenane prepared via orthogonal 2,9-disubstituted Cu(I)-bis(phenanthroline)s and platinum(II)–pyridine coordination driven self-assembly by Wu et al. (c) Reversible oxidation and reduction process of a radical [5]catenane reported by Stoddart and coworkers. Source: Reproduced with permission from Nguyen et al. [36]. Copyright 2018 Elsevier Inc.

with much higher DNA cleavage efficiency compared to its precursor components. This result further motivated the authors to investigate the interaction with the bacterial cells. The heterometallic complex showed superior antibacterial activity to the clinically isolated drug-resistant pathogens compared to the precursors. The enhanced antibacterial activity stems from the synergistic effect of heterometals in bacterium-binding and damaging activity as revealed by confocal observation of the 4′,6-diamidino-2-phenylindole/propidium iodide double stained cells.

Recently, Stoddart and coworkers have reported a viologen-based tetracosacationic [5]molecular necklace that confines five polycationic ring components in a nanospace (six positive charges per nm^3) (Figure 8.6a) [36]. As coulombic repulsive force prevents close proximity of the polycationic rings, they employed template-directed strategy based on radical–radical recognition, thereby turning the repulsive positively charged viologen components into attractive viologen radical dimers. In the presence of Cu(0) as a reducing agent, two bis-viologen axles containing terminal azide and alkyne groups, respectively, and an octamethylene spacer between the two viologen units were subjected to form two [3]pseudorotaxanes with cyclobis(paraquat-*p*-phenylene) bisradical dication rings. Subsequently, tandem click reactions between these two [3]pseudorotaxanes catalyzed by the *in situ* generated Cu(I) species, followed by oxidation with $NOPF_6$, resulted in the radial [5]catenane in 36% overall yield. Evidently, Cu powder served a dual role as a reducing agent and a catalyst in the synthesis of the radial [5]catenane. The radial [5]catenane was fully characterized by NMR spectroscopy and MS spectrometry as well as single-crystal XRD crystallography (Figure 8.6b). Reduction of tetracosacationic [24+] state of the radial [5]catenane by Zn dust produced dodecaradical dodecacationic [12(·+)] state, which could be reconverted to the original [24+] state through oxidation using $NOPF_6$ (Figure 8.6c). On the other hand, reversible electrochemical switching between the [24+] and [12(·+)] states was demonstrated without any change of the peak currents in cyclic voltammetry scans up to five cycles.

8.3 Main Chain Polycatenanes Composed of Covalent Macocyclic Building Block

In 2017, Rowan and coworkers first achieved the synthesis of linear poly[*n*]catenanes (*n* up to 27), molecular analogs of a "chain link," that paves the way to the design of polymeric materials with utmost flexibility but robustness [18]. In their strategy, metallosupramolecular polymers were prepared to obtain poly[*n*]catenanes. They synthesized a 68-membered macrocycle containing two 2,6-bis(*N*-alkylbenzimidazolyl)pyridine (Bip) ligands and a threading molecule (a macrocycle precursor) with two Bip ligands and preassembled these ditopic components at the 1 : 1 ratio in the presence of 2 equivalents of Zn^{2+} to afford linear supramolecular copolymers (Figure 8.7a). The preassembled structures composed of alternating links between macrocycles and threading molecules were subjected to undergo an olefin metathesis reaction at the terminal olefins of the threading molecules for the ring

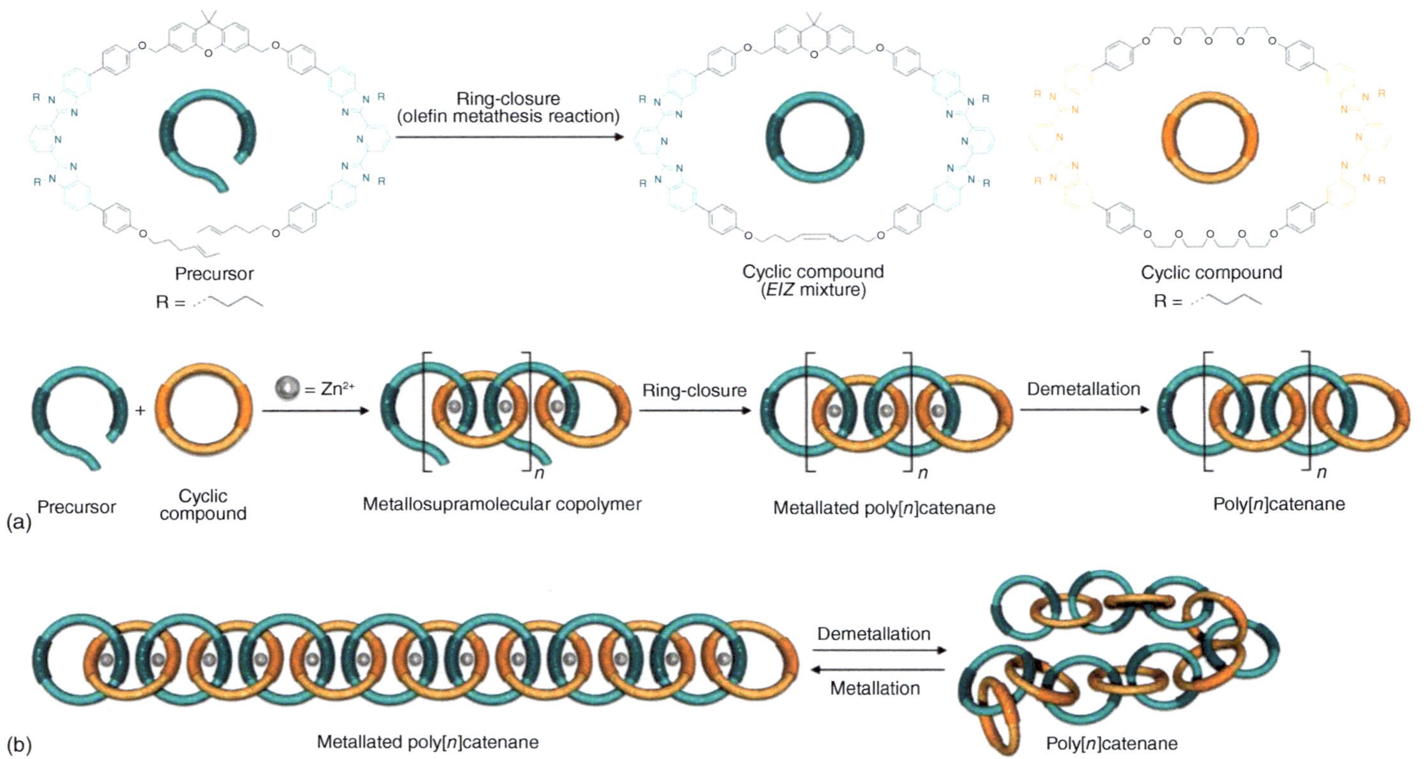

Figure 8.7 (a) Schematic illustration of Rowan' s strategy for the synthesis of poly[*n*]catenanes by assembling Bip-based macrocycles into a metallosupramolecular polymer followed by ring closure via olefin metathesis reaction. (b) Schematic representation of changes in the conformation of the poly[*n*]catenanes upon demetallation/metalation.

closure, leading to a full interlocked, polymeric chain link architecture. The alternating arrangement of the macrocycle and the threading molecule is enthalpically favored by the principle of maximal site occupancy. Finally, demetallation using tetrabutylammonium hydroxide gave pure organic interlocked poly[n]catenanes (n = 7–130, depending on the degree of branching) with number-average molecular weights of >20 000 g mol^{-1} with 75% yield. The poly[n]catenanes were further separated by preparative GPC to obtain four fractions, among which the main fraction (60%) contained mainly linear poly[7–27]catenanes (85%) and cyclic poly[4–7]catenanes, as confirmed by ^{1}H NMR, MALDI-TOF, and multi-angle light scattering experiments.

The conformational flexibility of the polymeric structure could be controlled by Zn^{2+} coordination with the Bip ligands of the polycatenanes (Figure 8.7b). Interestingly, an increase in the hydrodynamic radii of linear (from 3.9 to 6.6 nm) and branched (from 5.0 to 7.6 nm) polycatenanes was observed upon the addition of Zn^{2+} ions, while the cyclic one remained almost unaffected. This result was also corroborated well with all-atom molecular dynamics simulations of Zn^{2+} complexed linear poly[n]catenane. In addition, differential scanning calorimetry (DSC) analysis showed that the glass transition temperature (T_g) increased from 97 to more than 160 °C, suggesting a reduction of flexibility in the conformation of linear and branched polycatenanes upon Zn^{2+} coordination. Molecular dynamic simulation studies of poly[n]catenanes (n = <25) in the melt state revealed that the high flexibility of the poly[n]catenanes inhibits their entanglement like linear covalent polymers [37]. At the same time, their interlocked geometry does not allow π–π stacking interactions between the chains, thereby reducing cohesive energy density. As a result, the physical appearance of the poly[n]catenanes in the solid state differs significantly from conventional linear covalent polymers. For example, the poly[n]catenanes appear as solid powders, while the linear covalent polymers are often convertible to solid films.

8.4 Main Chain Nano-polycatenanes Composed of Non-covalent Building Block

Characterization of main chain poly[n]catenanes composed of large and flexible macrocycles is challenging because crystallization of such interlocked compounds is extremely difficult. Moreover, mixtures of poly[n]catenanes are intrinsically impossible to be crystallized. Proving the branching catenane structure is also not easy with conventional methods such as size exclusion chromatography and diffusion-ordered spectroscopy (DOSY). Recently, Yagai and coworkers reported nanoscale poly[n]catenanes by kinetically organizing a small molecule into toroidal supramolecular polymers with a radius of about 13 nanometers that enabled easy characterization with atomic force microscopy (AFM) [19]. The authors used a diphenyl naphthalene (DPN) appended barbituric acid derivative, which formed various topological supramolecular polymers depending on the supramolecular polymerization conditions (Figure 8.8a).

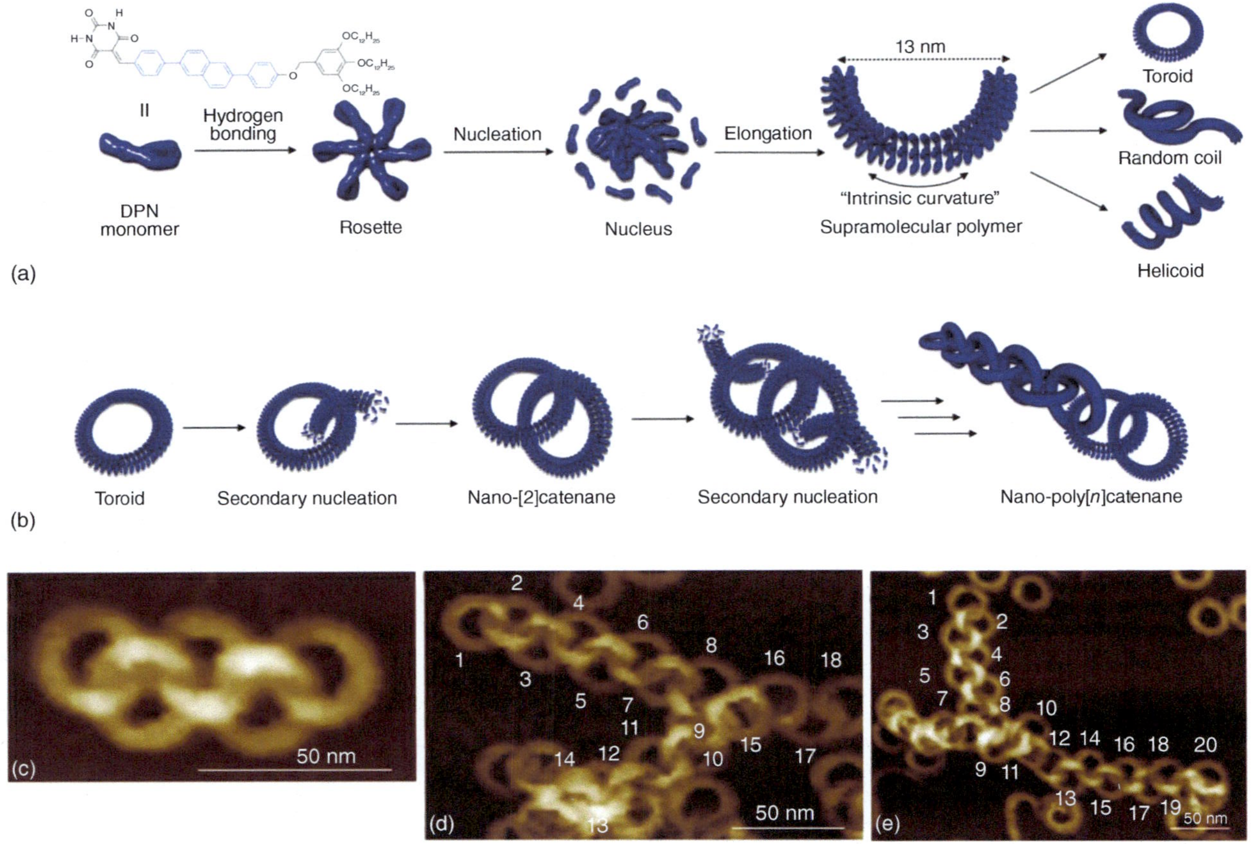

Figure 8.8 (a) Molecular structure of DPN monomer and schematic representation of the unique stacking of hydrogen-bonded cyclic hexamers to generate an intrinsic curvature. (b) Mechanism of nano-poly[*n*]catenane formation in a stepwise fashion reported by Yagai' s group. AFM images of nano-(c) [5], (d) [18], and (e) [22]catenanes. Source: Reproduced with permission from Datta et al. [19]. 2020 Springer Nature.

By applying the "injection" method, i.e. rapid mixing of the monomer solution in chloroform (good solvent) with methylcyclohexane (poor solvent), the authors initially found by AFM that the formation of a high population of toroids (44%). Interestingly, they found that a significant number of toroids (~150 out of 5000) were paired up into nano-[2]catenanes by mechanical bond, while nano-[3]catenanes were rarely observed [19]. The calculated yield of the interlocked toroids on the basis of monomer was found to be 1.6%, which surpassed the reported yield of Wasserman' s stochastic catenation [5, 6]. When the poor solvent was changed from methylcyclohexane to less polar linear alkanes such as *n*-hexane and *n*-octane, the yield of the nanocatenane jumped to about 5–7% and also enabled oligo-catenation up to nano-[6]catenane. The nano-[5]catenane was named "Nanolympiadane" (Figure 8.8c) to pay tribute to the Stoddart's "Olympiadane" [15].

Similar to the idea of preorganization of precursors by metal–ligand coordination template in Sauvage' s approach, the authors anticipated that the surface of the toroid formed initially, which is covered by alkyl chains, could induce nucleation of the remaining monomers through van der Waals interactions. The phenomenon of nucleation of monomers on the surface of a preformed aggregate is called "secondary nucleation," which could occur at the interior and outer surfaces with equal probability [38]. To confirm the secondary nucleation of monomers on the toroid surface, the authors designed a seeding experiment [39]. As the self-assembled interlocked nanostructures are composed of toroidal building blocks, toroids could be considered as seeds for secondary nucleation. Since the authors could not selectively obtain pure toroids by any polymerization protocols, they removed open-ended assemblies from the mixture of various supramolecular topologies, including toroids, by exploiting the higher thermal stability and smaller sizes of toroids (Figure 8.9). Then, the authors added the isolated toroidal solution to a hot

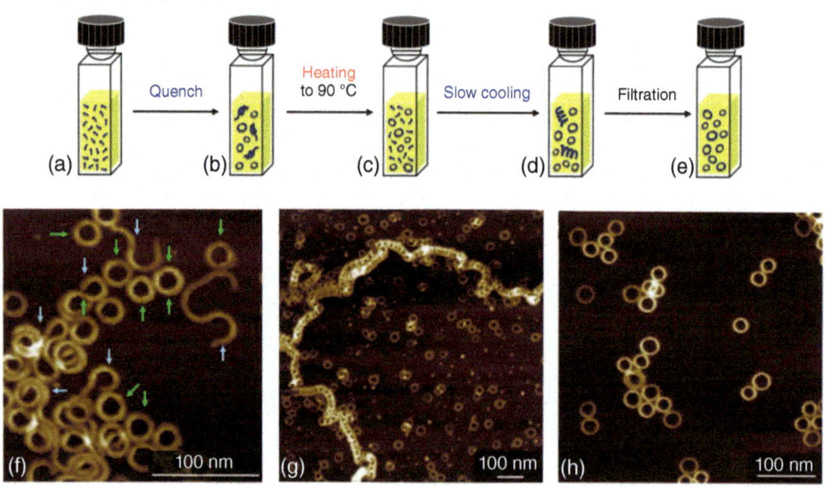

Figure 8.9 (a–e) Upper panel displays protocols for separating toroids from open-ended fibers. (f–h) Lower panel shows AFM images obtained from (b), (d), and (e) solutions. Green and blue arrows in AFM image (f) show toroids and open-ended fibers, respectively.

monomer solution and subsequently cooled down to slightly below the "primary" nucleation temperature of the monomers. Monitoring the supramolecular polymerization by UV–Vis spectroscopy unveiled that the nucleation of monomers was clearly accelerated in the presence of toroids. The secondary nucleation of monomers on the surface of toroids was also supported by means of a scaling method developed to analyze protein aggregation [38] and coarse-grained molecular dynamics simulation.

The "injection" protocol was modified further to achieve polycatenation by improving the efficiency of secondary nucleation. In contrast to Rowan' s polycatenane synthesis [18], the growth of self-assembled polycatenane should occur more in a "living" manner at the termini of catenanes (Figure 8.8b). Accordingly, instead of adding the monomer solution all at once to the poor solvent, which produces nano-catenanes at the same time, the monomer solution was injected portion-wise to elongate the nano-catenanes sequentially. In addition, such a portion-wise injection protocol could maintain a lower local concentration regime of monomers to favor secondary nucleation and reduce the production of randomly coiled fibers [40, 41]. After the trial of several combinations of the following parameters such as injection speed, number of injections and intervals, the addition of the poor solvent ten times at an interval of 1 s was found to be the best for the formation of nano-polycatenanes. Analysis of AFM images of the aggregates revealed that the modified protocol dramatically enhanced not only the yield of interlocked toroids but also the catenation number (Figure 8.8d,e). While the single injection experiment was able to produce nano-[6]catenane as the longest species, the 10-injections protocol succeeded in creating unprecedented nano-poly[22]catenane having a length of approximately 500 nm. A nano-poly[18]catenane containing a branched nano-[4]catenane as a part of the structure suggests that three supramolecular polymer chains can pass through the internal void space of a single toroid. The formation of such sterically challenging, branched structures also supported the contribution of secondary nucleation, which enabled the pre-organization of the monomers at the inner surface of the preformed catenanes.

8.5 Polycatenanes Composed of Metal–Organic Coordination Cages

In 1994, Fujita et al. exploited the hydrophobic effect in palladium(II)–pyridine coordination-driven self-assembly to obtain interlocked macrocycles. The yield of the interlocked macrocycles was very high (99%), but they were at equilibrium with the corresponding discrete macrocycles [42]. The authors found that the equilibrium shifted towards the interlocked species at high concentrations, while a dilute condition favored the discrete macrocycles. In contrast, platinum(II)-coordination-driven self-assembly resulted in a kinetically trapped state, which was shown by the concentration-independent ratios of discrete and interlocked species.

Followed by this work, Fujita discovered one-pot synthesis of a triply interlocked cage composed of trigonal ligand-based two discrete coordination cages in 61–65%

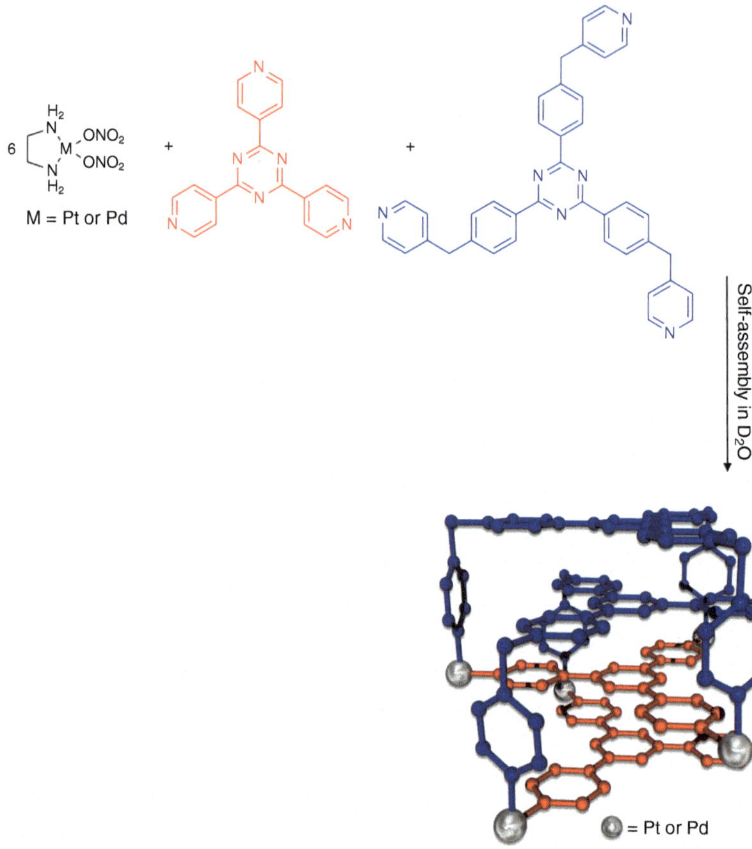

Figure 8.10 Fujita' s triply interlocked cage complex is formed by the self-assembly of four organic ligands with six square-planar-coordinating metal cations (Pd^{II} or Pt^{II}).

yield (Figure 8.10) [43]. Spontaneous self-assembly of two different tridentate ligands with three cis-blocked square-planar metal cations (Pd(II) or Pt(II)) in D_2O initially led to the formation of a kinetically distributed oligomeric mixture, which further transformed to the thermodynamically stable interlocked molecular cage upon heating at 100 °C for three days. Notably, this result was in clear contrast to their previous observation on the Pd(II)/Pt(II)–pyridine coordination-driven self-assembly, wherein interlocking of macrocycles occurred in 99% yield under kinetic conditions. The catenation of the cages was driven by π–π stacking interactions between four triazine-cored π-surfaces as revealed by XRD crystallographic analysis and an entropic contribution associated with the release of cavity-confined solvent molecules. Although the additional intermolecular stacking of the triazine-cored π-surfaces between catenanes leading to an infinite one-dimensional strand was observed in the solid state, polycatenation was not observed.

Followed by Fujita' s work, a few examples of similar three-dimensional [2]catenane were reported independently by Hardie and coworkers [44] and Kuroda and coworkers [45]. However, the construction of cage-based polycatenane has not been

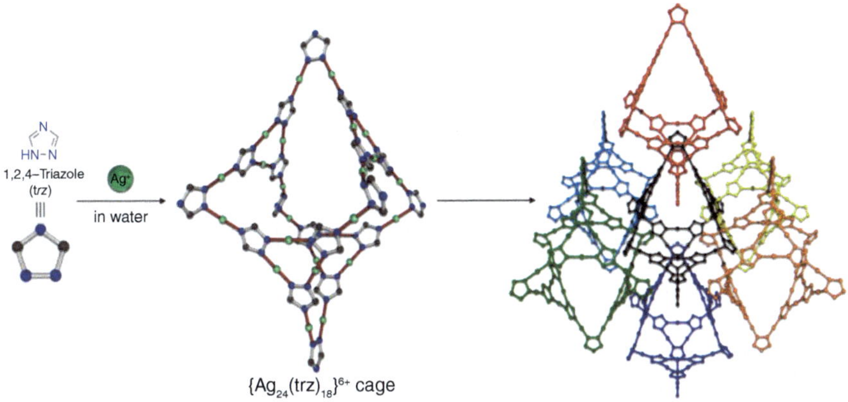

Figure 8.11 Synthesis of polycatenane based on metallocages reported by Lu and coworkers.

achieved until the report by Lu and coworkers [46]. Instead of using any tridentate ligand, a simple 1,2,4-triazole (trz) ligand was subjected to undergo hydrothermal reaction with trilacunary Keggin $[A\text{-}PW_9O_{34}]^{9-}$ anion and $Ag(O_2CCH_3)$ in water to produce a crystalline material $\{Ag_{26}(trz)_{20}\}[PW_{12}O_{40}]_2$ (Figure 8.11). Single-crystal XRD analysis of the material revealed that sextuple inter-catenation of discrete adamantane-like $\{Ag_{24}(trz)_{18}\}^{6+}$ nano-cages in three dimensions led to the formation of an infinite polycatenated architecture in which the Keggin polyoxometalate anions $[PW_{12}O_{40}]^{3-}$ worked as templates. In the absence of Keggin polyoxometalate anion, self-assembly of Ag^+ cations with 1,2,4-triazole afforded a two-dimensional [Ag(trz)], supporting a crucial role of the anions as templates in the polycatenation process.

Lu and coworkers reported one-dimensional and three-dimensional polycatenanes constructed by discrete tetrahedral cages [47]. Reaction of H_3tcba (H_3tcba = tri (4-carboxy-benzyl)amine) with $[NiL]Cl_2$ (L = 1,4,8,11-tetraazaundecane) in aqueous/dimethylformamide (DMF) mixture resulted in $\{[NiL]_4(Htcba)_4(H_2O)_2\}\cdot 35H_2O$ as needle-shaped pink crystals. On the other hand, treating H_3tcba with $[NiL](ClO_4)_2$ in methanol/acetonitrile mixture produced block-shaped octahedral purple crystals of $\{H[NiL]_6(tcba)_4(ClO_4)\}\cdot 32H_2O$. XRD study unveiled the existence of one-dimensional and three-dimensional polycatenated assemblies of discrete tetrahedral cages in the former and latter crystals, respectively (Figure 8.12). In the polycatenated architecture, the ClO_4^- anion occupied the center of the tetrahedral cages. This result is in contrast to the previous report of $\{Ag_{26}(trz)_{20}\}[PW_{12}O_{40}]_2$ three-dimensional polycatenanes in which $[PW_{12}O_{40}]^{3-}$ anions were found to reside in the cavities of the adamantane-like nanocages. The lattice water molecules of one- and three-dimensional polycatenanes could be removed under 120 and 130 °C in thermogravimetric analysis. The dynamic pores of the latter allowed reversible dehydration and rehydration, as confirmed by the XRD pattern. Furthermore, the dehydrated material could also show size-dependent adsorption/desorption behaviors of alcohols.

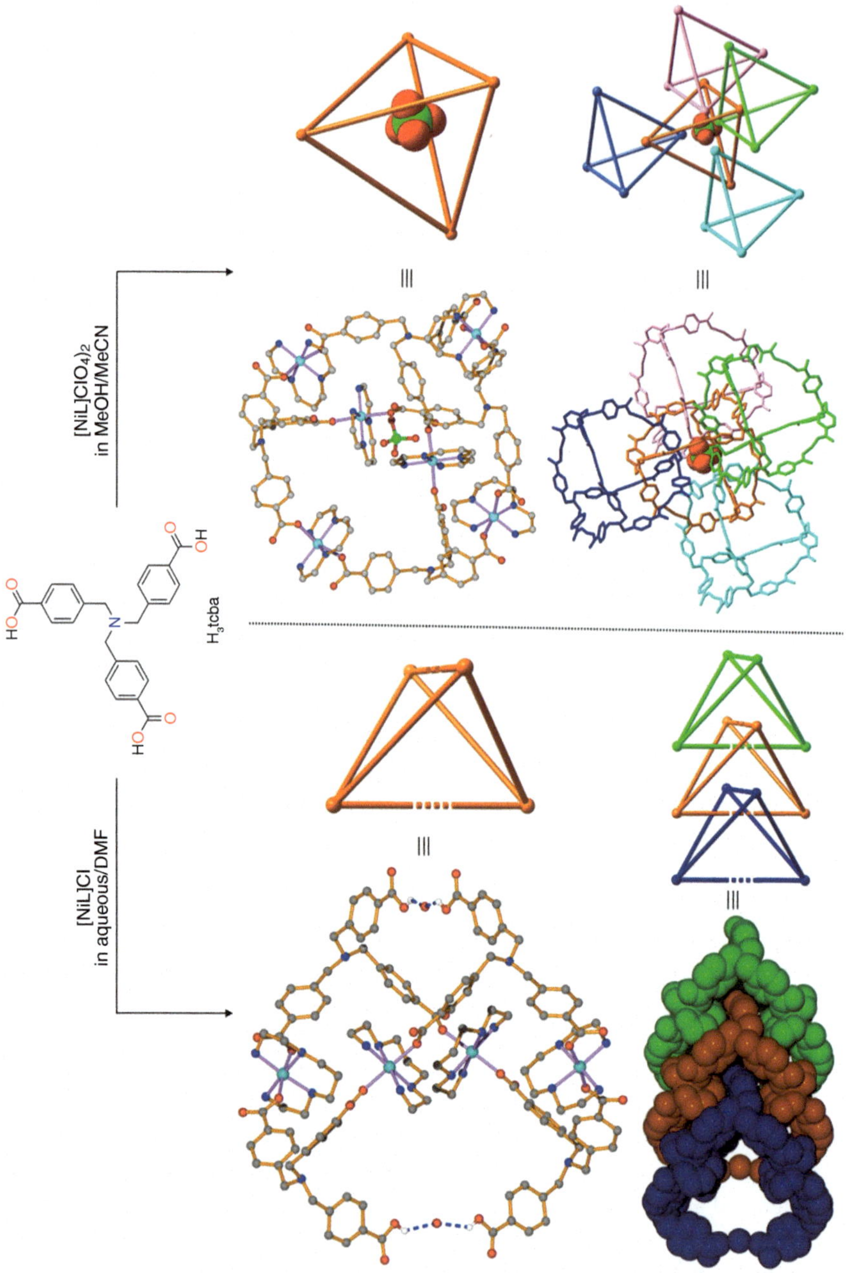

Figure 8.12 Schematic presentation of preparation of one-dimensional and three-dimensional polycatenated assemblies composed of discrete tetrahedral cages by Lu and coworkers. Source: Reproduced with permission from Jiang et al. [47]. Copyright 2012 Springer Nature Limited.

Dehnen and coworkers employed a tridentate ligand 2,4,6-tris(4-pyridyl)pyridine (ptpy) to prepare a strand-like polycatenane composed of icosahedral cages [48]. Coordination-driven self-assembly of the tridentate ligand with $ZnCl_2$ generated colorless hexagonal plates of $[(ZnCl_2)_{12}(pytpy)_8]_n \cdot xCHCl_3$ which upon characterization by single crystal XRD analysis revealed a linear chain of interlocked icosahedral cages with an inner volume of more than 2700 Å^3. Although face-to-face π–π interactions of the ligand cores were involved, similar to the reports by Fujita and others, the success of the polycatenation of cages, unlike their system, relies on the charge neutrality and adequate volume of the icosahedral cages.

Recently, Martí-Rujas and coworkers have reported the synthesis of a poly[n] catenane composed of interlocked $M_{12}L_8$ icosahedral nanometric cages under kinetic control [49]. They utilized tris-pyridyl benzene (TPB) ligand that has a relatively more electron-rich core compared to the ptpy ligand. Vigorous mixing of $ZnCl_2$ with TPB in a mixture of nitrobenzene and methanol led to the formation of a microcrystalline poly[n]catenane material $[(ZnCl_2)_{12}(TPB)_8]_n{\cdot}9(C_6H_5NO_2)$ in 58% yield, while slow crystallization process additionally produced a coordination polymer. Similar polycatenated materials were also obtained by using other aromatic solvents such as p-chlorotoluene, toluene, chlorobenzene, and 1-bromo-4-fluorobenzene instead of nitrobenzene. In contrast, chloroform could not produce nanocages, indicating that aromatic solvent molecules acted as a template in the formation of the nanocages. Density functional theory calculations suggested that catenation involving ligand aromatic–aromatic interactions is favored over the formation of isolated (monomeric) cage units as kinetic products according to the analysis of the packing energy in terms of monomeric and dimeric cages. High-resolution synchrotron X-ray study revealed that the material obtained under a kinetic condition is composed of interlocked $M_{12}L_8$ nanocages having six ordered and three disordered nitrobenzene guest molecules. Thermogravimetric analysis showed that the weakly bound disordered nitrobenzene guest molecules were released at lower temperature (160 °C) without any structural alteration of the cage, whereas the removal of ordered nitrobenzene molecules by heating occurred at much higher temperatures (c. 210 °C) and was accompanied by a collapse into another thermodynamically stable structure as confirmed by XRD experiment. The partially emptied nanocages could reabsorb the nitrobenzene guest molecules, while the original structure could not be recovered from the new phase obtained at 230 °C even after soaking in nitrobenzene for two days.

8.6 Poly[2]catenane

Poly[2]catenane is structurally different from poly[n]catenane in that the former can be synthesized by polymerizing a [2]catenane monomer unit. Sauvage, Müllen, and coworkers prepared a poly[2]catenane by polymerizing a Cu(I)-phenanthroline complex-based [2]catenane with two ester functions (Figure 8.13a) [50]. The [2]catenane monomer was synthesized by template-directed synthesis using Cu(I)-phenanthroline complexation. The ester groups were then

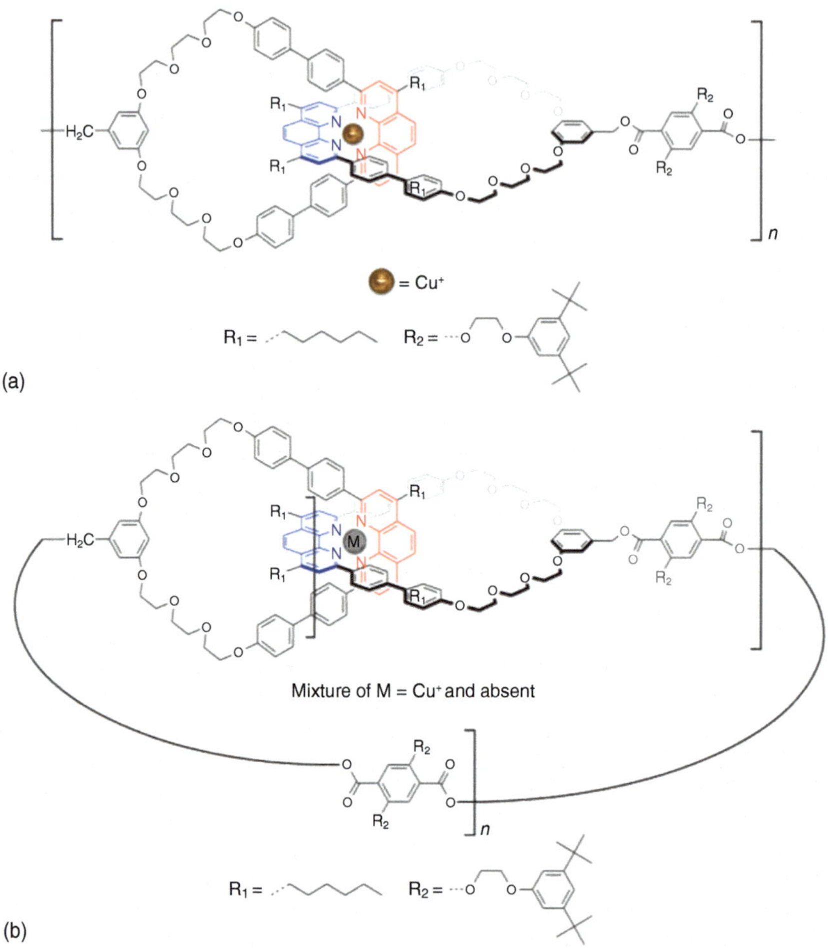

Figure 8.13 Poly[2]catenanes reported by (a) Sauvage and (b) Greets.

reduced by diisobutylaluminum hydride to produce the corresponding dibenzylic alcohol derivative, which was subjected to undergo polymerization reaction with a terephthalic acid derivative to afford poly[2]catenane. GPC analysis suggested that the polymers have weight-average molecular weight (M_w) of 1800 kDa and a number-average molecular weight (M_n) of 55 kDa. The structures were validated by ^{1}H and ^{13}C NMR, and MALDI-TOF MS. After several years, Kern, Sauvage, Geerts, and coworkers attempted to prepare similar poly[2]catenane using related monomer components, and they found that the copolymerization reaction using the demetalated [2]catenane resulted in the formation of cyclic oligo[2]catenanes owing to high flexibility of the demetallized [2]catenane unit (Figure 8.13b) [51]. The GPC analysis with indicated molecular weight ranges of M_w = 42–47 kDa and M_n = 22–25 kDa; number-average degree of polymerization (DPn) = 8–9.

Stoddart and coworkers prepared a bis[2]catenane precursor by donor–acceptor-based template-directed coupling reaction between a dicationic bis(bipyridinium) salt and a substituted *m*-xylylene dibromide followed by counterion exchange [52]. The polymerization of the bis[2]catenane with bis(4-isocyanatophenyl)methane resulted in a poly(bis[2]catenane)s (Figure 8.14a). GPC analysis revealed M_n of 45.0 kDa, from which DP_n was estimated to be 15. The research group also prepared a bis[2]catenane bearing a bipyridine motif which could be engaged to coordinate with CF_3SO_3Ag in acetonitrile to produce one-dimensional Ag^+-coordinated poly([2]catenane) without using covalent polymerization method [52, 53]. The M_n determined by GPC with protein standards was 150 kDa, which corresponds to DP_n of 40.

Using tetracationic cyclophane-aromatic crown ether motif, Grzybowski, Stoddart, and coworkers prepared a bistable side chain poly[2]catenane, which can show a redox controlled actuation. The alkyne-functionalized [2]catenane was employed to undergo click reaction in DMF at 60 °C with a methacrylate polymer with azide terminated side-chains (M_w = 55.0 kDa, M_n = 39.0 kDa, PDI = 1.4), which was obtained by atom transfer radical polymerization (Figure 8.14b) [54]. Monitoring the click reaction for three hours by ^1H NMR in DMF-d_7 confirmed 95% conversion. The M_n value of the polymer was determined to be 128 kDa by the end-group analysis using ^1H NMR. On the other hand, size exclusion chromatography coupled with multiangle light scattering indicated M_n and M_w values of 870.0 and 1300.0 kDa, respectively (PDI = 1.5), suggesting aggregation. Scanning electron microscopy (SEM) and dynamic light scattering (DLS) experiments revealed the formation of hollow spherical nanoparticles with sizes around 200 nm. The bistability of the employed [2]catenane stems from an equilibrium between two switchable isomers in which the ground-state co-conformation (GSCC) and the metastable-state co-conformation (MSCC) correspond to the encirclement within the tetrathiafulvalene (TTF) and the dioxynaphthalene units of the polyether macrocycle, respectively, by the tetracationic cyclophane ring. The GSCC and MSCC isomers coexist at 9 : 1 ratio at 298 K, and the latter are favored at high temperatures, as confirmed by variable temperature UV–Vis spectroscopy. The redox properties of the TTF unit enabled on/off switching of the bistable poly[2]catenane both chemically and electrochemically. Sequential chemical or electrochemical oxidation of the TTF unit to its radical dication in a reversible fashion triggered a Coulombic repulsion-powered circumrotation of the polyether macrocycle with respect to the tetracationic cyclophane ring of the [2]catenanes along the polymer side chains. Therefore, the switching properties of mechanically interlocked molecules under chemical or electrochemical control are retained after grafting into polymeric scaffolds and their corresponding secondary superstructures.

The mechanical bonds of catenanes have been found to be as strong as the covalent bonds along the polymer backbone against mechanical stimulation by Craig and coworkers. Using sonochemical polymer mechanochemistry, they successfully probed the relative mechanical strength of a multicatenane copolymer compared to its macrocyclic and linear counterparts [1, 55]. Specifically, polymers of *gem*-dichlorocyclopropanated-polybutadiene ($_g$DCC-PB) embedded with 5 mol% of catenane or other topological controls were synthesized by entropy-driven

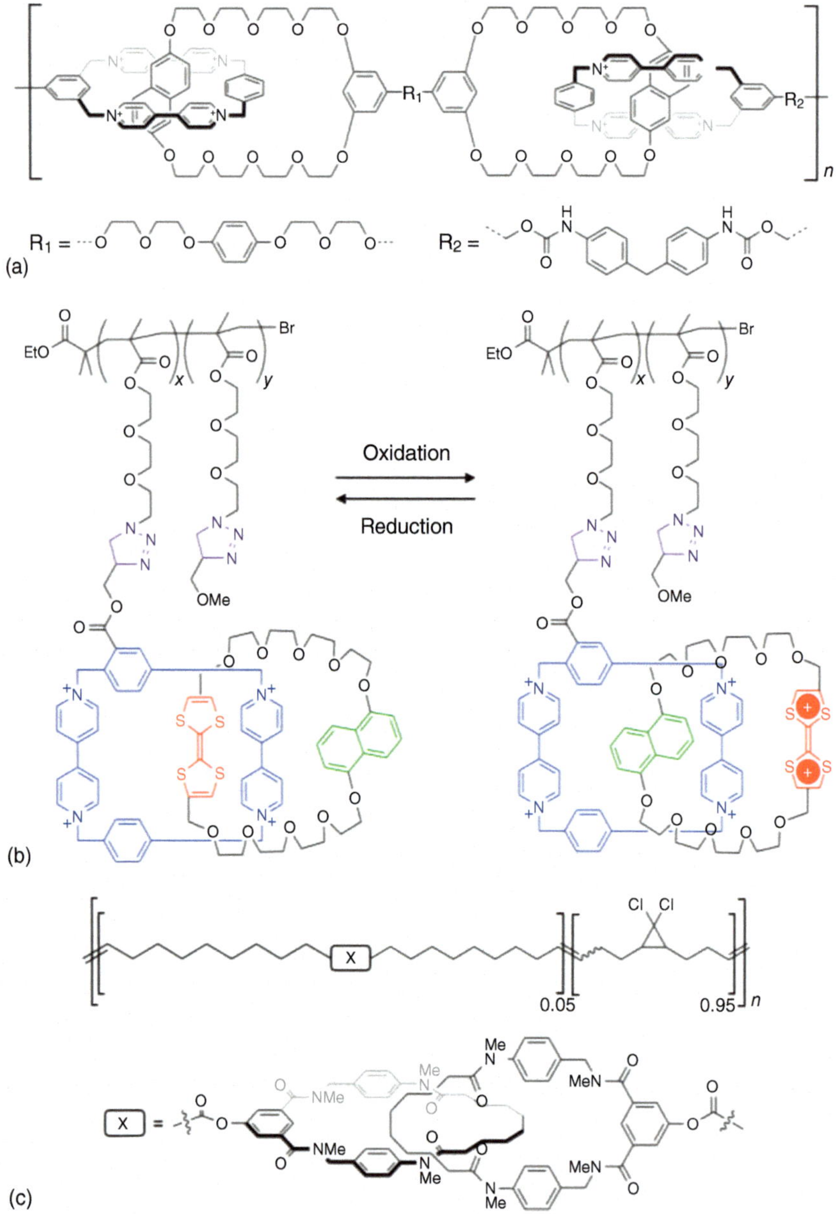

Figure 8.14 Poly[2]catenanes reported by (a) and (b) Stoddart and (c) Craig.

ring-opening metathesis polymerization (Figure 8.14c). The polymer solutions were subjected to pulsed ultrasound for four hours in an ice-water bath to break the polymer chains mechanically into shorter chains with the effective limiting molecular weights (M_{lim}), which did not undergo further changes in the time scale of hours. Estimation and comparison of $M_{lim,}$ which corresponds to the relative mechanical strength of each topology, revealed a narrow distribution of M_{lim} (28–30 kDa) for each polymer. When the limiting contour length (L_{lim}) for each polymer was considered instead of M_{lim} to assess the relative mechanical strength of topology, no significant difference was observed between the three systems.

Despite the similar strength of mechanically interlocked and covalent bonds, the incorporation of the former more flexible linkage at the molecular level may offer additional advantages by endowing the bulk materials with special dynamic and unique mechanical properties. Wu, Huang, and coworkers reported an adaptive polymer gel prepared by thiol-ene "click" reaction of a bisvinyl functionalized [2]catenane with a PEG derivative containing multiple thiol groups (PEG-SH) (Figure 8.15a) [56]. In the resulting polymer network, the [2]catenane comprising hydrogen-bonded macrocyclic rings served as crosslinkers that could be switched reversibly between "rigid" and "flexible" states by the virtue of stimuli-responsive property (temperature and pH) as characterized by ^{1}H NMR spectroscopy (Figure 8.15b). The rigid [2]catenane comprising hydrogen-bonded rings gave rise to a tough gel, which transformed to a soft state upon application of stimuli that can weaken or destroy the hydrogen bonds. The transformation of the gel from a tough to soft state upon changing pH from 7 to 1.7 was confirmed by the tensile modulus, which decreased from 166 to 70 kPa and the breaking strain, which increased from 0.77 to 1.35 mm mm^{-1} in rheological and mechanical tests. Notably, the gels crosslinked by [2]catenane units showed better mechanical performance than a control gel lacking [2]catenane crosslinker.

Xing and Shi reported metallosupramolecular polymers comprising not only host–guest interactions but also [2]catenane linkage in the main chain (Figure 8.16) [57]. The authors specifically designed a homoditopic [2]catenane-based guest monomer containing two cyano terminal groups and a heteroditopic host monomer bearing a pillar[5]arene on one end and a terpyridyl group on the other end. One-pot self-assembly of the host, guest and $Zn(ClO_4)_2 \cdot 6H_2O$ in $CDCl_3/CD_3CN$ (v/v = 1/1) at 1 : 2 : 1 molar ratio resulted in a linear supramolecular ternary polymer which was characterized by ^{1}H NMR, NOESY, DOSY, DLS, and viscosity measurements. The formation of supramolecular polymers with high molecular weight and a high degree of linear chain extension enabled isolation of a rod-like fiber and characterization by SEM. Furthermore, the supramolecular polymerization at high concentration of the components resulted in a stimuli-responsive organogel. Notably, the supramolecular polymer gel homoditopic [2]catenane-based guest exhibited higher viscoelasticity in the rheology experiment compared to a control supramolecular polymer gel which was obtained by replacing the [2]catenane unit with an alkyl chain.

Xing and Wang also employed the same [2]catenane motif to prepare a homoditopic benzo-21-crown-7 (B21C7)-linked crosslinker, which was further used to

(a)

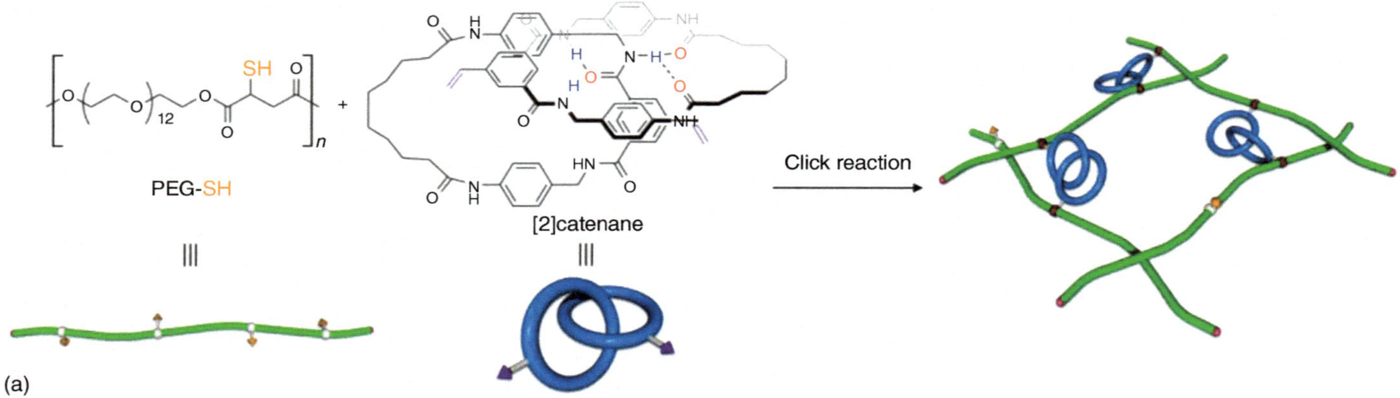

(b)

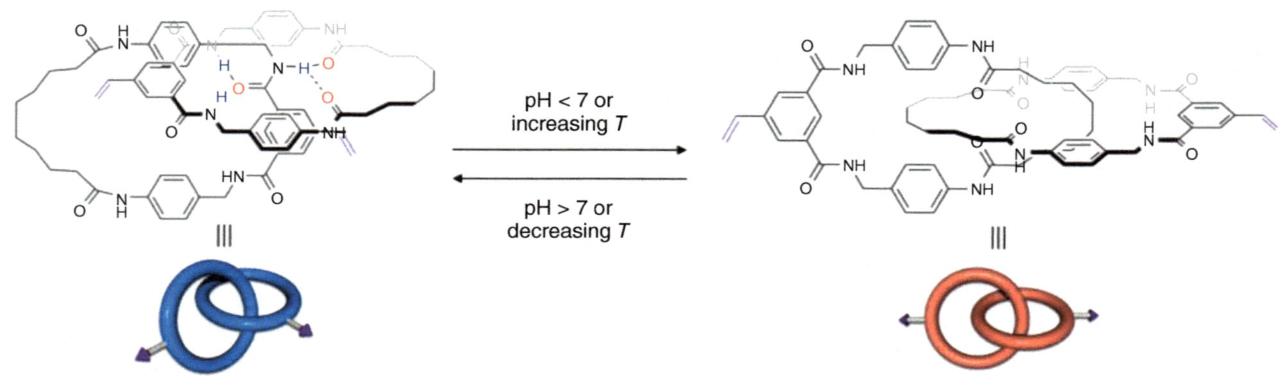

Figure 8.15 (a) Schematic illustration of the preparation of a polymer gel by thiol-ene "click" reaction of a bisvinyl functionalized [2]catenane with PEG-SH. (b) Structural change of the [2]catenane unit under external stimuli.

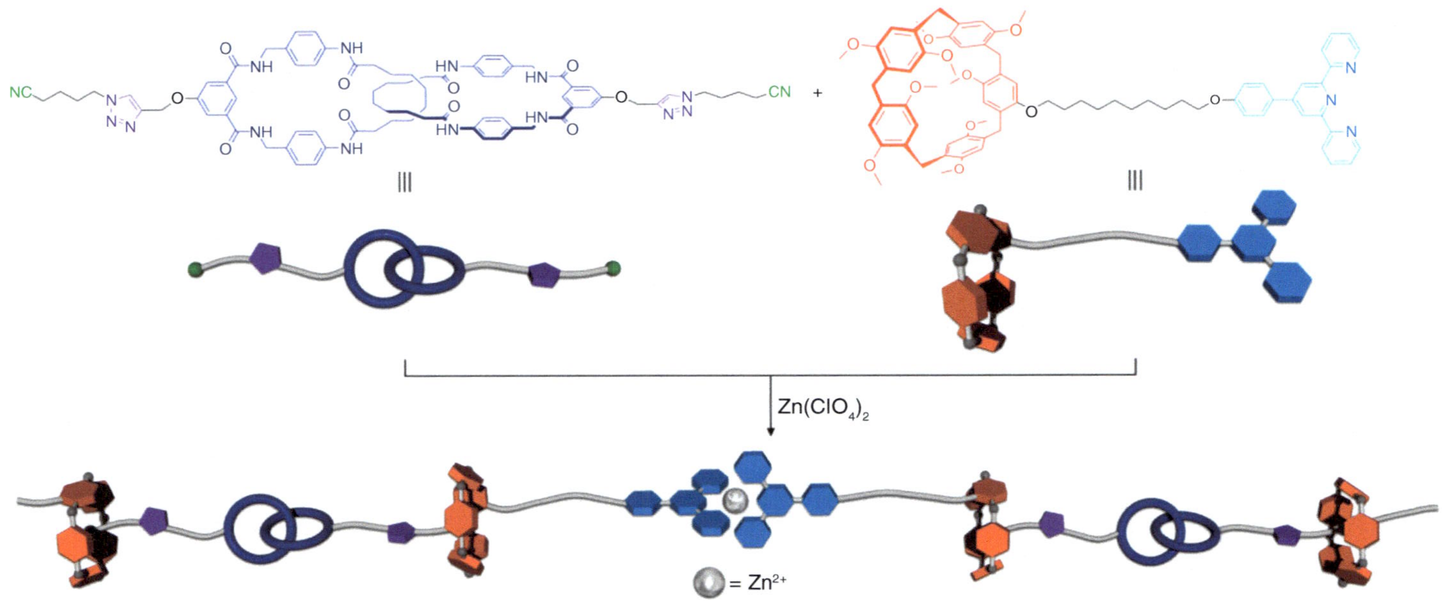

Figure 8.16 Cartoon representation of the formation of supramolecular main-chain poly[2]catenanes based on orthogonal host–guest interaction and metal ion coordination reported by Shi and colleagues.

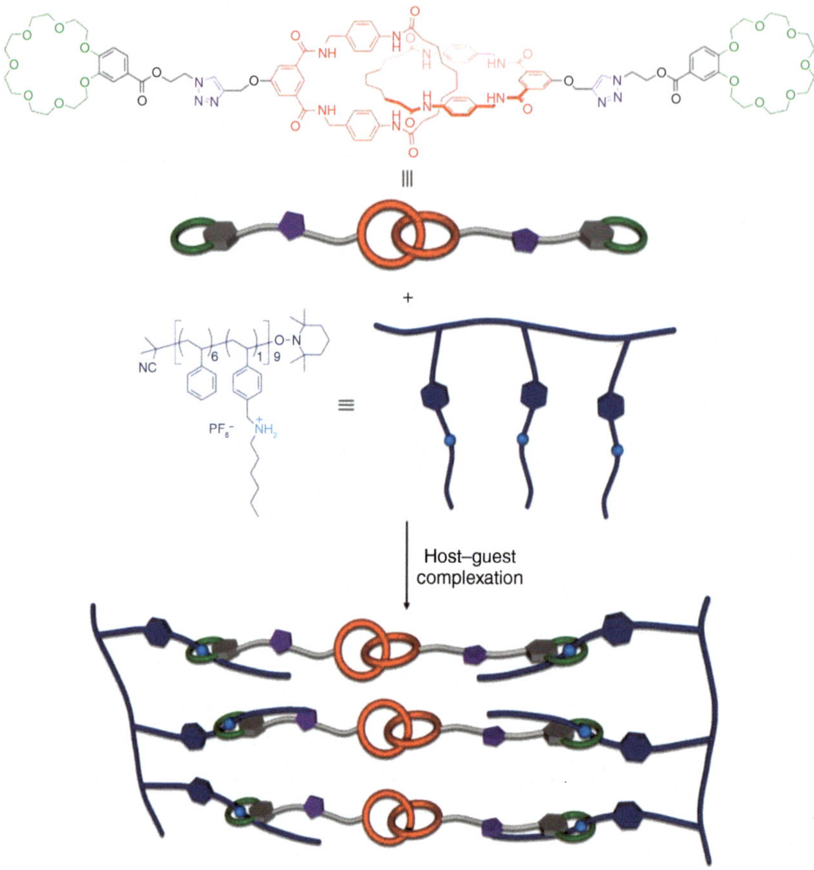

Figure 8.17 Schematic representation of the formation of a cross-linked poly[2]catenane supramolecular polymer through host–guest interactions of a homoditopic B21C7-linked crosslinker with a complementary secondary ammonium salt-functionalized graft polymer.

form a cross-linked poly[2]catenane supramolecular polymer through host–guest interactions with a complementary secondary ammonium salt-functionalized graft polymer (Figure 8.17) [58]. In $CHCl_3$, threading of the secondary ammonium salt of the polymer through the cavity of B21C7 produced [3] pseudorotaxane-type cross-linked network, which was characterized by ^1H NMR, NOESY, and SEM measurements. The cross-linked polymer exhibited superior viscoelastic properties compared to a control supramolecular polymer obtained from a homoditopic B21C7 cross-linker containing an alkyl chain spacer instead of the [2]catenane unit. Supramolecular polymerization at high crosslinker concentration (keeping the 1 : 1 molar ratio between B21C7 and secondary ammonium salt moieties) led to the formation of an organogel that exhibited thermo-, pH-, and chemo-responsive sol–gel transition behaviors. The critical gelation concentration was determined to be 45 and 10 mM for the mechanically interlocked guest and the polymer host, respectively.

8.7 Summary and Outlook

In this chapter, we featured a broad scope of polycatenanes covering poly[2]catenanes, radial poly[*n*]catenanes, and main chain linear poly[*n*]catenanes, in which the annular subunits are constructed by covalent, non-covalent and coordination bonds. It is now about four decades since Jean-Pierre Sauvage' s seminal report of the first synthesis of a metallo-catenane, which is unquestionably one of the greatest discoveries in modern chemistry and revolutionized the platform of scientific community to make breakthroughs at the interface of chemical topology, supramolecular chemistry, polymer chemistry, and materials science. The continuing development of the templated methodology enabled different synthetic strategies to realize diverse oligo- and polycatenanes. Even though many other interactions such as hydrogen bonds, donor–acceptor, π–π stacking, van der Waals and radical-pairing interactions have been employed in the synthesis of different types of oligo- and polycatenanes, it is still Sauvage' s template-directed strategy that plays as the basic underlying principle.

The main chain polycatenanes of high molecular weight remained unexplored until Rowan' s report because of issues such as inefficient interlocking and cyclization of multiple rings and difficulty in controlling the degree of catenation and even the nature of the mechanical connectivity. The recent endeavor by Yagai and coworkers has been able to advance polycatenanes further by using the potential self-assembly property of molecules instead of relying on complex synthetic methods. This is the first instance of the realization of a complex topological structure through molecular self-organization at the mesoscopic scale that enables easy characterization by AFM. An important appeal of this invention is that gaining control over the relative motion and position of the large constituent rings would lead us closer to the performance of cellular machinery. These structures with unique topologies also open new avenues for the fabrication of advanced organic photonics and electronics. Judicious collaboration of scientists from different disciplines should be directed toward utilization of polycatenanes as next-generation materials that can compete with the analogous covalently bonded polymers and biological counterparts.

Although a mechanical bond links the components together with similar bond strength a covalent bond does, the former provides an additional advantage of controlling molecular motion. This structural feature was elaborated by embedding [2]catenane as a repeating unit in a polymer chain wherein the impact of molecular motion manifested as superior bulk properties (mechanical and thermal) and enabled the fabrication of switchable functional polymeric materials.

Although significant progress in template-directed synthetic strategies and supramolecular chemistry has enabled the preparation of various types of catenated systems, many unmet issues are yet to be addressed for the full bloom of this research area. Control over the degree of cyclic oligomerization is important to avoid a widely dispersed mixture of molecular necklaces comprising different numbers of interlocked units. The discovery of self-assembled polycatenanes at the mesoscale is a vital step toward mimicking natural systems. However, the homogeneous surface of the elementary nanorings renders their catenation uncontrolled. Controlling interlocking patterns, topological isomerism, and relative motion of the rings are crucial not only for the exploration of enticing properties of these topologically exotic molecular assemblies but also for the fabrication of molecular machines operating at the mesoscopic scale. From the point of view of responsiveness of poly[2]catenanes, integrating multiple types of catenanes that can function orthogonally in a polymer chain would produce novel multi-responsive materials. Despite significant exploration of stimuli-responsive changes in the mechanical properties of polycatenated soft materials, a limited number of reports are known for crystalline materials composed of interlocked units. Recently, Sato, Aida, and coworkers have reported a [2]catenane-based metal–organic crystal that exhibited reversible deformation and recovery of its crystalline framework in response to an externally applied pressure [59]. The hydrogen bonding interaction associated with the [2]catenane units was modulated by solvent, which allowed switching between the mechanically rigid and elastic states. Further structural sophistication of the catenated building blocks could give rise to crystalline materials that can self-heal damage autonomously. We believe that the methodologies described in this book chapter will motivate chemists to develop such advanced oligo- and polycatenaned materials, which will enrich nanotechnology, supramolecular material science, and related biological fields.

References

1 Lee, B., Niu, Z., and Craig, S.L. (2016). The mechanical strength of a mechanical bond: sonochemical polymer mechanochemistry of poly(catenane) copolymers. *Angew. Chem. Int. Ed.* 55: 13086–13089.

2 Gil-Ramírez, G., Leigh, D.A., and Stephens, A.J. (2015). Catenanes: fifty years of molecular links. *Angew. Chem. Int. Ed.* 54: 6110–6150.

3 Pakula, T. and Jeszka, K. (1999). Simulation of single complex macromolecules. 1. Structure and dynamics of catenanes. *Macromolecules* 32: 6821–6830.

4 Rauscher, P.M., Rowan, S.J., and de Pablo, J.J. (2018). Topological effects in isolated poly[*n*]catenanes: molecular dynamics simulations and rouse mode analysis. *ACS Macro Lett.* 7: 938–943.

5 Wasserman, E. (1960). The preparation of interlocking rings: a catenane. *J. Am. Chem. Soc.* 82: 4433–4434.

6 Wasserman, E. (1962). Chemical topology. *Sci. Am.* 207: 94–100.

7 Stoddart, J.F. (2017). Mechanically interlocked molecules (MIMs)-molecular shuttles, switches, and machines (Nobel lecture). *Angew. Chem. Int. Ed.* 56: 11094–11125.

8 Schill, G. and Lüttringhaus, A. (1964). The preparation of catena compounds by directed synthesis. *Angew. Chem. Int. Ed. Engl.* 3: 546–547.

9 Dietrich-Buchecker, C.O., Sauvage, J.P., and Kintzinger, J.P. (1983). Une nouvelle famille de molecules: les metallo-catenanes. *Tetrahedron Lett.* 24: 5095–5098.

10 Hunter, C.A. (1992). Synthesis and structure elucidation of a new [2]catenane. *J. Am. Chem. Soc.* 114: 5303–5311.

11 Vögtle, F., Meier, S., and Hoss, R. (1992). One-step synthesis of a fourfold functionalized catenane. *Angew. Chem. Int. Ed. Engl.* 31: 1619–1622.

12 Johnston, A.G., Leigh, D.A., Pritchard, R.J., and Deegan, M.D. (1995). Structurally diverse and dynamically versatile benzylic amide [2]catenanes assembled directly from commercially available precursors. *Angew. Chem. Int. Ed. Engl.* 34: 1209–1212.

13 Caballero, A., Zapata, F., White, N.G. et al. (2012). A halogen-bonding catenane for anion recognition and sensing. *Angew. Chem. Int. Ed.* 51: 1876–1880.

14 Gilday, L.C., Lang, T., Caballero, A. et al. (2013). A catenane assembled through a single charge-assisted halogen bond. *Angew. Chem. Int. Ed.* 52: 4356–4360.

15 Amabilino, D.B., Ashton, P.R., Reder, A.S. et al. (1994). Olympiadane. *Angew. Chem. Int. Ed.* 33: 1286–1290.

16 Sauvage, J.-P. (2017). From chemical topology to molecular machines (Nobel lecture). *Angew. Chem. Int. Ed.* 56: 11080–11093.

17 Feringa, B.L. (2017). The art of building small: from molecular switches to motors (Nobel lecture). *Angew. Chem. Int. Ed.* 56: 11060–11078.

18 Wu, Q., Rauscher, P.M., Lang, X. et al. (2017). Poly[*n*]catenanes: synthesis of molecular interlocked chains. *Science* 358: 1434–1439.

19 Datta, S., Kato, Y., Higashiharaguchi, S. et al. (2020). Self-assembled poly-catenanes from supramolecular toroidal building blocks. *Nature* 583: 400–405.

20 Yagai, S., Kitamoto, Y., Datta, S., and Adhikari, B. (2019). Supramolecular polymers capable of controlling their topology. *Acc. Chem. Res.* 52: 1325–1335.

21 Sauvage, J.P. and Weiss, J. (1985). Synthesis of dicopper(I) [3]catenates: multiring interlocked coordinating systems. *J. Am. Chem. Soc.* 107: 6108–6110.

22 Bitsch, F., Dietrich-Buchecker, C.O., Khémiss, A.-K. et al. (1991). Multiring interlocked systems: structure elucidation by electrospray mass spectrometry. *J. Am. Chem. Soc.* 113: 4023–4025.

23 Langton, M.J., Matichak, J.D., Thompson, A.L., and Anderson, H.L. (2011). Template-directed synthesis of π-conjugated porphyrin [2]rotaxanes and a [4]catenane based on a six-porphyrin nanoring. *Chem. Sci.* 2: 1897–1901.

24 Okada, M. and Harada, A. (2003). Poly(polyrotaxane): photoreactions of 9-anthracene-capped polyrotaxane. *Macromolecules* 36: 9701–9703.

25 Higashi, T., Morita, K., Song, X. et al. (2019). One-pot synthesis of cyclodextrin-based radial poly[*n*]catenanes. *Commun. Chem.* 2: 78.

26 Whang, D., Park, K.-M., Heo, J., and Kim, K. (1998). Molecular necklace: quantitative self-assembly of a cyclic oligorotaxane from nine molecules. *J. Am. Chem. Soc.* 120: 4899–4900.

27 Roh, S.-G., Park, K.-M., Park, G.-J. et al. (1999). Synthesis of a five-membered molecular necklace: a 2+2 approach. *Angew. Chem. Int. Ed.* 38: 638.

28 Kim, K. (2004). Designed self-assembly of molecular necklaces using host-stabilized charge-transfer interactions. *J. Am. Chem. Soc.* 126: 1932–1933.

29 Dasgupta, S. and Wu, J. (2011). Template-directed synthesis of kinetically and thermodynamically stable molecular necklace using ring closing metathesis. *Org. Biomol. Chem.* 9: 3504–3515.

30 Li, S., Huang, J., Zhou, F. et al. (2014). Self-assembly of triangular and hexagonal molecular necklaces. *J. Am. Chem. Soc.* 136: 5908–5911.

31 Loeb, S.J., Tiburcio, J., Vella, S.J., and Wisner, J.A. (2006). A versatile template for the formation of [2]pseudorotaxanes. 1,2- Bis(pyridinium)ethane axles and 24-crown-8 ether wheels. *Org. Biomol. Chem.* 4: 667–680.

32 Loeb, S.J. and Wisner, J.A. (1998). A new motif for the self-assembly of [2]pseudorotaxanes; 1,2-Bis(pyridinium)ethane axles and [24]crown-8 ether wheels. *Angew. Chem. Int. Ed.* 37: 2838–2840.

33 Ye, Y., Wang, S.-P., Zhu, B. et al. (2015). Self-assembly of [3]catenanes and a [4]molecular necklace based on a cryptand/paraquat recognition motif. *Org. Lett.* 17: 2804–2807.

34 Huang, F., Switek, K.A., Zakharov, L.N. et al. (2005). Bis(m-phenylene)-32-crown-10-based cryptands, powerful hosts for paraquat derivatives. *J. Org. Chem.* 70: 3231–3241.

35 Wu, G.-Y., Shi, X., Phan, H. et al. (2020). Efficient self-assembly of heterometallic triangular necklace with strong antibacterial activity. *Nat. Commun.* 11: 3178.

36 Nguyen, M.T., Ferris, D.P., Pezzato, C. et al. (2018). Densely charged dodecacationic [3]- and tetracosacationic radial [5]catenanes. *Chem* 4: 2329–2344.

37 Rauscher, P.M., Schweizer, K.S., Rowan, S.J., and de Pablo, J.J. (2020). Thermodynamics and structure of poly[n]catenane melts. *Macromolecules* 53: 3390–3408.

38 Cohen, S.I.A., Vendruscolo, M., Dobson, C.M., and Knowles, T.P.J. (2012). From macroscopic measurements to microscopic mechanisms of protein aggregation. *J. Mol. Biol.* 421: 160–171.

39 Ogi, S., Sugiyasu, K., Manna, S. et al. (2014). Living supramolecular polymerization realized through a biomimetic approach. *Nat. Chem.* 6: 188–195.

40 Anwar, J., Khan, S., and Lindfors, L. (2015). Secondary crystal nucleation: nuclei breeding factory uncovered. *Angew. Chem. Int. Ed.* 54: 14681–14684.

41 Kondepudi, D.K. and Asakura, K. (2001). Chiral autocatalysis, spontaneous symmetry breaking, and stochastic behavior. *Acc. Chem. Res.* 34: 946–954.

42 Fujita, M., Ibukuro, F., Hagihara, H., and Ogura, K. (1994). Quantitative self-assembly of a [2]catenane from two preformed molecular rings. *Nature* 367: 720–723.

43 Fujita, M., Fujita, N., Ogura, K., and Yamaguchi, K. (1999). Spontaneous assembly of ten components into two interlocked, identical coordination cages. *Nature* 400: 52–55.

44 Westcott, A., Fisher, J., Harding, L.P. et al. (2008). Self-assembly of a 3-D triply interlocked chiral [2]catenane. *J. Am. Chem. Soc.* 130: 2950–2951.

45 Fukuda, M., Sekiya, R., and Kuroda, R. (2008). A quadruply stranded metallohelicate and its spontaneous dimerization into an interlocked metallohelicate. *Angew. Chem. Int. Ed.* 47: 706–710.

46 Kuang, X., Wu, X., Yu, R. et al. (2010). Assembly of a metal–organic framework by sextuple intercatenation of discrete adamantane-like cages. *Nat. Chem.* 2: 461–465.

47 Jiang, L., Ju, P., Meng, X.-R. et al. (2012). Constructions of two polycatenanes and one polypseudo-rotaxane by discrete tetrahedral cages and stool-like building units. *Sci. Rep.* 2: 668.

48 Heine, J., auf der Günne, J.S., and Dehnen, S. (2011). Formation of a strandlike polycatenane of icosahedral cages for reversible one-dimensional encapsulation of guests. *J. Am. Chem. Soc.* 133: 10018–10021.

49 Torresi, S., Famulari, A., and Martí-Rujas, J. (2020). Kinetically controlled fast crystallization of $M_{12}L_8$ poly-[n]-catenanes using the 2,4,6-tris(4-pyridyl)benzene ligand and $ZnCl_2$ in an aromatic environment. *J. Am. Chem. Soc.* 142: 9537–9543.

50 Weidmann, J.-L., Kern, Q.J.-M., Sauvage, J.-P. et al. (1996). Poly[2]catenanes containing alternating topological and covalent bonds. *Chem. Commun.* 1996: 1243–1244.

51 Weidmann, J.-L., Kern, J.-M., Sauvage, J.-P. et al. (1999). Poly[2]catenanes and cyclic oligo[2]catenanes containing alternating topological and covalent bonds: synthesis and characterization. *Chem. Eur. J.* 5: 1841.

52 Hamers, C., Raymo, F.M., and Stoddart, J.F. (1998). Main-chain and pendant poly([2]catenane)s incorporating complementary π-electron-rich and -deficient components. *Eur. J. Org. Chem.* 10: 2109–2117.

53 Hamers, C., Kocian, O., Raymo, F.M., and Stoddart, J.F. (1998). A poly(bis[2] catenane) containing a combination of covalent, mechanical, and coordinative bonds. *Adv. Mater.* 10: 1366–1369.

54 Olson, M.A., Braunschweig, A.B., Fang, L. et al. (2009). A bistable poly[2]catenane forms nanosuperstructures. *Angew. Chem. Int. Ed.* 48: 1792–1797.

55 Lee, B., Niu, Z., Wang, J. et al. (2015). Relative mechanical strengths of weak bonds in sonochemical polymer mechanochemistry. *J. Am. Chem. Soc.* 137: 10826–10832.

56 Xing, H., Li, Z., Wu, Z.L., and Huang, F. (2018). Catenane crosslinked mechanically adaptive polymer gel. *Macromol. Rapid Commun.* 39: 1700361.

57 Xing, H. and Shi, B. (2016). Supramolecular main-chain polycatenanes formed by orthogonal metal ion coordination and pillar[5]arene-based host–guest interaction. *Polym. Chem.* 7: 6159–6163.

58 Wang, W. and Xing, H. (2018). A novel supramolecular polymer network based on a catenane-type crosslinker. *Polym. Chem.* 9: 2087–2091.

59 Meng, W., Kondo, S., Ito, T. et al. (2021). An elastic metal–organic crystal with a densely catenated backbone. *Nature* 598: 298–303.

9

Synthesis, Properties, and Applications of Mechanically Interlocked Polymers

Leonie Braks and Ali Coskun

University of Fribourg, Department of Chemistry, Chemin du Musee 10, Fribourg, 1700, Switzerland

9.1 Introduction

In biological materials, unique properties often arise from a combination of molecular structure and hierarchical architecture. This combination can impart materials with exceptional properties, i.e. spider silk that is extremely strong, tough and elastic, while still being light-weight and consisting only of organic matter. The benefits of hierarchical design have been recognized and many new materials have been realized through bio-inspired design principles [1]. However, it still remains a challenge to mimic the hierarchical architectures synthesized by complex biochemical machinery. Supramolecular chemistry provides a bottom-up approach to the construction of complex structures because of the precise control of noncovalent interactions between different building blocks. Next to covalent and supramolecular bonds, a third type of bond exists: that is the mechanical bond. The mechanical bond connects molecules topologically, and they cannot be separated without breaking at least one of the covalent bonds of one of the connected parts. Some examples of molecules connected by a mechanical bond and the freedom of motions allowed by such bonds are shown in Figure 9.1.

Incorporation of the mechanical bond into materials can grant properties that are otherwise only found in biological materials, like extreme toughness and elasticity. The first claimed syntheses of [2]catenanes in 1960 by Wasserman [2], which was recently revisited by Leigh and coworkers [2, 3] and rotaxanes in 1967 by Harrison and coworker [4] were based on statistical or directed multi-step synthesis schemes, respectively. These procedures had ultralow yields, but since then the field has grown massively thanks to the introduction of template-directed synthesis. Via the use of passive [5] and active metal-templating [6], π–π stacking and donor–acceptor interactions [7], intermacrocycle hydrogen bonds [8, 9] and host–guest chemistry [10], the possible library of mechanically interlocked molecules (MIMs) expanded fast. In 2016, the Nobel Prize in Chemistry was awarded jointly to Jean-Pierre Sauvage, Sir J. Fraser Stoddart, and Bernard L. Feringa for their work on the design and synthesis of molecular machines based on MIMs and molecular rotors [11].

Mechanically Interlocked Materials: Polymers, Nanomaterials, MOFs, and more, First Edition.
Edited by Emilio M. Pérez.

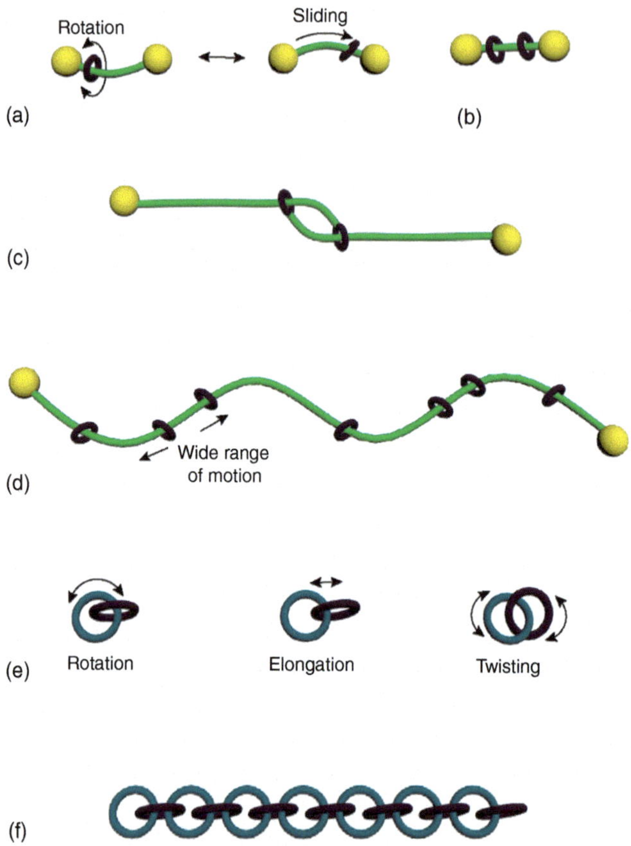

Figure 9.1 Graphical representation of mechanically interlocked molecules and polymers. (a) [2]Rotaxane. (b) [3]Rotaxane. (c) [c2]Daisy chain. (d) Main-chain polyrotaxane. (e) [2]catenane. (f) Poly[*n*]catenane.

Due to the dynamic nature of the mechanical bond, MIMs have some fundamentally different properties from their covalently bonded counterparts, especially regarding their mechanical properties and stimuli-responsiveness. However, to utilize and exploit the mechanical properties on a macroscopic scale, one needs to create polyrotaxanes or polycatenanes: polymers containing mechanical bonds (Figure 9.1d,f).

Polyrotaxanes are a fascinating class of MIMs that have attracted significant attention in recent years due to their unique properties and potential applications in various fields. Polyrotaxanes commonly consist of a linear polymer chain threaded through multiple cyclic molecules called "rings". The interlocked structure of polyrotaxane imparts novel properties such as high mechanical strength, tunable supramolecular properties, and controlled release behavior that make them attractive for a wide range of applications.

The synthesis of polyrotaxanes has been the subject of extensive research over the past few decades, and various methods have been developed to create polyrotaxanes using different polymers and rings [12–14]. These methods range from simple mixing of the components to more complex chemical reactions involving crosslinking and polymerization reactions.

Polyrotaxanes have been applied in various fields, such as materials science, electronics, and biomedical applications [15]. In materials science, the unique properties of polyrotaxanes have been exploited to develop new materials such as self-healing materials, stimuli-responsive materials, and high-strength fibers. In electronics, polyrotaxanes have been used as molecular wires, energy storage materials, and memory devices. In the biomedical field, polyrotaxanes have been used as drug delivery vehicles, tissue engineering scaffolds, imaging agents, and biosensors.

This chapter aims to provide an overview of the synthesis and applications of polyrotaxanes. We will discuss the various synthetic methods used to prepare polyrotaxanes, their unique properties, and their applications in different fields. We hope that this chapter will provide a comprehensive understanding of polyrotaxanes and inspire further research in this exciting area of science.

9.2 Synthesis and Physical Properties of Polyrotaxanes

Rotaxanes can be integrated or transformed into polymeric structures (Figure 9.2). Main-chain polyrotaxanes (Figure 9.2a) have the ring incorporated onto the polymeric backbone, either by having the rings around the polymer backbone or by including the ring in the polymer chain. The polymer properties for these

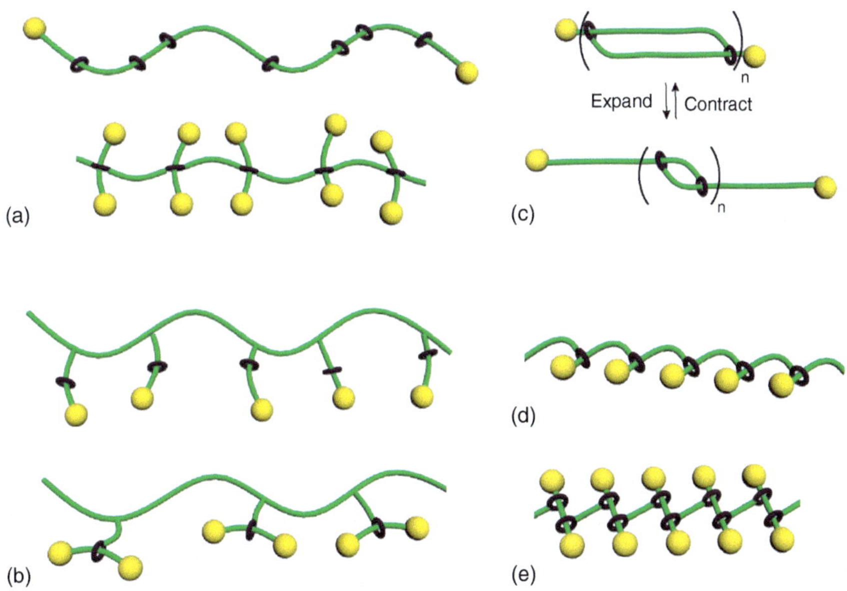

Figure 9.2 Graphical representations of different subclasses of polyrotaxanes. (a) Main-chain polyrotaxane. (b) Side-chain polyrotaxane. (c) [c2]Daisy chain. (d) Poly[2]rotaxane. (e) Poly[3]rotaxane.

types of structures depend strongly on the non-covalent interactions between the rings and the chains that influence the mobility or aggregation of the rings. Alternatively, rotaxane can be incorporated as side chains as shown in Figure 9.2b. Daisy chain- polyrotaxanes, also called artificial molecular muscles, (Figure 9.2c) are formed by covalently binding the ring and threads which gives rise to interesting macroscopic properties when the structure expands or contracts. Poly[2]rotaxanes (Figure 9.2d) are formed by connecting consecutive [2]rotaxanes and poly[3]rotaxanes (Figure 9.2e). These architectures can then be used to create more advanced materials with attractive mechanical properties.

9.2.1 Synthetic Strategies Towards Main-Chain Polyrotaxanes

Main-chain polyrotaxanes are among the most widely researched mechanically interlocked polymers and the most interesting ones for practical applications. Two main approaches in their synthesis can be distinguished by the starting compounds, which can either be a polymer chain and rings or (pseudo)polyrotaxanes that are polymerized. As shown in Figure 9.3a, a polyrotaxane can be synthesized by introducing polymerizable stoppers, into the [2]rotaxane. Naturally, this approach results in the formation of a polyrotaxane with high ring density, albeit with a limited sliding range for the ring components, owing to the presence of intermediate stoppers. This issue can be overcome by using a threading followed by stoppering approach, where the ring components are threaded onto a polymeric thread with reactive end groups, thus forming a pseudorotaxane, and subsequently capped with a stopper (Figure 9.3b). Since partial dethreading of the ring components during polymerization is inevitable in this approach, it naturally results in a lower ring density.

When starting from a polymer thread, a favorable enthalpic contribution is desired to increase the number of threaded rings onto the polymer chain. The most common of these approaches is the threading followed by stoppering approach as shown in Figure 9.3c where inclusion of the ring depends completely on the favorable formation of the inclusion complex. For the "clipping" method (Figure 9.3d), the polymer chain is first capped with bulky end groups and then mixed with the ring precursor, followed by a cyclization reaction, thus forming the ring *in situ*. The slipping approach (Figure 9.3e), on the other hand, involves the favorable inclusion complex formation along with the precise tuning of the size of the cap and the ring components so that threading is likely under synthetic conditions, and less favorable under application conditions.

Compared to traditional polymers, polyrotaxanes offer unique physical properties due to the non-covalent interactions between the chain and the threaded rings and the rings between themselves. For the systems with fixed rings that interact strongly with the backbone, a crystalline-like material is formed, but when the backbone and ring interact loosely, a more amorphous material is formed. The non-covalent interactions between the rings and backbone can be disturbed by changes of solvent [16], pH [17] or by modifying functional groups of the ring [18, 19] to tune the mechanical properties of the resulting polymer. Furthermore, the elastic modulus

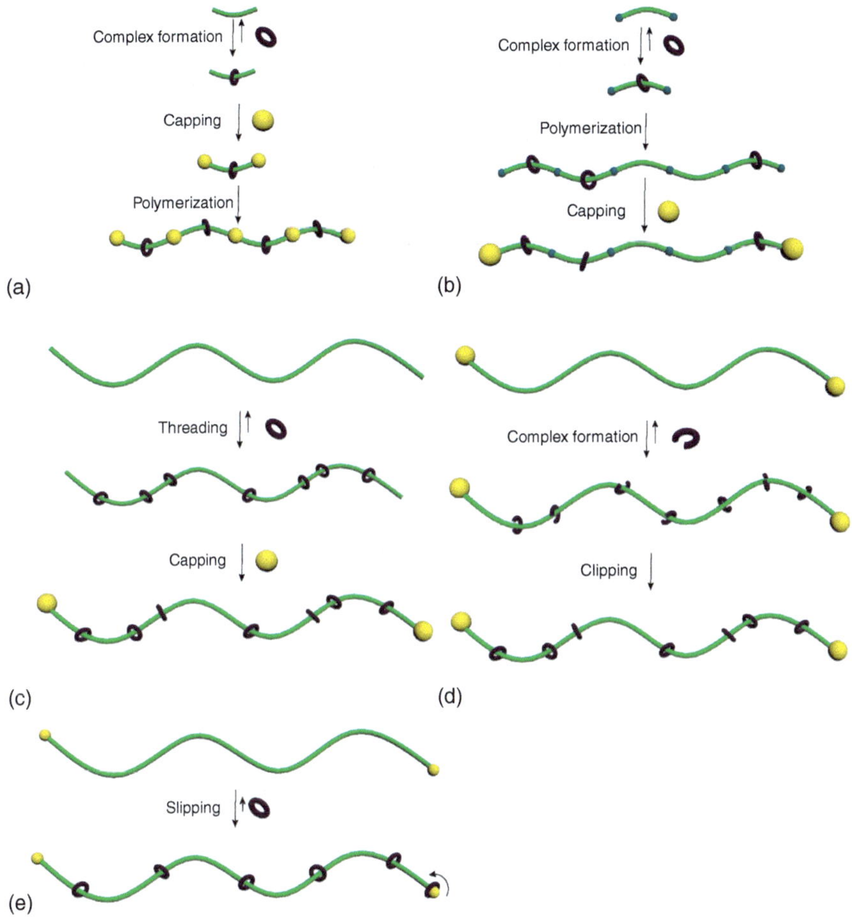

Figure 9.3 Graphical representation of synthetic strategies towards main-chain polyrotaxanes. (a) Capping followed by polymerization. (b) Polymerization followed by capping. (c) Threading followed by capping. (d) Clipping. (e) Slipping.

of polyrotaxane can be controlled by limiting the mobility of rings to certain parts of the polymer chain by using a block copolymer backbone. Under conditions where the macrocycles form a crystalline domain in the central block, fluid-like behavior is observed, but when the macrocycles are localized on the outer part, they allow for the bridging of separate polymer chains via the crystalline domain, thus resulting in slight elastic behavior [20].

9.2.2 Structural Diversity of Polyrotaxanes

Host–guest chemistry is the most attractive templating mechanism for the synthesis of polyrotaxanes, as it facilitates the assembly of multiple rings onto a polymeric thread. Figure 9.4 shows some examples of common rings used for the formation of inclusion complexes.

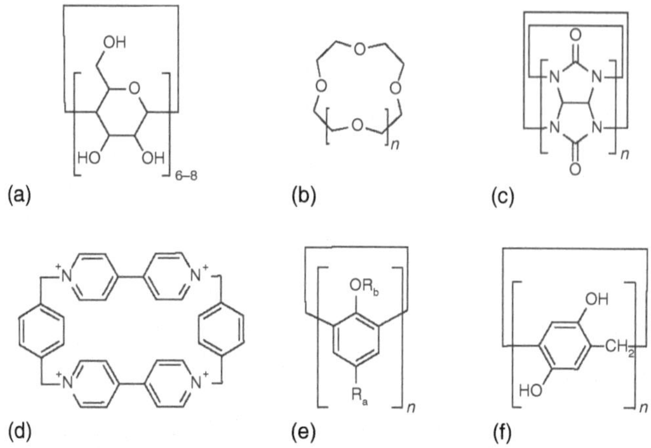

Figure 9.4 Common rings for polyrotaxane synthesis. (a) Cyclodextrin. (b) Crown ether. (c) Cucurbituril. (d) $CBQT^{4+}$. (e) Calixarene. (f) Pillarene.

Cyclodextrins (CDs, Figure 9.4a) are obtained by enzymatic cleavage of corn starch and contain six to eight D-glucose molecules connected to form a ring. While their cavity is hydrophobic, the presence of multiple -OH moieties renders the exterior of the ring hydrophilic. In polar solvents such as water, van der Waals and hydrophobic interactions drive the inclusion of linear polymers with matching radius into the core of CDs. Furthermore, dipole–dipole and charge-transfer interactions can also play a role in the binding of certain guests with CD [21]. By tuning the cavity size of the CD (α-, β-, γ-CD have progressively larger cavities) to the polymer thread and cap a multitude of polyrotaxanes can be synthesized.

Crown ethers (Figure 9.4b) of different sizes, have been shown to form inclusion complexes with various dialkylammonium salts based on hydrogen-bonding interactions [22]. Gibson et al. showed the inclusion complex formation of crown ethers with acrylate and methacrylate [23], acrylonitrile [24], and styrene [25] monomers and the subsequent polyrotaxane formation by radical polymerization of the [2]rotaxane precursor monomers. Furthermore, crown ethers containing one or more benzene rings can be synthesized easily, thus giving access to a large diversity of polyrotaxane structures like main-chain polyrotaxanes with the ring in the backbone [26], side-chain polyrotaxanes [27], and poly[3]rotaxanes [28].

Cucurbit[*n*]urils (CB[*n*], Figure 9.4c) have a hydrophobic cavity and can form ion–dipole interactions between the lone pairs on the oxygen with cationic species such as diaminoalkanes [29]. A polyrotaxane with a high density of CB[*n*] was synthesized using azide–alkyne click chemistry catalyzed by the cavity of the macrocycle, thus forming a perfect (one ring per binding site) main-chain polyrotaxane [30].

Cyclobis(paraquat-*p*-phenylene) ($CBQT^{4+}$, Figure 9.4d), which is highly electron deficient, can form inclusion complexes with electron-rich guest molecules based on donor–acceptor and hydrogen bonding interactions. $CBQT^{4+}$ can, for example, form inclusion complexes with electron-rich naphthalene derivatives, which were then used to create different polyrotaxane architectures by the polymerization of the corresponding pseudorotaxane using click-chemistry [31]. In these rotaxanes, either the ring or the electron-rich binding site can be chemically or electrochemically reduced or oxidized, respectively, to control the properties of the polymers; thus, they can be used to form stimuli responsive polymers [31, 32].

Pillar[*n*]arenes (Figure 9.4e) and calix[*n*]arenes (Figure 9.4f) are two other types of macrocycles known to form strong inclusion complexes with small molecules that have been studied less in relation to polyrotaxane synthesis up to now. Specifically, calixarenes can be easily tuned to form complexes with a variety of guests. In this direction, the hydroxy group on the lower rim of a calix[8]arene was modified with propylene to be suitable for threading onto a poly(dimethylsiloxane) backbone, which resulted in a stable polyrotaxane [33]. Pillar[*n*]arenes can form inclusion complexes with electron-deficient molecules such as bipyridinium salts [34], which has been used to synthesize polyrotaxane-containing pillar[5]arenes on a backbone of alkylviologen [35].

9.2.3 Cyclodextrin-Poly(ethylene glycol) Main-Chain Polyrotaxanes

The most commonly produced polyrotaxane consists of α-cyclodextrin (α-CD) rings and poly(ethylene glycol) (PEG) thread synthesized by threading followed by a stoppering approach. α-CD composed of six glucose moieties, is a cone-shaped macrocycle with an inner and outer diameter of the ring of respectively 0.45 and 0.67 nm as shown in Figure 9.5. In 1990, inclusion complex formation between the hydrophobic cavity of α-CD and poly ether backbone was demonstrated by Harada and Kamachi [10]. Two years later, the same group created a polyrotaxane containing many CDs by threading the α-CD rings onto the amine-functionalized PEG and then reacting this polypseudorotaxane with dinitrofluorobenzene [36]. The large synthetic advantage of PEG and α-CD based polyrotaxane is that the individual components are water-soluble, while the inclusion compound is not and can thus be precipitated out of the solution.

The yield of polyrotaxane synthesis was further improved to 91–98% by very efficient carboxylation of PEG end-groups (>98% yield), which after the capping reaction with adamantylamine gives polyrotaxane in high yield [37]. This rotaxane, which is now very commonly used in various applications, is shown in Figure 9.6.

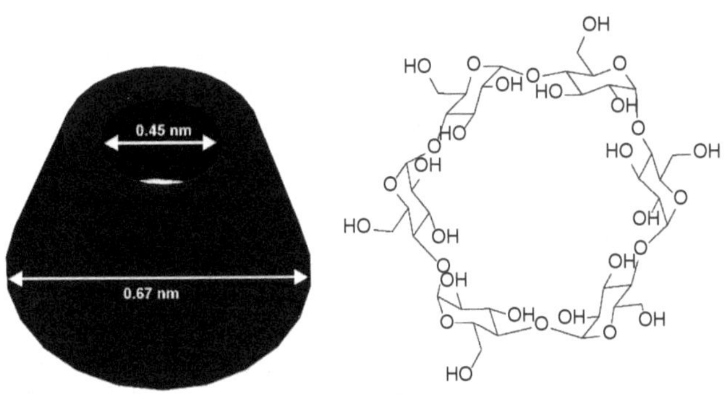

Figure 9.5 Schematic and molecular structure of α-cyclodextrin.

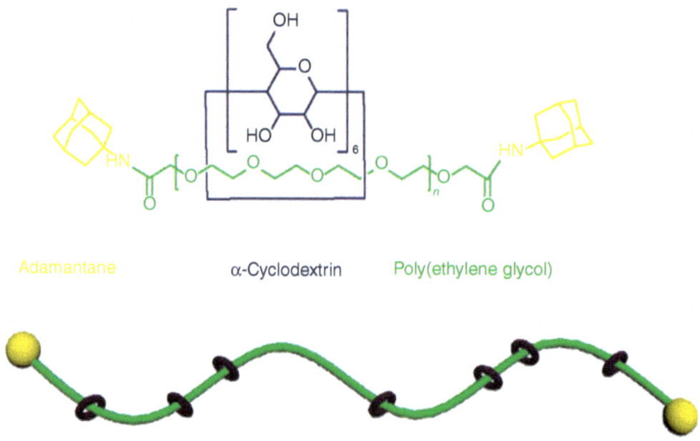

Figure 9.6 α-CD based PEG main-chain polyrotaxane.

For very short PEG chains, the threading density, calculated as the number of CDs over the number of binding sites (one CD per two ether groups), can go up to 100%. For PEGs larger than 10 000 g mol^{-1} inclusion is normally limited to approximately 30% because of the precipitation and aggregation of the chains at that threading density. The amount of α-CD on the PEG chain can be synthetically modified using different strategies, which is advantageous to control the physical properties of the resulting polyrotaxane. By varying the PEG/CD ratio, temperature and complexation time, the ring number on a 20 000 g mol^{-1} PEG could be tuned from 3 up to 125 [38]. Another approach describes adding dimethylsulfoxide (DMSO) to water prior to the capping step in order to dissociate some of the rings and thus obtain lower ring density in the resulting polyrotaxane [13]. Capping the pseudopolyrotaxane with shorter and longer PEG chains can also reduce the threading ratio [39]. Finally, modifying the reactive endgroup of the PEG during the threading step can drastically lower the coverage ratio to about 5% as it changes the favorability of complex formation [40].

A lower threading density increases the decomposition temperature and the mobility of the ring as CD crystallization is reduced [39]. This also improves the solubility of the unmodified polyrotaxane in solvents like dimethylformamide and water. The coverage ratio can have a strong impact on the glass transition temperature (T_g), as the rings restrict the free cooperative motion of polymer chains even at temperatures above T_g [41].

Without modification, CD rings aggregate on the polymer backbone due to the strong hydrogen bonding between CD moieties, which complicates processability and further functionalization. Unmodified PEG/CD-based polyrotaxanes are only soluble in DMSO, 1 M NaOH and some specifically designed high polar solvents like ionic liquids and dimethylacetamide/LiCl. Functionalization of the CDs after threading is challenging because the reaction must proceed in the aforementioned solvents, and conditions must be mild to prevent the cleavage of the stoppers and/or the backbone, as this leads to complete dissociation of the supramolecular complex.

For example, hydroxypropylation and methylation of the polar OH groups will render the polyrotaxane soluble in DMF and water. Another useful functionalization is the introduction of acetyl group, which enables solubility in common organic solvents like acetone, dichloromethane, and toluene, thus allowing for the use of common polymer characterization techniques and processing strategies like film casting and blending [42].

Main-chain polyrotaxane glasses can be formed by the methoxy ethylation of CD, which lowers T_g under room temperature. Films made of these kinds of polyrotaxanes show very different mechanical properties depending on the temperature. At temperatures above T_g both ring and chain are mobile, resulting in liquid-like polymer behavior. After the glass transition, the CD rings are immobilized, while the polymer chains are still highly mobile. At temperatures far below T_g substantial subrelaxation takes place, strongly increasing the viscoelastic properties of the glassy polymer material [43].

9.2.4 Polyrotaxane Networks

Slide-ring materials or topological gels can be synthesized by chemically crosslinking the threaded rings in solution. In 2001, the first slide-ring material was prepared by reacting a main-chain polyrotaxane composed of PEG and α-CD rings in 1 M NaOH with cyanuric acid as the crosslinker [44]. This approach gives a so-called figure-of-eight crosslink that functions as a pulley (Figure 9.7a). When tension is applied to a slide-ring gel, the figure-of-eight crosslinks can move freely along the polymer chain and thus decrease the mechanical tension on the polymer chain as shown in Figure 9.9.

To discuss the mechanical properties of polymer materials, stress–strain curves are often utilized. The shape of the stress–strain curve gives information about the stress-relaxation mechanism and also gives absolute information like the Young's modulus, yield strength and fracture. Figure 9.8 shows archetypical stress–strain curves of brittle polymers (R-shaped) and physically (J-shaped with hysteresis), chemically (S-shaped), and topologically (J-shaped) crosslinked gels. Conventional polymers such as polyacrylic acid exhibit nonlinear softening, resulting in characteristic R-shaped curves and materials that cannot recover their original shape [45]. In a chemically crosslinked gel (e.g. natural rubber), mechanical stress is dissipated by conformational change, until all the tension concentrates on the shortest chain resulting in a typical S-shaped curve with a relatively low yield strength. In contrast, a physical gel with dynamic crosslinks gives a J-shaped curve with large hysteresis, caused by the recombination of the crosslinks upon stress. These types of materials recover their original shape after the release of tension. Non-hysteresis J-shaped curves are often found in soft biological materials such as arterial walls [46] or spider silk [47] and signify materials that are very elastic at low extensions and very stiff at high extensions, usually with a high yield strength. The topological gel exhibits a J-shaped curve as well because of the freely movable crosslinks between the polymer chains. However, no hysteresis behavior is observed as the sliding motion is completely reversible. The remaining free rings also have a significant effect on the elasticity of the topological gel, as the uncrosslinked rings add resistance to the sliding ring motion when tension is applied, which is called "sliding elasticity" [48]. The free CD will be more unevenly distributed

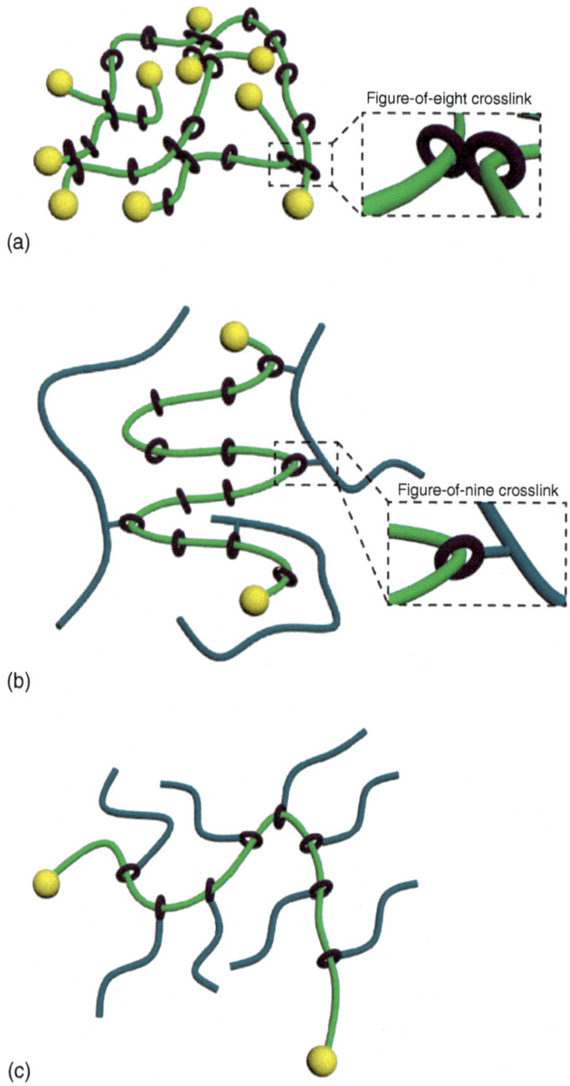

Figure 9.7 Graphical representation of different architectures of slide-ring materials. (a) Figure-of-eight slide-ring material. (b) Figure-of-nine slide-ring material. (c) Sliding graft copolymer.

after tensioning and thus create a counter force [49]. The combination of the pulley effect that equalizes the tension among polymer chains and the ring entropy that maximizes the distribution of the free rings on the polymer chain results in the observed low Young's modulus, negligible hysteresis and significant energy dissipation over a wide range of deformation rates. Readers interested in a more detailed explanation of the mechanical properties of slide-ring materials and how this can be modeled using the free junction model are referred to [50] or [51] for a quantitative description of the sliding elasticity of the free rings.

The J-shaped stress–strain curve thus underlies some of the interesting mechanical properties, which makes the slide-ring materials interesting for mechanically demanding applications. With low applied stress, the material extends massively,

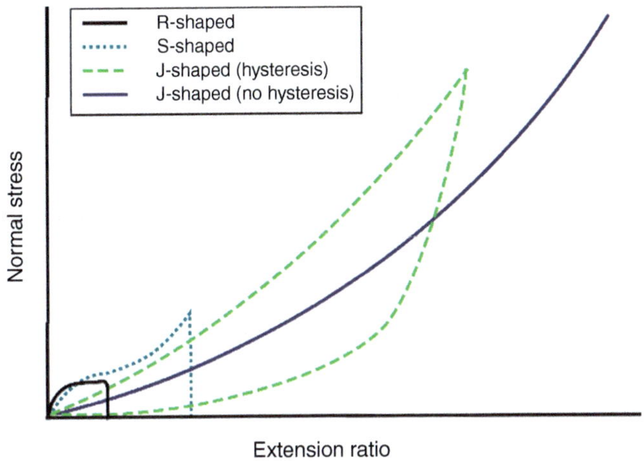

Figure 9.8 Typical stress–strain curves of polymers (R-shaped) and chemically (S-shaped), physically (J-shaped), and topologically (J-shaped) crosslinked networks.

which prevents crack formation. At higher stresses, closer to the breaking point, the network gets stiffer due to the high local concentration of rings, ensuring high breaking strength (Figures 9.8 and 9.9).

Using the aforementioned strategies to tune the ring and crosslinking density, different gels can be produced with very different stress–strain curves. Upon extension, uncrosslinked rings in densely threaded polyrotaxane agglomerate, which is entropically unfavorable, causing a high ring pressure and fragile polymers. On more sparsely threaded gels, the chain-sliding mechanism of the rings remains active for larger extensions, thus leading to the remarkable elastic behavior of slide-ring materials [40]. Most of the described research has been done in a (semi-)wet gel state of the polyrotaxane, where the presence of solvents partly enables the mobility of the rings. Surprisingly, the pulley-effect is also active in the solid state when the rings have been modified such that they do not agglomerate.

One such strategy to prevent ring agglomeration is by crosslinking polyrotaxane covalently with another polymer through the ring components, as shown in

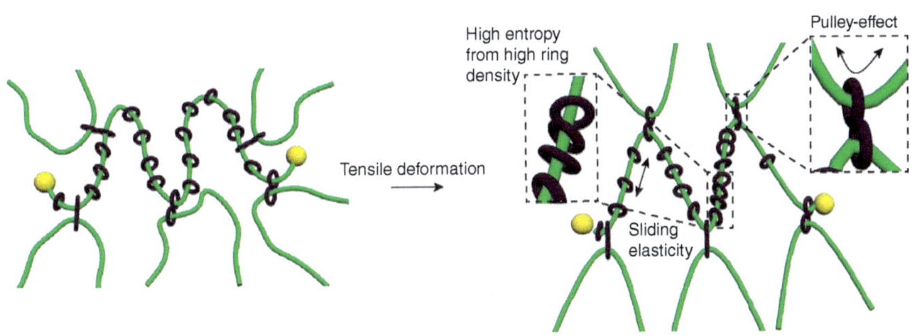

Figure 9.9 Schematic explanation of the mechanism of stress dissipation in slide-ring materials.

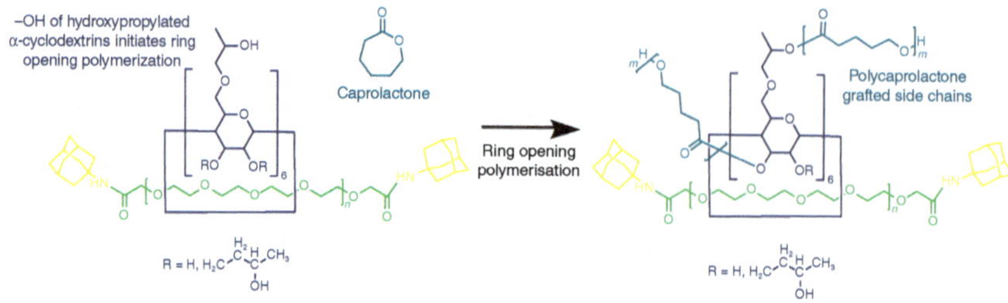

Figure 9.10 Polycaprolactone grafted polyrotaxanes as developed in [52]. Source: Adapted from Araki et al. [52].

Figure 9.7b, to obtain a figure-of-nine slide-ring material. This approach allows the retention of many of the material properties of the original polymer while providing superior elasticity and toughness.

Another important modification of polyrotaxanes is sliding graft copolymers, where side chains are grafted to the rings (Figure 9.7c). One of the most used polyrotaxane materials is the sliding graft copolymer developed by Ito, as shown in Figure 9.10 [52]. This polyrotaxane is based on the hydroxy propylated α-CD-PEG polyrotaxane, where pending alcohol groups catalyze the *in situ* ring-opening polymerization of caprolactone onto the CD. The polycaprolactone (PCL) chains improve the polymer processability, which can be film cast using solvents like chloroform and toluene. Additionally, it has a melting point of 60 °C in contrast to polyrotaxane, which does not melt before degradation. This polymer is now commercially available by Advanced Softmaterials Inc., enabling new application-oriented research in different fields. Further crosslinking this structure with a di-isocyanate linker creates a very flexible and tough network that can be solutioncast into a dry film with the same advantages as the previously discussed slide-ring gels.

9.2.5 Insulated Molecular Wires

The polyrotaxane principle is used in a different direction for the assembly of insulated molecular wires (IMW). This term is associated with conducting polymers that are interesting for molecular electronics applications with an extremely high ring coverage, thus giving a very rigid, rod-like structure, as in Figure 9.11. The protective sheath of rings endows the polymer with enhanced stability, solubility and even better electrical properties because crosstalk between different chains is prevented [53]. Alternative ways to get to this architecture are via polymer-wrapped π-systems or dendronized polymers [54].

The need for extremely high surface coverage triggered the development of different kinds of synthetic methods than previously described for the polyrotaxane networks, as the most favored threading followed by stoppering approach does not lead to the desired high coverage ratios.

CDs, such as α-CD or β-CD rings, are often used as rings in the synthesis of para-phenylene conducting polymers. This involves the Suzuki coupling reaction

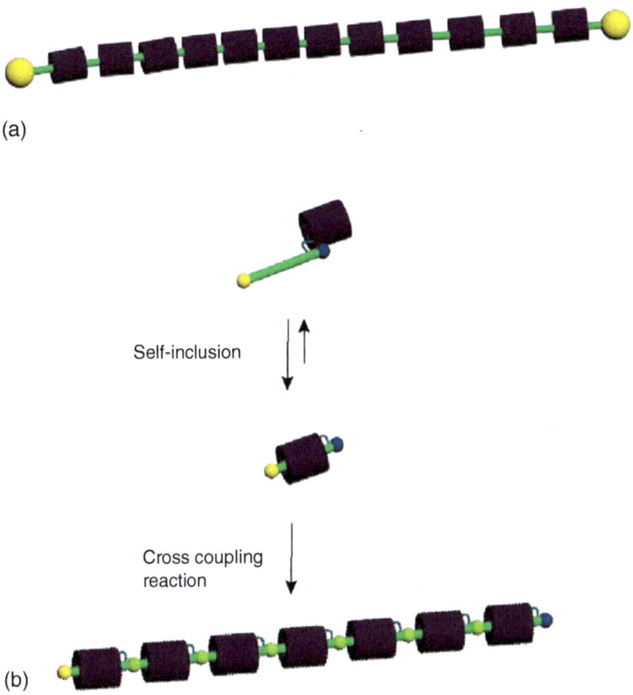

Figure 9.11 Cartoon representation of (a) Insulated molecular wire and (b) potential synthetic approach.

of conjugated monomers in water, along with a naphthalene-based stopper, using the threading followed by stoppering approach of Figure 9.3b [55]. The insulated wires showed improved processability as aggregation is prevented, and water could be used as a solvent, which allows for the incorporation of the IMW in polymer blends. More importantly the luminescence activity is enhanced and quenching by environmental impurities is reduced [56]. The photophysical properties, like emission energy and photoluminescent quantum yield and life time, can be tuned by the threading ratio and CD size [57] and work in solid-state applications [56].

The coverage ratio can be further increased by linking the CD rings covalently to the conducting polymer so that every monomer is sheathed [58]. The so-called self-inclusion strategy can be very efficient and, as shown in Figure 9.8b, leads to a densely covered chain. An even higher threading ratio (85%) was accessed by starting from a [2]rotaxane core made by self-inclusion, and subsequent polymerization gave a very regular poly(phenylene ethynylene) conducting polymer [59].

A large advantage of the insulation layer is that it can be functionalized to enable solubility in desired solvents without impacting the properties of the conjugated polymer. In this way, permethylation of α-CD covered acetylene-phenylene polymer allowed to give solubility in organic solvents for liquid solution processes [60]. By tuning the synthetic conditions and matching the CD ring size, different conducting polymers can thus be included, for example, β-CD+polyaniline [61], γ-CD+polysilane [62].

Next to CD-based sheaths, metal-based templating strategies are attractive because they provide a precise geometry and, thus, high control for the synthesis of IMWs. Additionally, the incorporated metal ions endow the system with high conductivities and potential environmental sensitivity for sensor applications [63].

9.3 Applications

The availability of cheap and scalable synthesis routes towards polyrotaxanes and materials that incorporate them, like slide-ring materials, has allowed for these materials to be researched for a plethora of applications. Most of these uses them for their excellent mechanical properties, like high elasticity and low Young's modulus, but also other properties emerging from the mechanical bond like stimuli-responsiveness, and protection from the environment.

9.3.1 Composite Materials

The most straightforward way to exploit polyrotaxane architecture is in elastomer materials, where the slide-ring motion and the pulley effect can greatly improve the desired properties. For these kinds of applications, polyrotaxane is usually mixed or crosslinked to some other polymer to create a composite. A slide-ring material based on PCL-modified polyrotaxane, when combined with suitable isocyanates and polyols, is used in scratch-resistant coatings. These types of coatings have high elasticity and flexibility, allowing them to return to their original shape and fill the gaps in the surface after the application of force. This means that small damages can be repaired in a matter of seconds, while also demonstrating strong adhesion to substrates, preventing crack and peel-off from the substrates. These properties are relatively independent of environmental factors such as temperature. Originally developed for use in the car industry, slide-ring materials have been applied to mobile devices, accessories, and more. By further functionalizing the PCL chains with acrylic or methacrylic residues, UV-curable scratch-resistant coatings have been developed [64].

By mixing 1% of PCL grafted polyrotaxane with polymethyl methacrylate (PMMA) and molding the mixture into membranes, both tensile and compressive properties are improved, improving the scratch resistance of the films. The authors attributed this effect to the enhanced molecular cooperativity between the PMMA chains even without any covalent bonding between the two polymers, although increasing the compatibility by end-functionalizing the PCL chains with methacrylate groups further improved the properties [65].

Compositing polyrotaxane with fillers can also improve material properties for different goals. For example, by using commercially available PCL grafted polyrotaxane (70%) with hexagonal boron nitride (hBN) as filler and using hexamethylene diisocyanate as a crosslinker, the resultant elastomer had almost double the tensile strength and Young's modulus of the unfilled polyrotaxane membrane, while also showing a higher extension than either the hBN or polyrotaxane on its own [66].

Another composite of chemically crosslinked polyrotaxane (35%) composite with graphene nanoplatelets manifested high thermal conductivity while still preserving the attractive flexible properties of polyrotaxane, which makes it an interesting material for flexible electronic devices [67].

9.3.2 Biomedical Applications

The absence of covalent bonds between the constituent parts of a polyrotaxane makes them interesting vehicles for drug or gene delivery in biological systems. By placing a linker between the thread and cap that can be cleaved upon exposure to certain external stimuli, the rings with physiological function will be freed, as schematically shown in Figure 9.12. The external stimuli can range from biological (e.g. enzymatic), chemical (e.g. pH), or physical (e.g. light). The advantage of polyrotaxanes over regular polymers is that cleavage on the thread results in the complete dissociation of the assembled structure and thus allows for full cargo delivery. The cargo can be small DNA or RNA fragments, proteins, drugs, etc.

An example of a polyrotaxane-based drug delivery system is shown in Figure 9.13. Hydroxypropyl-β-CD is a potential drug for Niemann-Pick disease as it can promote the excretion of accumulated cholesterol in the diseased cells. Administration of β-CD directly is challenging because of its cytotoxicity in the required concentrations, but using a polyrotaxane system enhances the therapeutic effect. The polyrotaxane masks the toxicity of hydroxypropyl-β-CD during administration,

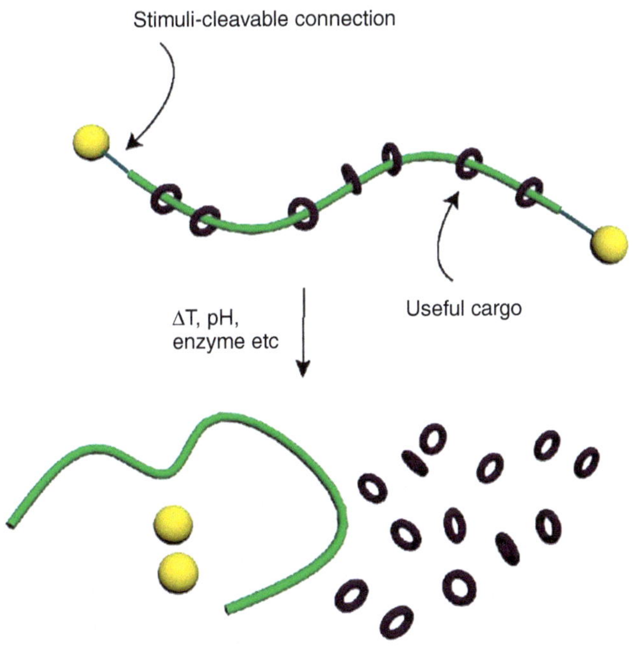

Figure 9.12 Schematic depiction of a stimuli responsive drug delivery system based on polyrotaxane.

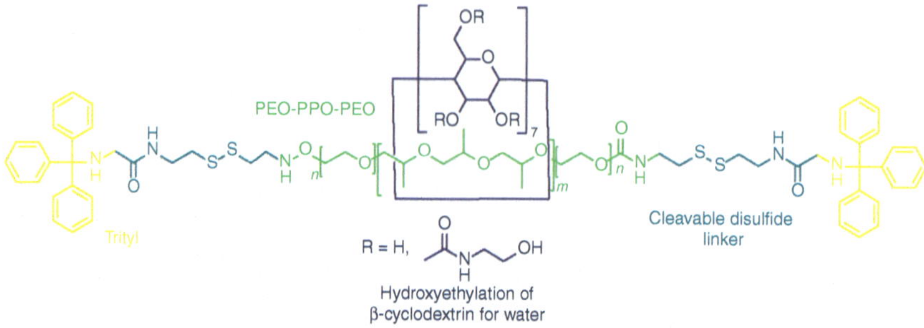

Figure 9.13 Polyrotaxane drug delivery system as described in Ref. [68]. Source: Adapted from Tamura and Yui [68].

prolongs the blood half-life and allows the accumulation in the targeted cellular structures. Once polyrotaxane is taken up in the intracellular environment of the targeted lysosomes, the disulfide linkages are cleaved, and β-CD is released [68]. The intracellular cleavage of disulfide linkages is also utilized for the delivery of plasmid DNA (pDNA) or small interfering RNA (siRNA). Polyrotaxanes with *N*,*N*-dimethylaminoethyl (DMAE)-group functionalized CD have strong electrostatic interactions with DNA/RNA, and the mobility of the ring allows for intimate contact. As shown before, once the polyrotaxane enters to the intracellular environment, a single cleavage leads to the dissociation of the entire complex. Polyelectrolyte complex formation with polyrotaxanes can also be used for protein delivery by adapting the CD functionalization [69].

The molecular mobility of polyrotaxanes offers an additional advantage for the multivalent interaction with receptor proteins. For example, maltose functionalized α-CD on PEG thread binds to concanavalin A receptor better than standard grafted polymers [71, 72]. This concept is schematically shown in Figure 9.14. This improved multivalent interaction can be exploited for the delivery of imaging agents to specific tissues or cells, for example, in the imaging of tumor tissues. α-CD functionalized with a biodegradable Gd-based contrast agents for magnetic

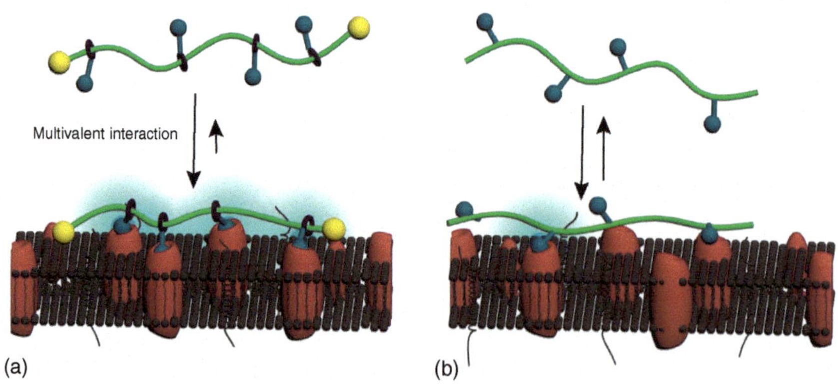

Figure 9.14 Schematic showing the effect of the mobile rings on multivalent binding of ligands on rings to membrane proteins.

resonance imaging (MRI) can be specifically delivered to breast tumor cells because of conjugated AS1311 aptamer. Accumulation of the contrast agent in the intracellular environment is again ensured by the cleavable disulfide bonds that release the functionalized CDs [70].

Polyrotaxanes integrated into a triblock copolymer have been extensively studied to tune the properties of biological surfaces. The terminal blocks are random copolymers of butyl methacrylate and phosphorylcholine that anchor the structure onto a hydrophobic surface and prevent non-specific biological responses. The threading density was varied, and CD was methylated to different degrees, where it was concluded that molecular mobility of the surface can be tuned using these parameters. As the cell behavior (adhesion, orientation, migration, and differentiation) is partly controlled by the extracellular matrix, polyrotaxane surfaces can play an important role in controlling cell growth by varying the mechanical, chemical, and topological properties of the surface [73].

Functionalization of CD ring with specific amino acid combinations that play a role in cell attachment helps to promote cell adhesion. The RGD-motif attached to the CD moieties using azide-click-chemistry promotes rapid initial cell adhesion due to the surface mobility [74, 75]. This block copolymer approach thus allows for tuning the polyrotaxanes surface by changing the copolymers, threading ratio and CD functionality. A follow-up study showed that by changing the functional group, stem cell differentiation toward different cell types could be induced [76].

9.3.3 Molecular Electronics

Conjugated polymers with an insulating sheath have the potential to further the field of organic electronics as they prevent π–π stacking interactions between chains and increase the stability toward solvents and temperature. Polyaniline with a dense β-CD insulating sheath has been used to fabricate a flexible resistive random-access memory (RRAM) device. The β-CD sheath protects the conductive core and makes it water-soluble, while retaining intrinsic flexibility. The hydrogen-bonding interaction between the CD rings and the conductive core allows for the doping of polyaniline by an applied electric field and the supramolecular architecture of the polymer is thus at the core of the device functionality. The RRAM device thus created shows fast response and good reliability, reproducibility, and stability. Due to its compatibility with solvents, the device can be printed and has high potential for wearable electronics applications [77].

The strong effect of non-covalent interactions on the polyrotaxane physical properties makes them interesting as sensor materials. For example, a pillar[5]arene-viologen polyrotaxane shows a color change upon heating, because the available thermal energy increases the molecular mobility of the rings, which stabilizes the photoinduced electron transfer and thus the absorption [35].

CB[*n*] rings on a conducting cationic poly(pyridyl vinylene) increase the solubility (and thus processability) and enhance the fluorescence of the backbone polymer due to the change of the micro-environment of the backbone. The fluorescence can then again be quenched by the addition of an electron-rich analyte like indole.

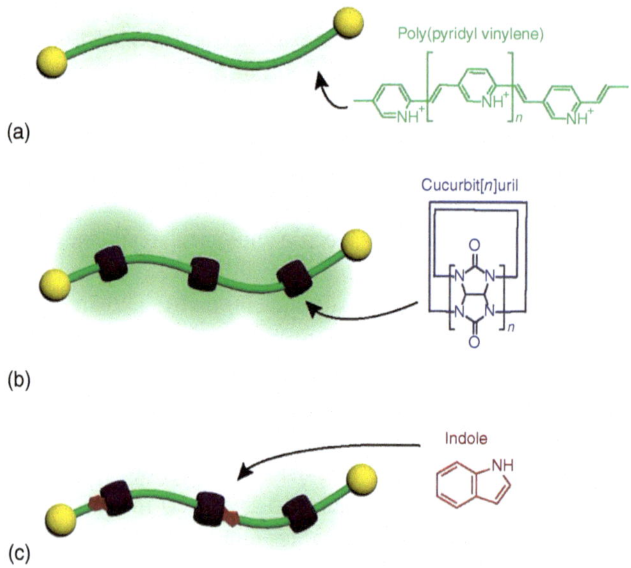

Figure 9.15 Graphical and molecular representation of (a) mildly fluorescent conducting polymer (b) strongly fluorescent polyrotaxane and (c) quenched polyrotaxane by analyte as in [78]. Source: Adapted from Willis-Fox et al. [78].

The fluorescence responsiveness is schematically shown in Figure 9.15. Due to the improved solubility these pseudopolyrotaxanes can be deposited as thin films on glass and can detect indoles in the air [78]. Similarly, thin films of rotaxane functionalized conjugated polymers could detect alcohol vapors by the attenuation of polymer fluorescence due to disruptive hydrogen bond formation [27].

Another strategy to implement polyrotaxanes for electronic applications is by mixing a functionalized PEG-α-CD based polyrotaxane with PEDOT-PSS. As shown in Figure 9.16 the PEG fragment enhances the water solubility for easier

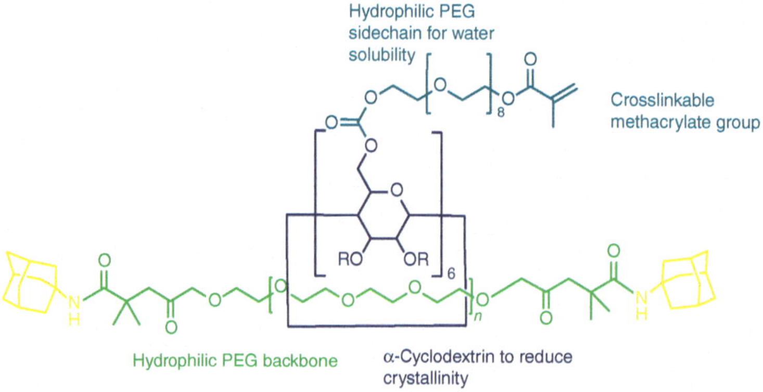

Figure 9.16 Polyrotaxane designed for flexible electronics applications as in [79]. Source: Adapted from Jiang et al. [79].

processing, and a crosslinkable methacrylate group enables photopatterning. PEG furthermore helps to promote the crystallization of PEDOT to promote higher conductivity while CD moieties prevent phase separation. The electrical conductivity of PEDOT-PSS is enhanced up to two orders of magnitude due to the presence of polyrotaxane, and an additional order of magnitude increase to ~2700 $S\,cm^{-1}$ after acid treatment, which was also retained upon stretching and relaxing. The created stretchable electrodes could be used to measure surface high-density electromyography (EMG) on moving muscles on an octopus's skin and in a rat's brainstem without any observable damage to the subjects [79].

9.3.4 Molecular Machines

Polyrotaxanes are synthetically limited by threads and rings that have strong non-covalent interactions to promote inclusion. Molecular pumps are envisioned to be able to drive rings away from equilibrium positions located on a thread or a ring. In 2015, Stoddart and coworkers used redox chemistry to thread two cyclobis(paraquat-*p*-phenylene) ($CBPQT^{4+}$) rings onto a thread by exploiting conformational rearrangement that favors complex formation between the stoppers and the ring. Upon oxidation, the stopper causes electrostatic repulsion of the ring while kinetically preventing dethreading, even though thermodynamically, the polymer is in an unfavorable high energy state [32].

This design was improved upon in 2020 to allow for up to ten rings to be threaded using chemical or electrochemical redox cycles. The structure of the developed rotaxane is shown in Figure 9.17. The central PEG chain in this case serves purely to collect the pumped ring as it does not have strong interactions with the $CBPQT^{4+}$ [80]. These kind of systems open up a new avenue for the development of fully organic energy storage devices due to the high amount of charge that can be stored in an out-of-equilibrium state.

In 2021, Leigh and coworkers used Fmoc-$OC_6H_4NO_2$ as a fuel to catalytically drive crown ether (24-C8) onto a thread of triazole and propylene units. Rings can be captured from the solution and kept on the polymer backbone as long as enough fuel is present to maintain the out-of-equilibrium state [81].

Controlled pump polyrotaxanes like these attached to a solid support can potentially store energy or sequence information when differently (fluorescently) labeled rings are used. Furthermore, when attached to vesicles or colloids, molecular

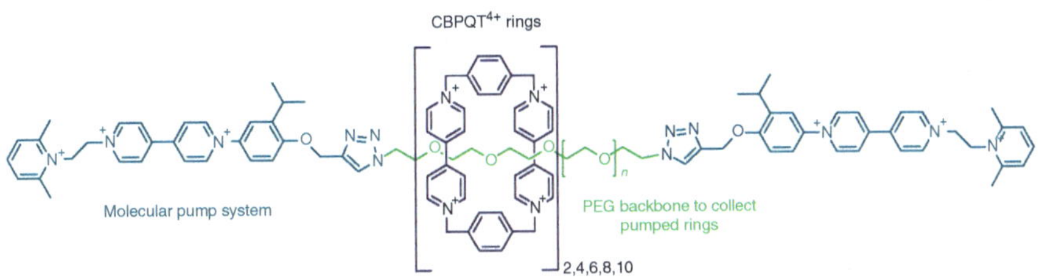

Figure 9.17 Molecular pump polyrotaxane based system as in [80]. Source: Adapted from Qiu et al. [80].

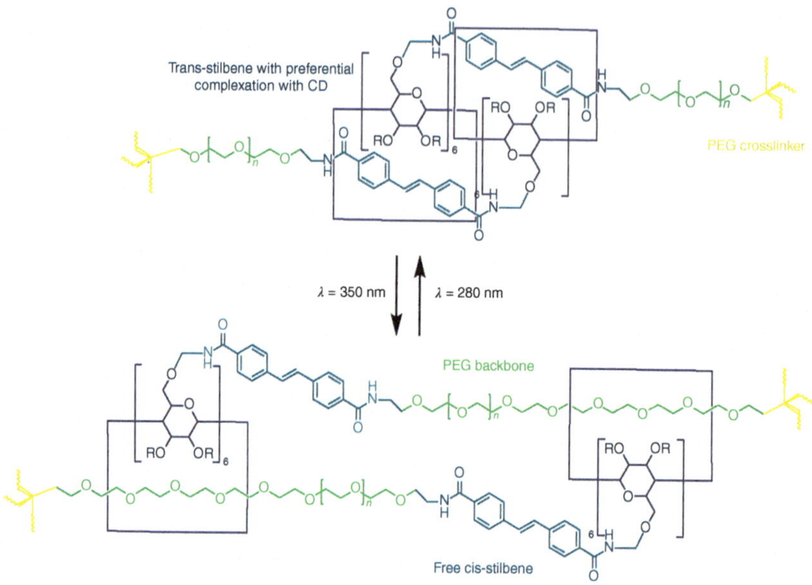

Figure 9.18 Artificial muscle based on polyrotaxane as in [82]. Source: Adapted from Ikejiri et al. [82].

pumps would enable to establish and maintain a concentration gradient between the enclosed space and the environment, analogous to natural systems [31].

[c2]Daisy-chain type polymers can be regarded as artificial muscles as they have access to contracted or expanded forms as the mechanical bonds slide past each other. By incorporating stimuli responsive binding sites, movement of the rings can be triggered on demand and macromolecular motion is initiated. An example is shown in Figure 9.18. Each α-CD was functionalized with stilbene, which forms an inclusion complex consisting of two CD and stilbene moieties. Polycondensation reaction with a four-arm PEG linker gives a crosslinked network structure. As stilbene binds stronger to α-CD in the trans conformation, switching between trans and cis state using irradiation with 350 or 280 nm light gives contraction and expansion of the dry gel. This gel can perform mechanical tasks like lifting something up to 15 times its own weight [82].

9.3.5 Batteries

Polyrotaxane networks have also found their way into the highly challenging field of battery materials. For example, silicon anodes show high potential in the battery field due to their high specific capacity. However, they experience such a massive volume change during cycling that conventional (crosslinked) polymeric binders cannot withstand it without fracturing the polymer anode composite.

In 2017, a polyrotaxane binder for silicon anodes based on PEG and α-CD was developed using a CDI-catalyzed crosslinking reaction of hydroxy propylated α-CD with the carboxylic acid groups of polyacrylic acid (PAA), as shown in Figure 9.19. The final network containing only 5 wt% polyrotaxane could withstand up to 390%

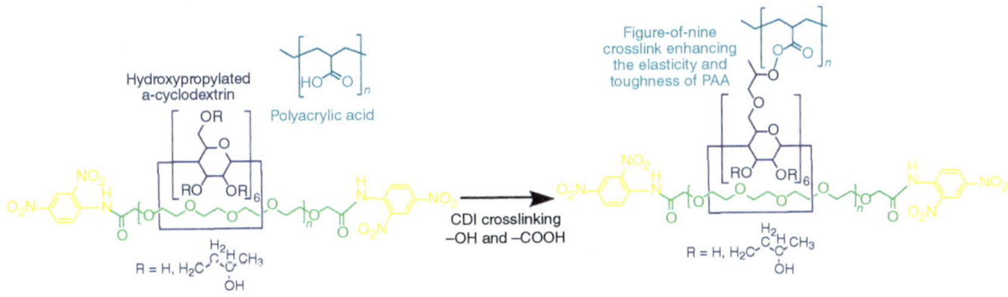

Figure 9.19 α-CD/PEG based polyrotaxane crosslinked with polyacrylic acid for binder for Si-anodes as in [45]. Source: Adapted from Choi et al. [45].

strain before breaking and showed excellent stress-recovery cycles. While conventional binders cannot maintain the electrode morphology during charge/discharge cycles, the newly designed anodes containing 5 wt% of polyrotaxane-PAA binder can effectively keep the pulverized particles together and maintain the electrode morphology upon repeated cycling. The improved properties of the binder network were attributed to the sliding motions of the rotaxane that could dissipate the stress equally based on the earlier described pulley effect [45].

In 2021, a derivative of polyrotaxane-PAA binder was introduced by replacing the covalent link between the slidable α-CD rings and PAA binder with dynamic boronic ester bonds. Using a higher fraction of polyrotaxane (17 wt%), the elongation at break for this crosslinked binder network is further increased by 552% and shows self-healing capabilities upon scratching. These properties led to an improved electrochemical performance compared to unmodified polyrotaxane-PAA binder, which is mainly attributed to the reversible bond formation of the boronic ester that self-heal upon cycling the battery [83].

Another highly demanding application for polymeric binders in batteries is the 3D lithium hosts. Carbon nanotubes can form a conducting percolating network that can take up Li, but are mechanically unstable because of the stress evolved during charge-discharge cycles. Again, a polyrotaxane-PAA binder, due to its pulley effect could maintain the structural integrity of the anode [83, 84].

Similarly, a mechanically interlocked network could also benefit from the dynamic motion of the sliding rings to disperse the stress created by Li stripping and deposition. The polyrotaxane, in this case, was based on the crosslinking of PMMS with a [2]rotaxane consisting of a functionalized crown ether (B21C7) on an ammonium-salt-based thread, as shown in Figure 9.20. The crosslinking can be done in situ on the Li metal interface to enable a thin and homogeneous coating layer. The crown ether favors Li ion mobility, and the slidable crosslinks give the network the required toughness and elasticity to withstand lithium dendrite growth [85].

The intrinsic molecular mobility of polyrotaxane materials can also be exploited in solid/gel electrolytes combined with lithium metal. The use of lithium metal as an anode for batteries is challenging because of uneven lithium deposition and consequential uncontrolled lithium dendrite growth. PEG is often used as a gel polymer electrolyte but suffers from insufficient ionic conductivity. α-CD/PEG-based polyrotaxane grafted with PCL show improved room temperature ionic conductivity of more than 10^{-4} S cm^{-1} compared to 10^{-6} S cm^{-1} for PEG. The high conductivity is

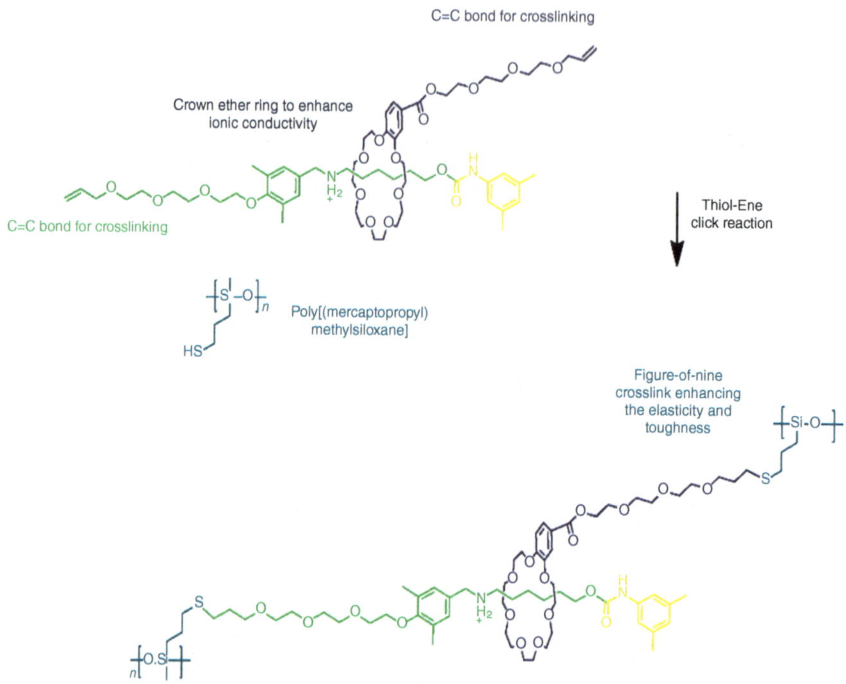

Figure 9.20 Mechanically interlocked network for lithium metal batteries based on polyrotaxane as in [85].

attributed to the fast-molecular motion of the PCL side chains. A larger ring size (changing from α-CD to γ-CD) provides a further increase in ionic conductivity as it allows more free motion. Polyrotaxane also demonstrates increased interfacial stability with lithium metal giving improved cycling performance [86].

A solid polymer electrolyte based on crosslinked PEG and 18-crown ether-6 (18C6) allowed for an increase in the ionic conductivity of PEG by a factor of 30 compared to the crosslinked PEG electrolyte, from 1.12×10^{-5} to 3.48×10^{-4} $S\,cm^{-1}$. In this case, the rotaxane is not used for its mechanical advantages (as crosslinking is not done on the rings), but the crown ether promotes Li^+ mobility in the electrolyte. Moreover, the crystallization of PEG that would negatively affect the Li^+ mobility is prevented by the presence of the rings and side reactions with electrodes are reduced. Together, this results in a battery that cycles stably with a Li-metal anode for 250 hours at room temperature [87].

9.4 Conclusion and Outlook

To conclude, polyrotaxanes and polyrotaxane-based materials show a number of advantages in comparison to conventional polymers. Firstly, the presence of the ring(s) in the polyrotaxane gives rise to unique supramolecular properties such as self-assembly, which can be exploited for various applications in materials science and drug delivery. Secondly, polyrotaxanes exhibit high mechanical strength due to their mechanically interlocked structure, which makes them ideal for use in materials that require high tensile strength and durability. Furthermore, the polyrotaxane architecture can be synthesized using a wide range of rings and polymers, allowing a high degree of versatility in their properties and applications. Even based on the same ring and polymer chain, the system is highly tunable, as aside from the usual way of modifying polymer properties by altering the length of the polymer chain, also the size of the ring and the ring density can be altered. This allows for precise control over their mechanical and supramolecular properties. Finally, the most abundant polyrotaxane consisting of PEG and CD has been shown to be biocompatible, making them a promising material for use in drug delivery and biomedical applications.

The versatility and tunability of polyrotaxanes allow for precise control over their properties, making them a material suitable for a wide range of applications:

- Composite materials: The unique interlocked structure of polyrotaxanes offers several advantages in composite materials, including enhanced mechanical properties (toughness, durability, elasticity) and improved adhesion.
- Molecular wires: Polyrotaxanes can be used as molecular wires to transport electrical charge in electronic devices. The interlocked structure of the polyrotaxane can provide high conductivity, making it a promising material for use in molecular electronics.
- Energy storage devices: Polyrotaxanes can be used as an electrolyte or binder material in energy storage devices such as batteries. The high mechanical strength and unique supramolecular properties of polyrotaxanes can provide better stability compared to conventional polymers, while the high molecular mobility can provide high conductivity.
- Sensors: Polyrotaxanes can be used as sensing materials in electronic sensors. The presence of the ring in the polyrotaxane can act as a recognition site for specific target molecules, allowing for highly selective and sensitive detection.
- Memory devices: Polyrotaxanes can be used as memory devices due to their ability to form reversible crosslinks between the polymer chain and the ring. This property can be exploited for the development of high-density data storage devices.
- Drug delivery systems: Polyrotaxanes can be used as drug or imaging agent delivery vehicles due to their biocompatibility and ability to bind drugs to the ring. The delivery can be targeted through the coupling of a ligand matched with the target receptor, allowing accumulation in relevant body parts, because of multivalent interaction allowed by polyrotaxane. The release of the drug can be controlled by stimuli-responsive bond cleavage in the intracellular environment, providing targeted and sustained drug delivery.
- Tissue engineering: Polyrotaxanes can be used as scaffolds in tissue engineering due to their high mechanical strength and biocompatibility. The polyrotaxane scaffold can provide a three-dimensional structure for cells to grow and differentiate, leading to the development of functional tissue.

As research into polyrotaxane-based materials continues, several key trends and future directions are emerging. The development of new and more complex polyrotaxanes and polyrotaxane-based materials is expected to continue. This includes the development of polyrotaxanes based on different ring systems than CDs as well as the incorporation of various functional groups into the polyrotaxane structure to impart new and unique properties. Also, the use of polyrotaxanes as building blocks for the development of supramolecular materials is an area of active research, which includes the use of polyrotaxanes for the development of supramolecular polymers, supramolecular gels, and other advanced materials with unique mechanical and physical properties. By incorporating stimuli-responsive groups into the polyrotaxane structure, researchers have also been exploring the development of materials that can respond to external stimuli such as temperature, pH, and light.

Overall, the future outlook for polyrotaxane-based materials is positive, with numerous potential applications in a wide range of fields. As researchers continue to develop and explore the unique properties of polyrotaxanes, new and exciting opportunities for these materials are expected to emerge.

References

1 Wang, Y., Naleway, S.E., and Wang, B. (2020). Biological and bioinspired materials: structure leading to functional and mechanical performance. *Bioact. Mater.* 5 (4): 745–757.

2 Wasserman, E. (1960). The preparation of interlocking rings: a catenane1. *J. Am. Chem. Soc.* 82 (16): 4433–4434.

3 Baluna, A.S., Galan, A., Leigh, D.A. et al. (2023). In search of Wasserman's catenane. *J. Am. Chem. Soc.* 145 (17): 9825–9833.

4 Harrison, I.T. and Harrison, S. (1967). The synthesis of a stable complex of a macrocycle and a threaded chain. *J. Am. Chem. Soc.* 89 (22): 5723–5724.

5 Dietrich-Buchecker, C.O., Sauvage, J.P., and Kintzinger, J.P. (1983). Une nouvelle famille de molecules: les metallo-catenanes. *Tetrahedron Lett.* 24 (46): 5095–5098.

6 Aucagne, V., Hänni, K.D., Leigh, D.A. et al. (2006). Catalytic "click" rotaxanes: a substoichiometric metal-template pathway to mechanically interlocked architectures. *J. Am. Chem. Soc.* 128 (7): 2186–2187.

7 Ashton, P.R., Goodnow, T.T., Kaifer, A.E. et al. (1989). A [2] catenane made to order. *Angew. Chem. Int. Ed. Engl.* 28 (10): 1396–1399.

8 Vögtle, F., Meier, S., and Hoss, R. (1992). One-step synthesis of a fourfold functionalized catenane. *Angew. Chem. Int. Ed. Engl.* 31 (12): 1619–1622.

9 Hunter, C.A. (1992). Synthesis and structure elucidation of a new [2]-catenane. *J. Am. Chem. Soc.* 114 (13): 5303–5311.

10 Harada, A. and Kamachi, M. (1990). Complex formation between poly(ethylene glycol) and α-cyclodextrin. *Macromolecules* 23 (10): 2821–2823.

11 Leigh, D.A. (2016). Genesis of the nanomachines: the 2016 Nobel Prize in chemistry. *Angew. Chem. Int. Ed.* 55 (47): 14506–14508.

12 Arunachalam, M. and Gibson, H.W. (2014). Recent developments in polypseudorotaxanes and polyrotaxanes. *Prog. Polym. Sci.* 39 (6): 1043–1073.

13 Araki, J. and Ito, K. (2007). Recent advances in the preparation of cyclodextrin-based polyrotaxanes and their applications to soft materials. *Soft Matter* 3 (12): 1456–1473.

14 Hoyas Pérez, N. and Lewis, J.E.M. (2020). Synthetic strategies towards mechanically interlocked oligomers and polymers. *Org. Biomol. Chem.* 18 (35): 6757–6780.

15 Hart, L.F., Hertzog, J.E., Rauscher, P.M. et al. (2021). Material properties and applications of mechanically interlocked polymers. *Nat. Rev. Mater.* 6 (6): 508–530.

16 Lin, Q., Hou, X., and Ke, C. (2017). Ring shuttling controls macroscopic motion in a three-dimensional printed polyrotaxane monolith. *Angew. Chem. Int. Ed.* 56 (16): 4452–4457.

17 Lin, Q., Li, L., Tang, M. et al. (2018). Rapid macroscale shape morphing of 3D-printed polyrotaxane monoliths amplified from pH-controlled nanoscale ring motions. *J. Mater. Chem. C Mater.* 6 (44): 11956–11960.

18 Inomata, A., Sakai, Y., Zhao, C. et al. (2010). Crystallinity and cooperative motions of cyclic molecules in partially threaded solid-state polyrotaxanes. *Macromolecules* 43 (10): 4660–4666.

19 Chen, Z., Aoki, D., Uchida, S. et al. (2016). Effect of component mobility on the properties of macromolecular [2]rotaxanes. *Angew. Chem. Int. Ed.* 55 (8): 2778–2781.

20 Uenuma, S., Maeda, R., Takahashi, S. et al. (2016). Self-assembled structure of polyrotaxane consisting of β-cyclodextrin and poly(ethylene oxide)-block-poly(propylene oxide)-block-poly(ethylene oxide) triblock copolymer in bulk system. *Chem. Lett.* 45 (8): 991–993.

21 Liu, L. and Guo, Q.-X. (2002). The driving forces in the inclusion complexation of cyclodextrins. *J. Incl. Phenom. Macrocycl. Chem.* 42 (1): 1–14.

22 Ashton, P.R., Glink, P.T., Stoddart, J.F. et al. (1996). Self-assembling [2]- and [3]rotaxanes from secondary dialkylammonium salts and crown ethers. *Chem. – A Eur. J.* 2 (6): 729–736.

23 Gibson, H.W., Bryant, W.S., and Lee, S.-H. (2001). Polyrotaxanes by free-radical polymerization of acrylate and methacrylate monomers in the presence of a crown ether. *J. Polym. Sci. A Polym. Chem.* 39 (12): 1978–1993.

24 Nagapudi, K., Leisen, J., Beckham, H.W., and Gibson, H.W. (1999). Solid-state NMR investigations of poly[(acrylonitrile)-rotaxa-(60-crown-20)]. *Macromolecules* 32 (9): 3025–3033.

25 Gibson, H.W., Engen, P.T., and Lee, S.-H. (1999). Synthesis of poly[(styrene)-rotaxa-(crown ether)]s via free radical polymerization. *Polymer (Guildf)* 40 (7): 1823–1832.

26 Lee, Y.-G., Koyama, Y., Yonekawa, M., and Takata, T. (2010). Synthesis of main-chain-type polyrotaxanes by new click polymerization using homoditopic nitrile N-oxides via rotaxanation–polymerization protocol. *Macromolecules* 43 (9): 4070–4080.

27 Kwan, P.H., MacLachlan, M.J., and Swager, T.M. (2004). Rotaxanated conjugated sensory polymers. *J. Am. Chem. Soc.* 126 (28): 8638–8639.

28 Oku, T., Furusho, Y., and Takata, T. (2003). First poly[3]rotaxane synthesized through the noncovalent step-growth polymerization of a homoditopic dumbbell compound and a macrocycle with a reversible thiol–disulfide interchange reaction. *J. Polym. Sci. A Polym. Chem.* 41 (1): 119–123.

29 Whang, D., Jeon, Y.-M., Heo, J., and Kim, K. (1996). Self-assembly of a polyrotaxane containing a cyclic “bead” in every structural unit in the solid state: cucurbituril molecules threaded on a one-dimensional coordination polymer. *J. Am. Chem. Soc.* 118 (45): 11333–11334.

30 Aprahamian, I., Miljanic, O.Š., Dichtel, W.R. et al. (2007). Clicked interlocked molecules. *Bull. Chem. Soc. Jpn.* 80 (10): 1856–1869.

31 Seale, J.S.W., Feng, Y., Feng, L. et al. (2022). Polyrotaxanes and the pump paradigm. *Chem. Soc. Rev.* 51 (20): 8450–8475.

32 Cheng, C., McGonigal, P.R., Schneebeli, S.T. et al. (2015). An artificial molecular pump. *Nat. Nanotechnol.* 10 (6): 547–553.

33 Resmerita, A.-M. and Farcas, A. (2014). Polyrotaxane based on polydimethylsiloxane and lower rim modified calix[8]arene derivative. *Rev. Roum. Chim.* 59: 459–465.

34 Ogoshi, T., Kanai, S., Fujinami, S. et al. (2008). para-Bridged symmetrical pillar[5]arenes: their Lewis acid catalyzed synthesis and host–guest property. *J. Am. Chem. Soc.* 130 (15): 5022–5023.

35 Ogoshi, T., Nishida, Y., Yamagishi, T., and Nakamoto, Y. (2010). High yield synthesis of polyrotaxane constructed from pillar[5]arene and viologen polymer and stabilization of its radical cation. *Macromolecules* 43 (17): 7068–7072.

36 Harada, A., Li, J., and Kamachi, M. (1992). The molecular necklace: a rotaxane containing many threaded α-cyclodextrins. *Nature* 356 (6367): 325–327.

37 Araki, J., Zhao, C., and Ito, K. (2005). Efficient production of polyrotaxanes from a-cyclodextrin and poly(ethylene glycol). *Macromolecules* 38 (17): 7524–7527.

38 Fleury, G., Brochon, C., Schlatter, G. et al. (2005). Synthesis and characterization of high molecular weight polyrotaxanes: towards the control over a wide range of threaded α-cyclodextrins. *Soft Matter* 1 (5): 378–385.

39 Kobayashi, Y., Nakamitsu, Y., Zheng, Y. et al. (2018). Control of the threading ratio of cyclic molecules in polyrotaxanes consisting of poly(ethylene glycol) and α-cyclodextrins. *Chem. Commun.* 54 (51): 7066–7069.

40 Kato, K., Okabe, Y., Okazumi, Y., and Ito, K. (2015). A significant impact of host–guest stoichiometry on the extensibility of polyrotaxane gels. *Chem. Commun.* 51 (90): 16180–16183.

41 Kato, K., Ohara, A., Yokoyama, H., and Ito, K. (2019). Prolonged glass transition due to topological constraints in polyrotaxanes. *J. Am. Chem. Soc.* 141 (32): 12502–12506.

42 Araki, J. and Ito, K. (2006). Polyrotaxane derivatives. I. Preparation of modified polyrotaxanes with nonionic functional groups and their solubility in organic solvents. *J. Polym. Sci. A Polym. Chem.* 44 (21): 6312–6323.

43 Kato, K., Mizusawa, T., Yokoyama, H., and Ito, K. (2015). Polyrotaxane glass: peculiar mechanics attributable to the isolated dynamics of different components. *J. Phys. Chem. Lett.* 6 (20): 4043–4048.

44 Okumura, Y. and Ito, K. (2001). The polyrotaxane gel: a topological gel by figure-of-eight cross-links. *Adv. Mater.* 13 (7): 485–487.

45 Choi, S., Kwon, T.W., Coskun, A., and Choi, J.W. (2017). Highly elastic binders integrating polyrotaxanes for silicon microparticle anodes in lithium ion batteries. *Science* 357 (6348): 279–283.

46 Claes, E., Atienza, J.M., Guinea, G. V, Rojo, F.J., Bernal, J.M., Revuelta, J.M., and Elices, M. (2010). Mechanical properties of human coronary arteries. *2010 Annual International Conference of the IEEE Engineering in Medicine and Biology*, 3792–3795.

47 Blackledge, T., Summers, A., and Hayashi, C. (2005). Gumfooted lines in black widow cobwebs and the mechanical properties of spider capture silk. *Zoology (Jena)* 108: 41–46.

48 Kato, K., Yasuda, T., and Ito, K. (2013). Viscoelastic properties of slide-ring gels reflecting sliding dynamics of partial chains and entropy of ring components. *Macromolecules* 46 (1): 310–316.

49 Kato, K., Yasuda, T., and Ito, K. (2014). Peculiar elasticity and strain hardening attributable to counteracting entropy of chain and ring in slide-ring gels. *Polymer (Guildf)* 55 (10): 2614–2619.

50 Ito, K. (2007). Novel cross-linking concept of polymer network: synthesis, structure, and properties of slide-ring gels with freely movable junctions. *Polym. J.* 39 (6): 489–499.

51 Mayumi, K., Tezuka, M., Bando, A., and Ito, K. (2012). Mechanics of slide-ring gels: novel entropic elasticity of a topological network formed by ring and string. *Soft Matter* 8 (31): 8179–8183.

52 Araki, J., Kataoka, T., and Ito, K. (2008). Preparation of a "sliding graft copolymer", an organic solvent-soluble polyrotaxane containing mobile side chains, and its application for a crosslinked elastomeric supramolecular film. *Soft Matter* 4 (2): 245–249.

53 Brovelli, S. and Cacialli, F. (2010). Optical and electroluminescent properties of conjugated polyrotaxanes. *Small* 6 (24): 2796–2820.

54 Frampton, M.J. and Anderson, H.L. (2007). Insulated molecular wires. *Angew. Chem. Int. Ed.* 46 (7): 1028–1064.

55 Taylor, P.N., O'Connell, M.J., McNeill, L.A. et al. (2000). Insulated molecular wires: synthesis of conjugated polyrotaxanes by Suzuki coupling in water. *Angew. Chem. Int. Ed.* 39 (19): 3456–3460.

56 Cacialli, F., Wilson, J.S., Michels, J.J. et al. (2002). Cyclodextrin-threaded conjugated polyrotaxanes as insulated molecular wires with reduced interstrand interactions. *Nat. Mater.* 1 (3): 160–164.

57 Brovelli, S., Latini, G., Frampton, M.J. et al. (2008). Tuning intrachain versus interchain photophysics via control of the threading ratio of conjugated polyrotaxanes. *Nano Lett.* 8 (12): 4546–4551.

58 Terao, J., Tsuda, S., Tanaka, Y. et al. (2009). Synthesis of organic-soluble conjugated polyrotaxanes by polymerization of linked rotaxanes. *J. Am. Chem. Soc.* 131 (44): 16004–16005.

59 Terao, J., Tanaka, Y., Tsuda, S. et al. (2009). Insulated molecular wire with highly conductive π-conjugated polymer core. *J. Am. Chem. Soc.* 131 (50): 18046–18047.

60 Terao, J. (2011). π-conjugated molecules covered by permethylated cyclodextrins. *Chem. Rec.* 11 (5): 269–283.

61 Yoshida, K., Shimomura, T., Ito, K., and Hayakawa, R. (1999). Inclusion complex formation of cyclodextrin and polyaniline. *Langmuir* 15 (4): 910–913.

62 Okumura, H., Kawaguchi, Y., and Harada, A. (2002). Complex formation between poly(dimethylsilane) and cyclodextrins. *Macromol. Rapid Commun.* 23 (13): 781–785.

63 Buey, J. and Swager, T.M. (2000). Three-strand conducting ladder polymers: two-step electropolymerization of metallorotaxanes. *Angew. Chem. Int. Ed.* 39 (3): 608–612.

64 Noda, Y., Hayashi, Y., and Ito, K. (2014). From topological gels to slide-ring materials. *J. Appl. Polym. Sci.* 131 (15): 40509.

65 Molero, G., Tsai, C.-Y., Liu, C. et al. (2021). Mechanical and scratch behaviors of polyrotaxane-modified poly(methyl methacrylate). *J. Appl. Polym. Sci.* 138 (42): 51237.

66 Inoue, K., Goto, T., Ito, T. et al. (2021). Boron nitride with high zeta potential via plasma processing in solution for preparation of polyrotaxane composite. *J. Phys. D Appl. Phys.* 54 (42): 425202.

67 Iida, M., Goto, T., Mayumi, K. et al. (2021). Fabrication of polyrotaxane and graphene nanoplate composites with high thermal conductivities. *Polym. Compos.* 42 (10): 5556–5563.

68 Tamura, A. and Yui, N. (2014). Lysosomal-specific cholesterol reduction by biocleavable polyrotaxanes for ameliorating Niemann-Pick type C disease. *Sci. Rep.* 4: 4356.

69 Tamura, A. and Yui, N. (2014). Threaded macromolecules as a versatile framework for biomaterials. *Chem. Commun.* 50 (88): 13433–13446.

70 Zu, G., Cao, Y., Dong, J. et al. (2019). Development of an aptamer-conjugated polyrotaxane-based biodegradable magnetic resonance contrast agent for tumor-targeted imaging. *ACS Appl. Bio. Mater.* 2 (1): 406–416.

71 Ooya, T., Eguchi, M., and Yui, N. (2003). Supramolecular design for multivalent interaction: maltose mobility along polyrotaxane enhanced binding with concanavalin A. *J. Am. Chem. Soc.* 125 (43): 13016–13017.

72 Ooya, T., Utsunomiya, H., Eguchi, M., and Yui, N. (2005). Rapid binding of concanavalin A and maltose–polyrotaxane conjugates due to mobile motion of α-cyclodextrins threaded onto a poly(ethylene glycol). *Bioconjug. Chem.* 16 (1): 62–69.

73 Chen, L., Yan, C., and Zheng, Z. (2018). Functional polymer surfaces for controlling cell behaviors. *Mater. Today* 21 (1): 38–59.

74 Kakinoki, S., Seo, J.-H., Inoue, Y. et al. (2015). Mobility of the Arg-Gly-Asp ligand on the outermost surface of biomaterials suppresses integrin-mediated mechanotransduction and subsequent cell functions. *Acta Biomater.* 13: 42–51.

75 Seo, J.-H., Kakinoki, S., Inoue, Y. et al. (2013). Inducing rapid cellular response on RGD-binding threaded macromolecular surfaces. *J. Am. Chem. Soc.* 135 (15): 5513–5516.

76 Seo, J.-H., Kakinoki, S., Yamaoka, T., and Yui, N. (2015). Directing stem cell differentiation by changing the molecular mobility of supramolecular surfaces. *Adv. Healthc. Mater.* 4 (2): 215–222.

77 Zhou, J., Feng, H., Sun, Q. et al. (2022). Resistance-switchable conjugated polyrotaxane for flexible high-performance RRAMs. *Mater. Horiz.* 9 (5): 1526–1535.

78 Willis-Fox, N., Belger, C., Fennell, J.F. Jr., et al. (2016). Threading the needle: fluorescent poly-pseudo-rotaxanes for size-exclusion sensing. *Chem. Mater.* 28 (8): 2685–2691.

79 Jiang, Y., Zhang, Z., Wang, Y.X. et al. (2022). Topological supramolecular network enabled high-conductivity, stretchable organic bioelectronics. *Science* 375 (6587): 1411–1417.

80 Qiu, Y., Song, B., Pezzato, C. et al. (2020). A precise polyrotaxane synthesizer. *Science* 368 (6496): 1247–1253.

81 Amano, S., Fielden, S.D.P., and Leigh, D.A. (2021). A catalysis-driven artificial molecular pump. *Nature* 594 (7864): 529–534.

82 Ikejiri, S., Takashima, Y., Osaki, M. et al. (2018). Solvent-free photoresponsive artificial muscles rapidly driven by molecular machines. *J. Am. Chem. Soc.* 140 (49): 17308–17315.

83 Xie, Z.H., Rong, M.Z., and Zhang, M.Q. (2021). Dynamically cross-linked polymeric binder-made durable silicon anode of a wide operating temperature Li-ion battery. *ACS Appl. Mater. Interfaces* 13 (24): 28737–28748.

84 Yoo, D.J., Elabd, A., Choi, S. et al. (2019). Highly elastic polyrotaxane binders for mechanically stable lithium hosts in lithium-metal batteries. *Adv. Mater.* 31 (29): 1901645.

85 Zhao, J., Hong, M., Ju, Z. et al. (2022). Durable lithium metal anodes enabled by interfacial layers based on mechanically interlocked networks capable of energy dissipation. *Angew. Chem.* 134 (52): e202214386.

86 Imholt, L., Dörr, T.S., Zhang, P. et al. (2019). Grafted polyrotaxanes as highly conductive electrolytes for lithium metal batteries. *J. Power Sources* 409: 148–158.

87 Ding, P., Wu, L., Lin, Z. et al. (2023). Molecular self-assembled ether-based polyrotaxane solid electrolyte for lithium metal batteries. *J. Am. Chem. Soc.* 145 (3): 1548–1556.

Index

Mechanically Interlocked Materials: Polymers, Nanomaterials, MOFs, and more, First Edition.
Edited by Emilio M. Pérez.

m

n